NUCLEAR AND CONDENSED MATTER PHYSICS

Related Titles from AIP Conference Proceedings

To learn more about these titles, or the AIP Conference Proceedings Series, please visit the webpage
http://www.aip.org/catalog/aboutconf.html

NUCLEAR AND CONDENSED MATTER PHYSICS

VI Regional CRRNSM Conference

Palermo, Italy   14–15 October 1999

EDITOR

Antonino Messina
University of Palermo, Italy

Melville, New York
AIP CONFERENCE PROCEEDINGS ■ 513

Editor:

Antonino Messina
University of Palermo
Department of Physical and Astronomical Sciences
Via Archirafi 36
I-90123 Palermo
ITALY

E-mail: messina@fisica.unipa.it

L.C. Catalog Card No. 00-100717
ISBN 1-56396-929-7
ISSN 0094-243X
Printed in the United States of America

CONTENTS

CONDENSED MATTER PHYSICS

NUCLEAR PHYSICS

MISCELLANEOUS

PREFACE

This book contains 102 scientific contributions presented at the *VI Conferenza Scientifica Regionale del Comitato Regionale per le Ricerche Nucleari e di Struttura della Materia* (VI Scientific Conference on Nuclear and Condensed Matter Physics) held in Palermo, Italy, from 14 to 15 October 1999.

The Conference was attended by 144 physicists, most of them belonging to the Sicilian Universities of Palermo, Catania and Messina.

The meeting took place at the historical and beautiful *Circolo Ufficiali* in the heart of Palermo, generously provided by the Sicilian Military Authority.

The Conference consisted of invited talks and poster contributions presented during five morning and afternoon sessions. The contributions submitted to the Scientific Committee of the Conference have been refereed thanks to the generous help of numerous colleagues.

The scientific interest of the Conference, in my opinion, is witnessed both by the quality of the papers appearing in this volume and by the enthusiastic participation of many young physicists. This should rekindle the interest of the Sicilian Government toward the *Comitato Regionale per le Ricerche Nucleari e di Struttura della Materia*, hopefully leading, in the near future, to a substantial increase of the research funds in the areas of Nuclear and Condensed Matter Physics.

I wish to thank the colleagues Gaetano Cubiotti, Sandro Giuliano and Lorenzo Cordone for their help. I am particularly indebted to Marcello Lattuada for continuous encouragement and advise.
I am grateful for financial support from the sponsors and exhibitors listed in the following pages.
Special thanks are due to Major General Bruno Loi and Colonel Paolo Frasconà for their active interest in the Conference. I would like to acknowledge the pleasant and comfortable atmosphere created by Maresciallo Cardaci and Mr. Palmisano at the *Circolo Ufficiali* during the Conference.
Finally I would like to acknowledge the essential commitment to the organisation of the Conference of Anna Napoli, Eugenio Vitrano, Sabrina Maniscalco and Rosanna Migliore.

Antonino Messina

Sponsors

Presidenza della Regione Sicilia
Comando Militare Autonomo della Sicilia
Università degli Studi di Palermo
Ente per le Nuove tecnologie, l'Energia e l'Ambiente (ENEA)
Istituto Nazionale di Fisica della Materia (INFM)
Laboratori Nazionali del Sud (LNS)
Società Italiana di Fisica (SIF)
Facoltà di Scienze Matematiche Fisiche e Naturali di Catania
Facoltà di Ingegneria di Catania
Banca Popolare S. Angelo
Assoindustria
Osservatorio Astronomico di Palermo
Dipartimento di Scienze Fisiche ed Astronomiche di Palermo
Centralgas
Compaq
Devil Disk s.r.l.
Gestetner - Sistemi per ufficio - Palermo
La Fondiaria Assicurazioni S.p.A. - Agenzia Generale di Palermo
Makers s.r.l.

Exhibitors

Bollati Boringhieri
DEA Librerie Internazionali

Scientific Committee

Lorenzo Cordone

Vice-Chairman of Comitato Regionale
per le Ricerche Nucleari e di
Struttura della Materia (CRRNSM), Italy

Gaetano Cubiotti

Department of Physics, Messina, Italy

Marcello Lattuada

Department of Physics, Catania, Italy

Antonino Messina

Department of Physical and Astronomical Sciences, Palermo, Italy

Local Committee

Chairman

Antonino Messina

Department of Physical and Astronomical Sciences, Palermo, Italy

Treasurer

Eugenio Vitrano

Department of Physical and Astronomical Sciences, Palermo, Italy

Members

Anna Napoli

Department of Physical and Astronomical Sciences, Palermo, Italy

Magg. Gen. Bruno Loi

Comando Militare Autonomo della Sicilia, Italy

Secretary

Sabrina Maniscalco

Department of Physical and Astronomical Sciences, Palermo, Italy

CONDENSED MATTER PHYSICS

Hysteretic Behavior of Microwave Second Harmonic Emission by Superconductors in the Critical State

A. Agliolo Gallitto, M. Guccione and M. Li Vigni

Unità INFM di Palermo and Dipartimento di Scienze Fisiche e Astronomiche
via Archirafi 36, I-90123 Palermo, Italy

Abstract. We report experimental results on microwave second harmonic emission in different types of conventional and high-T_c superconductors in the critical state. We show that the peculiarities of the second harmonic signal strongly depend on the type of superconductors.

INTRODUCTION

It is well known that the time dependence of the magnetization of type II superconductors (SC), exposed to dc and ac magnetic fields, is strongly influenced by the motion regime of fluxons. It has been shown that superconducting samples which are in the critical state developed by a static magnetic field, when submitted to an intense em field, exhibit a magnetization vector containing Fourier components at harmonic frequencies of the driving field[1]. Bean had extensively studied harmonic emission of type II SC in the critical state at low frequencies[1]. According to his model, only odd harmonic emission has been detected. However, it has been shown that SC in the critical state, exposed to intense mw fields, exhibit odd as well as even harmonic emission[2]. A model has been elaborated, which well accounts for the peculiarities of second harmonic (SH) signal detected in ceramic $YBa_2Cu_3O_7$. However, recent experiments performed in different SC have shown that the peculiarities of the SH signal depend on the type of superconductor.

We report a detailed study of SH emission at mw frequencies in different types of conventional and high-T_c SC in the critical state. A comparison of the experimental results in different samples will be done. We will show that some peculiarities of the SH signals cannot be justified by the models reported in the literature.

EXPERIMENTAL RESULTS AND DISCUSSION

Experiments have been performed in crystals of $YBa_2Cu_3O_7$ (YBCO) and $Ba_{0.6}K_{0.4}BiO_3$ (BKBO), and in Nb powder, at T = 4.2 K. The sample is placed in a

CP513, *Nuclear and Condensed Matter Physics,* edited by A. Messina

bimodal cavity, resonating at the two frequencies ω and 2ω, with $\omega/2\pi \approx 3\,GHz$, in a region in which the magnetic fields $H(\omega)$ and $H(2\omega)$ have maximal intensity. The ω-mode of the cavity is fed by a pulsed triode oscillator, with pulse repetition rate of 200 pps and pulse width of $\sim 1\,\mu sec$, giving a maximal peak power of $\sim 1\,kW$. The harmonic signals radiated by the sample are detected by a superheterodyne receiver[3]. SH signals have been investigated as a function of the static field H_0, at different input power levels. All the measurements have been performed with $\mathbf{H_0} \parallel \mathbf{H}(\omega) \parallel \mathbf{H}(2\omega)$.

Fig.1(a) shows the SH signal intensity as a function of H_0 in a BKBO crystal. The arrows point out the sweep direction of H_0. The SH signal is more intense at decreasing than at increasing fields. The hysteresis height is roughly independent of the magnetic field range spanned up to fields of the order of 10 kOe. On the contrary, it strongly depends on the input power level; this dependence is shown in Fig.1(b).

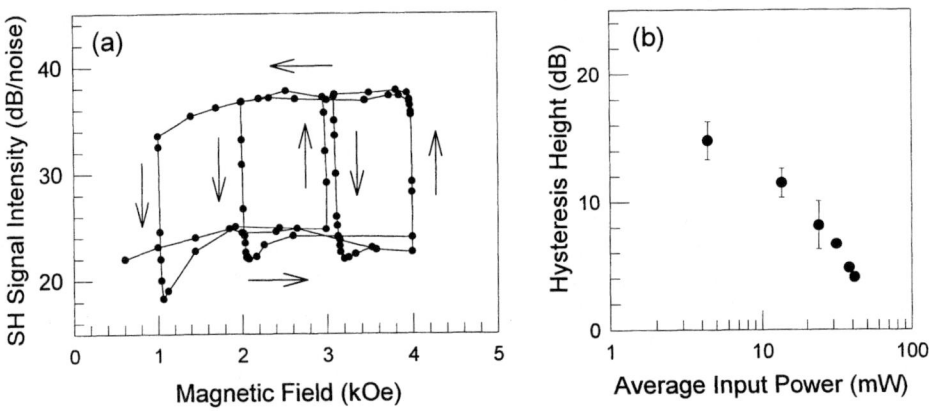

FIGURE 1. (a) SH signal intensity as a function of H_0 in a crystal of BKBO. $T = 4.2\,K$; average mw power $\approx 13.5\,mW$. **(b)** Power dependence of the hysteresis height. $H_0 = 2\,kOe$; $T = 4.2\,K$.

Fig.2 shows the SH signal intensity as a function of H_0 in a YBCO crystal with the c-axis parallel to $\mathbf{H_0}$. The signal does not exhibit hysteresis, within the experimental accuracy, except for the presence of enhanced dips in a narrow range of fields from the value of H_0 at which the inversion of the field sweep direction is operated. The same result has been obtained no matter what the input power level and the magnetic field range spanned. Similar results have been obtained in ceramic YBCO[2,3].

In Fig.3 we report the SH signal intensity as a function of H_0 in Nb powder. As one can see, the signal shows a hysteretic behavior. However, in this case, the hysteresis height depends on H_0. The field dependence of the hysteresis height is shown in the inset. Measurements performed at different values of the input power have shown that in this sample the hysteresis height is roughly independent of the input power level.

All results reported in the Figs.1-3 have been obtained after the samples were exposed to high fields, higher than the lower critical field. The same results have been obtained in zero-field-cooled samples, after the first run to high fields, and in field-

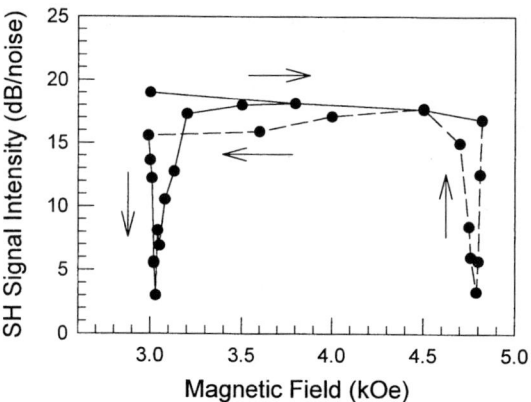

FIGURE 2. SH signal intensity as a function of H_0 in a sample of YBCO crystal with the c-axis parallel to H_0. Continuous and dashed lines refer to results obtained at increasing and decreasing fields, respectively. T = 4.2 K, input power ≈ 13.5 mW.

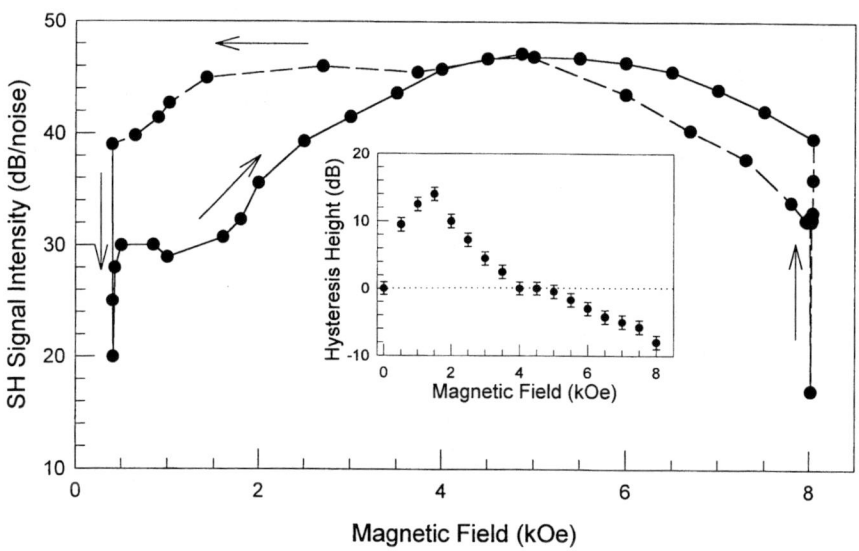

FIGURE 3. SH signal intensity as a function of H_0 in Nb powder. Continuous and dashed lines refer to results obtained at increasing and decreasing fields, respectively. The inset shows the H_0-dependence of the hysteresis height. T = 4.2 K, input power ≈ 1.5 mW.

cooled samples. During the first field sweep, virgin samples of BKBO and Nb exhibit an intense low field SH signal, which vanishes irreversibly when the samples are exposed to high fields. The low field signal can be ascribed to nonlinear processes in weak links[3,4]. After exposition to high fields, the trapped magnetic flux decouples the inter-grain Josephson junctions and the low-field signal disappears irreversibly. On the other hand, at low temperatures one can reasonably hypothesize that the SC, which have been exposed to high fields, are in the critical state.

Bean[1] has extensively studied nonlinear emission of type II SC in the critical state. From the Bean model, which assumes that the fluxon lattice follows the em field variations, only odd harmonic generation is expected. Nevertheless, at mw frequencies SH signals have been detected in both conventional and high-T_c SC in the critical state. A model has been elaborated in which a rectification process of the mw field operated by SC in the critical state is hypothesized[2]. Because of the rigidity of the fluxon lattice, the induction field inside the sample does not follow the variation of an em field oscillating at high frequencies, except when the variation involves motion of fluxons in the surface layers of the sample. It has been supposed that, for "direct critical state" developed by increasing magnetic fields, the induction flux in the sample does not vary during the positive semi-period of the mw field, while it does during the negative semi-period. The opposite occurs when decreasing fields develop the "reverse critical state". The model elaborated on this hypothesis predicts SH signals independent of the value and the sweep direction of H_0, in agreement with the experimental results obtained in YBCO[2]. The occurrence of the deeps has been ascribed to the fact that the inversion of the magnetic field sweep can remove the critical state and a finite variation of H_0 is needed to develop a new critical state.

The hysteretic behavior of SH signals detected in Nb and BKBO samples is not accounted for by the model discussed in Ref. 2. This is so because the model has been elaborated by supposing the complete rectification of the mw field, with a symmetric response of fluxons in direct and reverse critical state. On the contrary, the results in Nb and BKBO samples suggest that the response to the mw field of the fluxon lattice in the direct and reverse critical state is different. Indeed, SH generation is more enhanced when the fluxons are forced by the applied field to decrease their concentration than when they are forced to draw near each other.

We think that the SH generation is still ascribable to a rectification process operated by SC in the critical state. However, it could be hypothesized that the rectification is not complete. So, the mw field variation may modify the distribution of the flux inside the sample during the full wave period, but in an uneven way. The hysteretic behavior could arise from a different time response of the fluxons when they go in or out the sample surface. This might be related to surface barrier effects[5]. Further, the time response of the fluxon lattice could be different in different types of SC.

REFERENCES

1. Bean, C. P., *Rev. Mod. Phys.* **36**, 31 (1964).

2. Ciccarello, I., Fazio, C., Guccione, M., and Li Vigni, M., *Physica* **C 159**, 769 (1989).

3. Agliolo Gallitto, A., Ciccarello, I., Guccione, M., Li Vigni, M., and Persano Adorno, D., "Microwave Response of High-T_c Superconductors" in *Pair Correlations in Many Fermion Systems*, Edited by V. Z. Kresin, Plenum Press, New York (1998) pp. 111-131.

4. Agliolo Gallitto, A., Guccione, M., and Li Vigni, M., *Physica* **C 309**, 8 (1998).

5. Kugel, K. I., Mamsurova, L.G., Pigalskiy, K. S., Rakhmanov, A. L., *Physica* **C 300**, 270 (1998).

Defects Induced By Gamma Irradiation In Silica

S. Agnello, R. Boscaino, M. Cannas and F.M. Gelardi

INFM and Department of Physical and Astronomical Sciences, University of Palermo
Via Archirafi, 36, I-90123 Palermo (ITALY)

Abstract. We report an electron spin resonance study of the defects induced by γ irradiation in silica. By comparing various materials and by studying the defect growth as a function of the accumulated γ dose, interesting hints are found on the defect origin.

I. INTRODUCTION

Point defects in SiO_2 have been extensively investigated in a wide variety of silica materials but the nature of many of them is still not univocally determined and constitutes matter of controversy [1-5]. Point defects can be induced either by the manufacturing process or by external agents (drawing, heat treatments and particle or ionizing radiation). In particular, it is well known that γ irradiation can induce or convert defects so that information can be obtained on the generation mechanism and the structure of a given point defect by monitoring its concentration as a function of the accumulated γ dose (growth kinetics).

In this paper, we report the Electron Spin Resonance (ESR) study of the growth kinetics of the E' center [6] and three structures with field split of 1.36, 7.4 and 11.8 mT respectively, induced by γ irradiation in commercial silica of natural type I, II, and synthetic type III, IV [7]. From the comparison of the growth kinetics hints are found on the origin and the relations existing among these defects. Moreover, interesting suggestions on the defect nature are obtained by comparing the effects of the irradiation on the various silica types.

II. EXPERIMENTAL RESULTS

Our experiments were performed in samples of commercial origin [8,9] of the four standard silica types [7]: Natural dry (Type I): Infrasil 301 (I301), silica EQ (EQ906), (EQ912) and Puropsil (QPA); Natural wet (Type II): Herasil (H1), (H3) and Homosil (HM); Synthetic wet (Type III): Suprasil (S1), (S311); Synthetic dry (Type IV): Suprasil (S300). The samples of the various materials, slab shaped with dimensions ($5\times5\times1$ mm^3), were γ irradiated at room temperature in a ^{60}Co source accumulating doses in the range from 5×10^{-4} Mrad to 10^3 Mrad. ESR measurements were done at

CP513, *Nuclear and Condensed Matter Physics*, edited by A. Messina
© 2000 American Institute of Physics 1-56396-929-7/00/$17.00

T = 300 K with a spectrometer (Bruker EMX) working at 9.8 GHz, and with modulation field at 100 kHz.

The signal of the E' center was detected in the various irradiated samples using a microwave power P_{mw}=0.8 µW and a field modulation amplitude H_m=0.01 mT. As reported in fig.1a, this signal is characterized by the typical powder-like line shape with principal g values of 2.0018, 2.0006, 2.0003 [6]. In both natural and synthetic silica samples, the E' concentration increases in an initial dose range and reaches a constant intensity at larger γ doses (fig.1b). The dose range where the E' concentration increases is characteristic of a given silica samples and differs by even one order of magnitude from a sample to another. The same observation applies to the maximum spin concentration whose value ranges from 10^{16} to 2×10^{17} centers/cm^3. In fig.1a we report a structure partially overlapped with the E' line. This structure was revealed in all our samples (P_{mw}=50 µW, H_m=0.07 mT) and is characterized by three poorly resolved peaks with a maximum field split of 1.36 ± 0.03 mT. This structure shows the same growth kinetics as the E' center even if it begins to be detectable at doses higher than the E', as reported in fig.1b for EQ906, and its saturated concentration is lower than the E' one.

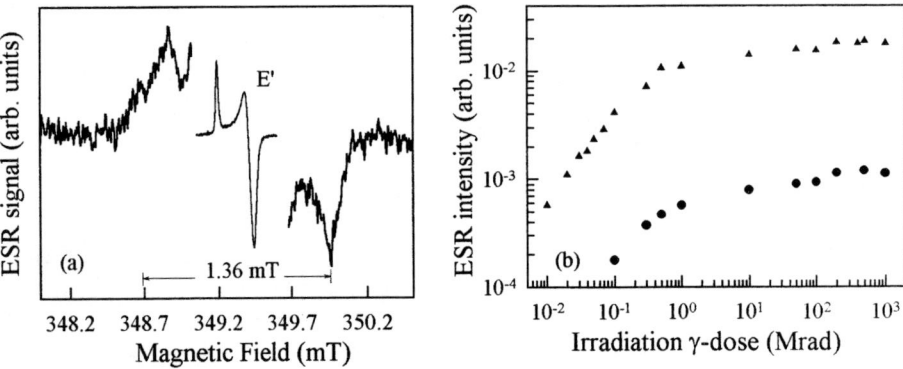

FIGURE 1. (a) Typical ESR derivative signals of E' and 1.36 mT structures. (b) Growth kinetics of the E' (triangles) and 1.36 mT (circles) structures in the sample EQ906.

Another structure, observed in all the irradiated samples is a doublet split by 7.36 ± 0.03 mT asymmetrically centered on the E' line (fig.2a). The characteristic growth kinetics of this doublet is reported in fig.2b, where the concentration increases with γ irradiation following a square root law. Moreover, no dose saturation has been detected for the intensity of this signal in all the samples up to a dose of 10^3 Mrad. As reported in fig.2a, we observed also a second doublet split by 11.83 ± 0.03 mT (P_{mw}=3.2 mW, H_m=0.4 mT) around the E' line. This doublet is present only in natural silica of type I and II and has a growth kinetics different from the E' centers. In fact, the doublet intensity increases up to a dose of ~0.1 Mrad, very lower (at least one order of magnitude) than the doses requested to complete the growth of the E' centers.

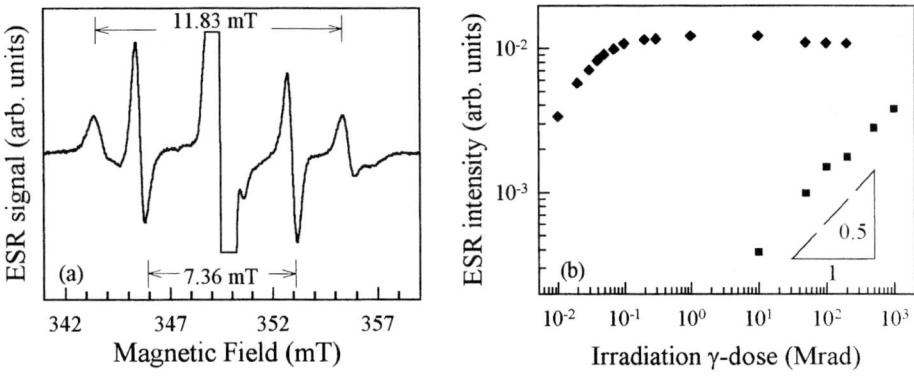

FIGURE 2. (a) Typical ESR derivative signals of the 7.36 mT and 11.8 mT doublets, the central line is the E' line distorted and saturated by microwave power. (b) Growth kinetics of the intensities of the 11.8 mT (diamonds) and 7.36 mT (squares) doublets in the sample EQ906.

III. DISCUSSION

By γ irradiating various samples of the four commercial silica types, we evidenced the generation of four main ESR structures: the well known E' center; a structure with maximum field split of 1.36 mT; and two doublets split by 7.36 and 11.8 mT respectively. The comparison of the irradiation effects in the various materials allows us to obtain interesting suggestion on the defects origin. The study of the growth kinetics shows that the E' center concentration reaches a constant value after prolonged irradiation both in natural and synthetic silica. From this observation we hypothesize that in our samples the γ rays convert some precursor of the E' defect already present in the as-grown material. Since differences are observed among the various samples both for the dose range of increasing defects concentration and for the maximum concentration, we infer that various precursors are present and are characterized by different conversion efficiencies. In each sample we observe that the 1.36 mT structure reproduces the growth kinetics of the E' centers and so it can be interpreted as an hyperfine structure of the E' signal. The 1.36 mT structure could arise from the interaction between the E' electron spin and a nuclear spin in its vicinity, probably an H atom, as proposed by several structural models reported in literature [10,11]. Since H atoms in our samples could arise from OH groups, the signal of the 1.36 mT structure should depend on the OH content, but we did not observe any particular feature in samples having OH concentration differing of some orders of magnitude.

As regards the 7.4 mT doublet, commonly referred as H(I) center [12], its kinetics is very different from that of E' center. The sublinear growth suggests a complex mechanism of generation. Moreover, the absence of saturation can be considered as an evidence of the direct action of the γ rays on the unperturbed matrix. According to the model attributing the H(I) center to an Si atom bonded to one H and two O atoms [10,

9

12, 13], it is possible to hypothesize that the γ rays act on a silicon normally bonded to four O atoms breaking two Si-O bonds and successively one H atom is linked to the Si atom [14].

Finally for the 11.8 mT doublet, usually named H(II) center [12], the saturation growth kinetics here reported indicates that it arises from a precursor activated by the irradiation. Moreover, the observation of the H(II) only in natural silica supports its attribution to an extrinsic defect. Our observations suggest that the H(II) structure involves a substitutional Ge atom bonded to one H and two oxygen atoms [12, 13, 15].

ACKNOWLEDGMENTS

The authors wish to thank Prof. E Calderaro for taking care of the γ irradiation in the irradiator IGS-2 of Dipartimento di Ingegneria Nucleare, Palermo, Italy. This work is a part of a National research project supported by Ministero Italiano della Ricerca Scientifica e Tecnologica, Roma, Italy.

REFERENCES

1. Devine, R.A.B. and Arndt, J., *Phys. Rev. B*, **39**, 5132-5138 (1989).
2. Nishikawa, H., Nakamura, R., Tohmon, R., Ohki, Y., Sakurai, Y., Nagasawa, K. and Hama, Y., *Phys. Rev. B*, **41**, 7828-7834 (1990).
3. Griscom, D.L., *J. Ceram. Soc.Jpn.*, **99**, 899-916 (1991).
4. Galeener, F.L., Kerwin, D.B., Miller, A.J., Mikkelsen J.C.Jr., *Phys. Rev.B*, **47**, 7760-7779 (1993).
5. Imai, H. and Hirashima, H., *J. Non-Cryst. Solids*, **179**, 202-213 (1994).
6. Weeks, R.A. and Nelson, C.M., *J. Appl. Phys.*, **31**, 1555-1558 (1960).
7. Hetherington, G., Jack, K.H. and Ramsay, M.W., *Physics & Chemistry of Glasses*, **6**, 6-15 (1965).
8. Heraeus Quartzglas, Hanau, Germany, Catalogue POL-0/102/E;
9. Quartz&Silice, Nemours, France, Catalogue OPT-91-3.
10. Tsai, T., Griscom, D.L., *J. Non-Cryst. Solids* **91**, 170-179 (1987).
11. Li, J., Kannan, S., Lehman, R.L. and Sigel, G.H.Jr., *Appl. Phys. Lett.*, **64** 2090-2092 (1994).
12. Radtsig, V.A. and Bobyshev, A.A., *Phys. Stat. Sol.(B)*, **133** 621-627 (1986).
13. Vitko, J.Jr., *J. Appl. Phys.*, **49**, 5530-5535 (1978).
14. Agnello, S., Boscaino, R., Cannas, M., Gelardi, F.M., Leone, M., *Nucl. Instrum. Methods* (in press).
15. Agnello, S., Boscaino, R., Cannas, M., Gelardi, F.M., Leone, M., *J. Non-Cryst. Solids*, **232-234**, 323-328 (1998).

Periodically Driven Noisy Nonlinear Physical Systems

N. V. Agudov[1] and B. Spagnolo

Istituto Nazionale per la Fisica della Materia, Unità di Palermo
and
Dipartimento di Energetica ed Applicazioni di Fisica, Università di Palermo,
viale delle Scienze,I-90128, Palermo, Italy

Abstract. Noise-induced nonequilibrium phenomena in nonlinear systems have recently attracted a great deal of attention in a variety of contexts. Noise can induce a number of interesting phenomena in periodically driven nonlinear dynamical systems, such as stochastic resonance (SR) and noise-enhanced stability (NES). We report here some new results concerning these noise-induced effects. Specifically: (i) the investigation of the SR phenomenon for a bistable system described by a piece-wise linear potential with arbitrary height and width ; (ii) the analysis of noise enhanced stability phenomenon in a piecewise linear potential in adiabatic approximation.

Noise-induced effects in nonlinear physical systems have recently shown to be useful to describe the dynamics of far from equilibrium systems. When a time-periodic external field is present the stochastic processes describing the systems are non-stationary stochastic processes. Such processes occur in many fields of physics, chemistry and biology and have attracted a lot of interest in the recent past. Typical problems are the kinetics of chemical reaction, where external periodic fields help the system to overcome energetic barriers and thus enhance the reaction rate, and the Josephson junctions, where external microwave fields enhance the thermal noise induced transitions between zero voltage states (superconducting) and a dissipative state [1].

The noise is not simply a nuisance which perturbs the observation and creates disorder, but rather noise might make possible new states and forms of behaviour which do not appear in a noise-free limit. Phenomena like noise-induced transitions, noise-enhanced stability (NES), resonant activation and stochastic resonance (SR) [2–5] are good examples in which the noise actually participates in the creation of ordered states or is responsible for surprising phenomena through its interaction with the nonlinearities of the system. This paper deals with two noise-induced

---

[1] Permanent address: Radiophysical Department, State University of N.Novgorod, 23 Gagarin Ave., Nizhny Novgorod, 603600, Russia.

CP513, *Nuclear and Condensed Matter Physics,* edited by A. Messina
© 2000 American Institute of Physics 1-56396-929-7/00/$17.00

phenomena in periodically driven physical systems, specifically: the SR and NES phenomena.

STOCHASTIC RESONANCE

For description of dynamical system the model of overdamped Brownian motion is used:

$$\dot{x} = -\frac{\partial U(x,t)}{\partial x} + \xi(t),$$ (1)

where $\xi(t)$ is the white Gaussian noise with zero mean and $\langle \xi(t)\xi(t+\tau) \rangle = 2q\delta(\tau)$, and $U(x,t)$ is a periodically driven potential profile

$$U(x,t) = \Phi(x) - xs(t)A\cos\Omega t, \quad A \ll q/L$$ (2)

where L is a characteristic size of the system.

We use the linear response theory and obtain the spectral characteristics $S(\omega)$ of the output signal $x(t)$ as follows

$$S(\omega) = \left(1 - \frac{a^2}{2D^{(0)}}\right) Q^{(0)}(\omega) + \frac{a^2}{2}\delta(\omega - \Omega)$$ (3)

where $D^{(0)}$ is the variance of the stochastic process $x(t)$, $Q^{(0)}$ is the spectral density of fluctuations which gives the platform in the spectrum and the second term is the δ-spike. We use quasistationary approximation, i. e. the frequency of the input signal is assumed to be less than the relaxation rate to the global equilibrium distribution. We obtain therefore the analytical expression of the Signal to Noise Ratio (SNR) as a function of the variance of the process $x(t)$ and of the correlation time $\tau_c^{(0)}$ of the system

$$R = \frac{a^2}{2S(0)} = \pi \left[\left(\frac{2q^2}{A^2 D^{(0)}} - 1\right)\tau_c^{(0)}\right]^{-1}.$$ (4)

where

$$\tau_c^{(0)} = \frac{1}{D^{(0)}} \int_0^\infty B^{(0)}[\tau]d\tau.$$ (5)

and $B^{(0)} = < x(t)x(t+\tau) > - < x(t) >^2$ is the covariance function of $x(t)$ without periodical driving.

The Eq.(4) allows us to study the dependence of the SNR on the widht and the height of the potential barrier. We consider a bistable steplike potential profile and we find that the SNR is greater for the narrow (or wide) and low barrier, while it is lower for the intermediate values of the width and for greater height.

NOISE ENHANCED STABILITY

The Mean First Passage Time of a Brownian particle moving in potentials fields normally decreases with noise intensity growth according to the Kramers formula. Recently the dependence on the noise intensity of the MFPT for metastable and unstable systems was revealed to have resonance character in different physical systems [3,5,6]. This effect was named by Mantegna and Spagnolo [3]: Noise Enhanced Stability, and gives a nonmonotonic behaviour of the average escape time as a function of the noise intensity. NES phenomenon is observed in different physical context: (i) in the intermittent chaotic behaviour of one-dimensional maps [7,8] (ii) in a model potential useful to describe first order phase transition [9], and (iii) in a tunnel diode [3].

We study the noise-driven escape of an overdamped Brownian particle in a piecewise linear potential profile in order to understand the NES effect. This potential profile can be useful to describe physical systems which are unstable or metastable in the absence of noise. We find that the average escape time of the unstable system can be increased by the external noise, contrary to what one might have expected. This means that the noise can modify the stability of the system in a counter-intuitive way.

We consider an overdamped Brownian particle moving in a piece wise linear potential $U(x,t)$ modulated by a periodic force

$$
U(x,t) = \begin{cases} -(k + A cos\omega_s t)x, & x < 0 \\ -Axcos(\omega_s t), & 0 < x < L \\ -(k + A cos\omega_s t)(x - L), & L < x < B \end{cases} , \tag{6}
$$

where B is the boundary and k is the slope of the two marginal linear intervals of the potential profile.

To calculate the MFPT we need an adiabatic condition in order that the particle can see a quasi-stationary potential. To do this we make the following steps: (i) we calculate the exact MFPT for the three kinds of the piece-wise linear potential profile for arbitrary initial condition and final position with an absorbing barrier; (ii) we apply the following adiabatic condition: $T_{x_i,x_i+\Delta x_i} \ll T_s = 2\pi/w_s$ or $T_i \approx T_s/N$ where $N \gg 1$ and T_i is the partial MFPT that the particle spends to go from the initial condition x_i to the absorbing barrier $x_i + \Delta x_i$; (iii) we obtain the total MFPT as the sum of the N different partial values $T_{x_i,x_i+\Delta x_i}$:

$$
T = \sum_{i=1}^{N} T_i \tag{7}
$$

and for different values of the noise intensity.

The MFPT T_i for potential profile $U(x,t)$ (Eq.(6)) with different initial conditions and absorbing barriers can be calculated by the following formula

$$
T(x_0, D) = \frac{1}{q} \int_{x_0}^{x_o+\delta x} e^{u(\zeta)/q} \int_{-\infty}^{\zeta} e^{-u(\chi)/q} d\chi d\zeta. \tag{8}
$$

The results of our theoretical calculation for $T_s = 100$ are shown in the following figure, where we reported the average escape time as a function of the noise intentensity. Noise enhanced stability is observed for more than two decade in the semilog plot of Fig. 1.

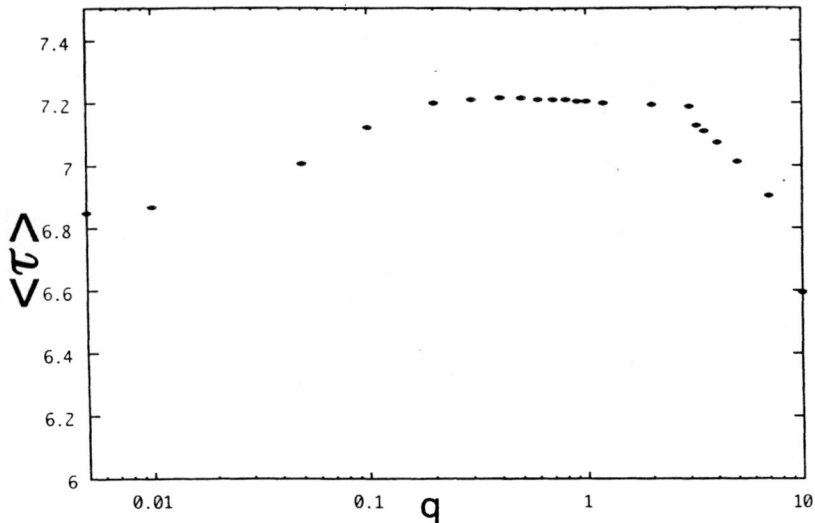

FIGURE 1. The average escape time as a function of the noise intensity.

The NES phenomenon is quite general and it is observed in various physical systems described by different kinds of the potential profiles.

Acknowledgements

This work has been supported by INFM and by MURST.

REFERENCES

1. P. Jung. *Phys. Rep.*, **234**. 175 (1993).
2. R. N. Mantegna and B. Spagnolo, *Phys. Rev. E*, **49**, R1792 (1994).
3. R. N. Mantegna and B. Spagnolo. *Phys. Rev. Lett.*, **76**, 563 (1996).
4. L. Gammaitoni. P. Hanggi, P. Jung and F. Marchesoni, *Rev. Mod. Phys.*, **70**. 223 (1998).
5. R. N. Mantegna and B. Spagnolo, *J. Phys. IV*, **8**, Pr6-247 (1998).
6. N. V. Agudov. *Phys.Rev.E* **57**. 2618 (1998).
7. J. E. Hirsch. B. A. Huberman. and D. T. Scalapino, *Phys. Rev. A* **25**, 519 (1982).
8. R. Wackerbauer.*Phys.Rev.E* **58**. 3036 (1998).
9. I. Dayan. M. Gitterman. and G. H. Weiss. *Phys. Rev. A* **46**. 757 (1992).

Tuning the Emission Spectra of Pyrromethene Doped Polycarbonate by Light Confinement

M. Allegrini, A. Arena, E. Cefali, C. Pace, S. Patanè and G. Saitta

Istituto Nazionale per la Fisica della Materia (INFM)
Dipartimento di Fisica della Materia e Tecnologie Fisiche Avanzate, Università di Messina

Abstract. Spectrophotometry and photoluminescence (PL) are used to investigate the optical properties and photodegradation of thin films of pyrromethene 580 dispersed in polycarbonate (PC). The films spin-coated on glass act as asymmetric planar waveguides and excited with blue light yield a linearly polarized emission whose colour changes according to the film thickness.

INTRODUCTION

Nowadays considerable research interest concentrates on the synthesis and characterization of luminescent materials, aimed at developing low-cost, easy processable active media to be used in optically-pumped thin-films lasers [1]. The principle on which such kind of devices works, is that the PL, isotropically emitted by the film, is partially confined at the interfaces between the material and lower refraction index surrounding media [2]. Light is allowed to travel back and forth through the material: if the film edges are sharp enough, the reflection at the edge provides the feedback mechanism needed in order to obtain oscillation and a highly directional, linearly polarized beam, can be obtained as output. As an evidence of the strong modifications the luminescence undergoes upon waveguiding, here we examine the optical behaviour of thin films of pyrromethene 580 (supplied by Exciton) dispersed in polycarbonate, deposited on glass.

RESULTS AND DISCUSSION

The absorption and emission spectrum of a diluted solution of pyrromethene in ethanol and of a solid solution of pyrromethene in PC are shown in fig. 1a. The solid state absorption spectrum closely resembles the ethanol solution spectrum, but for a rigid red shift of ~ 8 nm, due to self-absorption. A shoulder at about 490 nm and a peak at approximately 525 nm are responsible of the orange colour of the pyrromethene solid solutions. Excited with blue light the film isotropically emits an intense green PL, spectrally described by a ~50 nm wide band centred at about 545 nm. The excited state mean lifetime of the chromophores, measured using a multifrequency modulation technique is of the order of 10 ns both in ethanol and in polymer matrix, suggesting that neither the electron energy levels nor the deactivation mechanism of the dye-laser are affected by the interaction with the PC host.

CP513, *Nuclear and Condensed Matter Physics,* edited by A. Messina
© 2000 American Institute of Physics 1-56396-929-7/00/$17.00

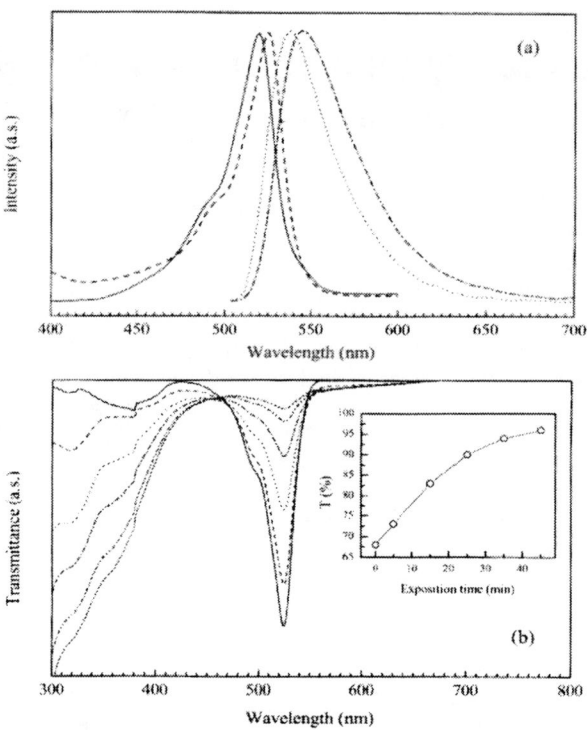

FIGURE 1. (a) absorption and excitation spectra of pyrromethene in ethanol (solid and pointed lines) and of pyrromethene in PC, spin-coated on glass (dashed and pointed-dashed lines); (b) transmittance spectra of a thin film of pyrromethene in PC as spin-coated and after exposure to UV light. The insert shows the transmittace at 525 nm vs the irradiation times.

At high dye molecules concentration, the pyrromethene solid solution in PC undergoes bleaching, observable under prolonged exposure of the films to continuos UV light sources. To avoid this solid-state dye-lasers based on pyrromethenes are pumped using pulsed laser light of longer wavelength [3]. As an example of the pyrromethene photodegradation, fig. 1b shows the trasmittance of a film of pyrromethene dispersed in PC as spin-coated and after different times of irradiation with a 100 W Xenon lamp. The unexposed film transmittance is characterized by a strong structure positioned at about 525 nm. The transmittance at 525 nm decreases progressively as the exposure time increases, approaching 95% after 45 minutes of irradiation: concurrentely an absorption tail, ascribed to the host matrix degradation, grows in the UV. Such results indicates that the material may be used as write-once media and suggests the possibility to use optical lithography to design gratings on the film surface or to obtain light confinement by selective bleaching.

The PL spectra of a number of films of pyrromethene in PC, spin-coated on glass, whose thickness ranges between several hundreds of nanometers and a few microns, are shown in fig. 2. The spectral lineshape of the emitted light, collected at

90° out of the surface normal, differs with respect to the PL peculiar of pyrromethene. Further, one can observe that acting on the film thickness, the luminescence maximum

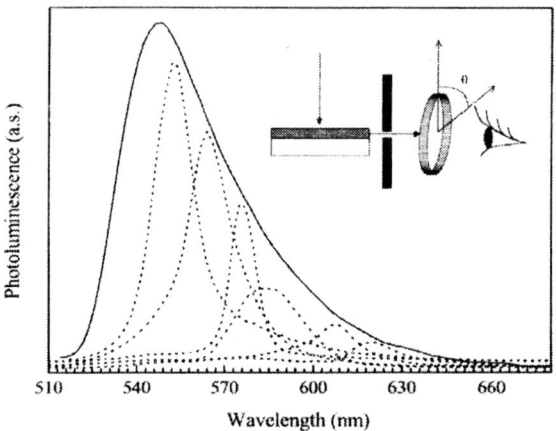

FIGURE 2. Luminescence spectrum of a film of pyrromethene dispersed in PC (solid). The dotted lines represent the arbitrarily scaled results of luminescence measurements, performed on films of different thickness, using the set-up described in the top of the figure. The linearly polarized and diaphragmed emission is collected at 90° out of the surface normal.

can be tuned between 545 and 625 nm. To explain this, one must take into account that as the polymer index of refraction (of the order of 1.58 in the red) is higher than those of the surrounding media (n=1 for air and n=1.5 for glass), the luminescence is partially confined in the polymer plane. Interference phenomena, the incidence of which depends on the film thickness, affect the emitted light. Accordingly, despite the luminescence of pyrromethene is unpolarized and peaked at approximately 545 nm, the PL observed along the film edge has a high linear polarization degree, is narrowed with respect to the non-waveguided emission and its maximum spans from the green to red according to the film thickness. As the film thickness increases the waveguided emission moves toward the red and the spectrum resolves in a higher number of structures. Such result is supported by theory, according to which the thicker is the film, the larger is the number of allowed waveguided modes. The waveguided emission spectra of a ≈ 800 nm thick film, at various polarization angles θ are shown in fig. 3. One can observe the presence of a very narrow structure, only nine nm wide, centred at about 576 nm. To support the hypothesis that such peak originates from waveguided luminescence, the insert in the top of fig. 3 shows that the relationship between the intensity of the peak and the polarization angle θ is quite well described by the $\cos^2\theta$ law, representing the theoretical dependence of a waveguided mode intensity upon the polarization angle. The intensity and the lineshape of the waveguided PL are affected by the nature of the film edges: the sharper are the film edges, the higher is the linear polarization degree and the lower is the observed band-broadening. The linear polarization degree calculated for the nine nm wide emission peak of fig. 3 is of the order of 86 %.

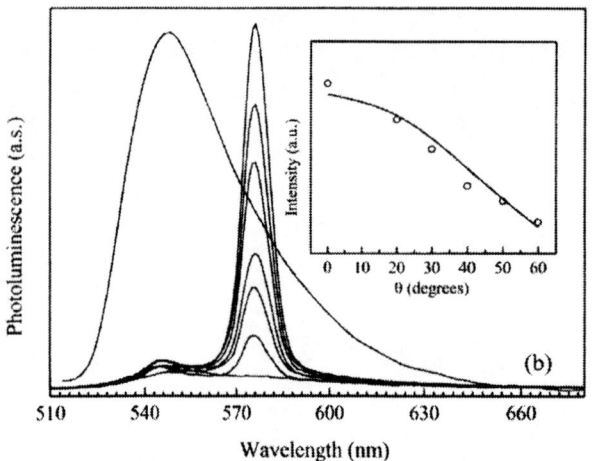

FIGURE 3. The wide PL band centred at 545 nm is the unpolarized emission peculiar of pyrromethene. The PL curves peaked at 580 nm are waveguided emission spectra, collected on a ~800 nm thick film, at different polarization angles. The insert in the top of the figure shows the experimental dependence of the PL intensity at 580 nm vs θ (open circles) and the theoretical $\cos^2\theta$ law.

In conclusion, we have shown how the PL spectra of films based on pyrromethene can be tuned between 545 nm and 625 nm using waveguiding. The partial light confinement allows narrow and linearly polarized emission band to be obtained. Such results can be improved using pulsed laser light as excitation and seem promising in view of using pyrromethene-PC as active medium in thin film lasers.

ACKNOWLEDGMENTS

This work is supported by INFM through the PAIS project and by the Regione Sicilia through CRRNSM.

REFERENCES

1. V.G. Kozlov, G. Parthasarathy, P.E. Burrows, S.R. Forrest, Y. You, and M.E. Thomson, K., *Appl. Phys. Letters* **73**, 144 (1998).

2. X. Long, A. Malinowski, D.C.D. Bradley, M. Inbasekaran and E.P. Woo, *Chem. Phys. Lett.* **272**, 6 (1997).

3. Y. Assor, Z. Burshtein and S. Rosenwalks, J. Opt. Soc. Am., 4914 (1998).

Cryoelectrophoresis:
Painless Administration of Drugs
through a Suitable Association
of Thermal and Electrical Techniques

A. Aloisi[1,4], M. Matera[5], R. Potenza[2,3,4], G. Santoro[1] and
C. Tuve'[2,3,4]

1) *AUSL n.3 – National Sanitary Service, Catania, Italy*
2) *Istituto Nazionale di Fisica Nucleare, Sezione di Catania, Italy*
3) *Centro Siciliano di Fisica Nucleare e di Struttura della Materia,Catania, Italy*
4) *Physics Department, University of Catania, Italy*
5) *Institute of Pharmacology – Medical School – University of Catania, Italy*

Abstract. A new painless technique, able to introduce drugs into the body deeply and locally, and the related equipment are presented here. This technique uses electricity for transport of the drugs, which are contained in an iced piece of water mixture that constitutes the active electrode of the dedicated equipment, which generates oscillating currents. It is explained why the association of thermal and electrical stimulation of the skin allows to overcome the disadvantages of the standard iontophoresis. Some clinical interesting results are presented.

THE TECHNIQUE

A dedicated power supply, which produces oscillating currents with non-zero mean value, is connected to an active electrode, which consists of an iced piece of water mixture, containing the drug to be administered and excipients, when needed. The return electrode consists of a large surface conductive rubber sheet, covered with a suitable conductive gel. The intensity and frequency of the current can be varied by the operator. This simple device, covered by international patent, is today used to introduce drugs inside human body to treat local diseases(1). The electrodes are to be placed directly in contact with the skin, possibly on both sides of the region to be treated. Ice must be brought to melt completely moving it gently always in contact with the skin, while the current passes through the thin surface liquid layer that forms during melting and through the interposed tissues down to the return electrode. This device associates together at least three techniques which help to obtain painless local and deep injections of drugs, as better explained below, in the next sections:
i)refrigeration of the skin at the point of application, which avoids burnings

CP513, *Nuclear and Condensed Matter Physics,* edited by A. Messina
© 2000 American Institute of Physics 1-56396-929-7/00/$17.00

and has a vasoconstrictor action with consequent reduction of blood flow in the derma;

ii) electrical stimulation of the skin, which increases its permeability to water and solutes by microporation(2);

iii) electrical dragging of the drugs brought in direct contact with the skin without interposed sponges or other containing devices, which often (if not always) represent very hard obstacles interposed along the way of the drug. So, this new technique overcomes the two main physiological disadvantages of the iontophoresis (see i) and ii)) and the main mechanical one (see iii)).

EXPERIMENTS ON NEW ZEELAND RABBITS

They were done at the Institute of Pharmacology of the Faculty of Medicine of the University of Catania and will be published elsewhere(3). We refer here on what has been already reported to the 12th and the 13th International Congresses of WCPT(1,4). The concentrations of aminopropylone (AMP), an antinflammatory drug, mol. wt. M=302,68 d, have been measured at different deepness inside the knee from epidermis down to the synovial membrane in rabbits treated with this drug in cryoelectrophoresis at the knee, and compared with those measured in blood, urine and feces. AMP levels were determined with High Power Liquid Chromatography (HPLC). Perfectly similar results were obtained for identical treatment with ^{35}S-marked mucopolysaccarides of mol. wt. ranging from 5 to 10 kd. The details of the measurements are given in ref.s(3,4) . Here we report as an example the level of concentrations of AMP 3 hours after the application: Epidermis =18.2 ± 2.9, Fascia = 13.5 ± 2.4, Capsule=9.9 ± 3.1, Synovial Membrane = 7.2 ± 1.9, Plasma = 0.27 ± 0.15, Faeces = 0, Urine = 0. The measured value show that while the systemic concentrations, represented by those in blood and excreta, are very low, the local concentrations are sensitively high and, as a matter of fact, about 100 times the systemic ones also at a relatively long time from the administration.

PHYSICS OF DRUG TRANSPORT IN CRYOELECTROPHORESIS

The drug, initially dissolved in water, separates about completely from the ice when the solution is iced, remaining dispersed in the solid mixture. The electrolytes dissociate in water, but don't remain so when brought to icing. In addition, the ion mobility in ice becomes very low, so that the conductance of the mixture is also very low, down to few μA/V, as our recent measurements have confirmed(4). In all clinical cryoelectrophoretic applications, on the contrary, currents are about 100 to 1000 times higher, driving to the conclusion that currents pass essentially through the liquid film that originates during the fusion of the ice, when it is placed in contact with hotter surfaces. In this liquid film, which goes from the metallic core of the active electrode to the skin, the water solution of the electrolyte is restored and the electric current can be efficiently transmitted. When the electric current passes through this film of melting ice and through the living tissues interposed between the active electrode and the return one, the transport of drug can occur for at least three mechanisms: i) diffusion; ii) electro-osmosis; iii) electrophoresis. Only the last mechanism needs electrolytic dissociation of the drug: the other ones allow also neutral molecules to pass. The quantity m of drug passing through the skin can be estimated by $m=fp[1+d(i)](M/zF)<i>t$, where f is the relative ionic fraction of the active drug (of mol. wt. M), d(i) is the quantity of drug passing in form of neutral molecules (which can depend on i), $p \leq 1$ is a coefficient taking into account the skin permeability for the different components of the solution, $<i>$ is the mean value of the current and t is the length of the application. When only electrolytes of comparable strength are dissolved in water, then m is about proportional to the molar concentration $c \sim f$, as early works on iontophoresis had recognized. So standing the things, the correct dosage of the drug for cryoelectrophoresis must be stated differently than in the standard systemic administration. In effect, the administered drug doesn't depend on the quantity dissolved in solution, but only on the electric charge $q=<i>t$ passed in the circuit and on the relative molar fraction f.

PHYSIOLOGY OF DRUG TRANSPORT

The association of thermal and electrical techniques in Cryoelectrophoresis shows its main advantages when the physiology of the transport is examined. First of all, as already said, also at relatively high currents (5 – 10 mA), the refrigeration caused by the ice avoids burnings of the skin, which were, on the contrary, a serious drawback of the standard iontophoresis. Furthermore, cold exercises a vasoconstrictor action which avoids that blood flow can drag away the drug from the derma while it is penetrating toward deeper tissues, and this allows the deep penetration of the drug and the consequent reduction of systemic dispersion. The AC component of the applied voltage can reinforce the vasoconstriction through the stimulation of the surface terminations of the autonomous nervous system, but its main action seems to be the production of the big increase of the skin permeability through the microporation (2) mentioned above.

CLINICAL APPLICATIONS OF CRYOELECTROPHORESIS

The main clinical applications of Cryoelectrophoresis well tested up to now are in the fields listed in Table 1 and a representative specimen of the cases treated in Catania laboratories is reported. It is seen that the percentage of success has been never less than 80%.

TABLE 1. Clinical Applications of Cryoelectrophoresis.

Field of applicat.	Disease	#cases	Success
Beauty Medicine	Cellulitis	>2,000	95 %
Urology	Induratio poenis	28	86 %
Neurorheumatology	Carpal tunnel syndr.	46	80 %
Neurorheumatology	Shoulder Periarthritis	40	100 %
Neurorheumatology	Facial palsy	3	100 %
Pain Therapy	Various	15	80 %
Antinflam. Th	Various	>200	98 %
Sport Accidents	Various	>100	85 %
Anaesthesiology	Bladder resection	2	100 %
Anaesthesiology	Phlebitis and varices	40	In progr.
Infectious diseases	Osteomyelitis	1	Full succ.

CONCLUSIONS

Due to its physical and physiological foundations, Cryoelectrophoresis looks promising to be the most advanced technique for treatment of local diseases, also in fields and cases not presented here. The most appreciable applications seem now those which allow to use of corticosteroids in high local concentrations (with neglegible systemic ones) and local therapy in cases of deficient blood circulation in the region of disease.

REFERENCES

1. Aloisi A. et al., *Proceedings of the 13th International Congress of World Confederation of Physical Therapy (W.C.P.T.), 1999, May 23rd-28th, Yokohama, Japan,* edited by The Japanese Physical Therapy Association, Tokyo, Japan, 1999, pag. 69;
2. Weaver J. C., *Proceedings of Electromed99 International Symposium, 1999, Apr. 12th-14th, Norfolk,VA, USA:IEEE Trans. on Plasma Sci.* (in press);
3. Aloisi A., Matera M., Potenza R., Santoro G., and Tuve' C., Submitted to *Physical Therapy*;
4. Matera M.et al., *Proceedings of the 12th International Congress of World Confederation of Physical Therapy (W.C.P.T.), 1995, June 25th–30th, Washington, DC, USA,* edited by The American Physical Therapy Association, Washington,DC, 1995, pag. 38.

Correlation of Structural and Electrical Transport Properties in Hydrogenated Silicon Films

F. Barreca[†], E. Fazio[†], F. Neri[†1], S Trusso[‡] and C. Vasi[‡]

(†)Istituto Nazionale per la Fisica della Materia and Dipartimento di Fisica della Materia e Tecnologie Fisiche Avanzate, Università di Messina, S.ta Sperone 31, 98166 Messina, Italy.
(‡)Istituto di Tecniche Spettroscopiche del CNR, V. La Farina 237, 98123 Messina, Italy.

Abstract. In this work we report on the correlation between the structural properties, i.e. crystalline/amorphous phase ratio and grain size, and the electrical transport properties of hydrogenated silicon thin films. The samples were deposited by means of pulsed laser ablation of a high purity silicon target in presence of hydrogen gas. Infrared spectroscopy measurements showed a monohydride preferential incorporation at the lower hydrogen pressures. The Raman spectroscopy studies of the TO phonon line suggest that crystallinity and hydrogenation of the films, deposited at room temperature, can be properly adjusted as a function of the deposition parameters. The temperature dependence of both the dark and the photo electrical conductivity shows a thermally activated behaviour, which is strictly related to the silicon microstructure and to the hydrogen content and bonding configuration.

INTRODUCTION

Two are the main goals which amorphous silicon based research is nowadays dealing with: the improvement of the materials performance and stability and the lowering of the deposition process temperature. It's well known that properties of a-Si:H rapidly deteriorates under illumination as a consequence of the Stabler-Wronsky effect, on the other hand good electronic properties can be obtained at substrate temperature in the 250-350 C range. In the last year very stable a-Si:H films were obtained by dilution of silane by hydrogen in chemical vapor decomposition based deposition processes. The presence of hydrogen during the deposition process was found to induce the growth of micro-crystalline silicon grains embedded in the amorphous phase. Moreover the amorphous/microcrystalline ratio was found to depend upon the hydrogen/silane one [1–3]. Due to the presence of energetic species in the emitted plume, the laser ablation deposition technique offers

---

1) e-mail: neri@ortica.unime.it

CP513, *Nuclear and Condensed Matter Physics,* edited by A. Messina
© 2000 American Institute of Physics 1-56396-929-7/00/$17.00

several advantages over the more conventional ones. Typically, a material of comparable quality can be grown at a lower substrate temperature [4]. In this work we report on the electronic properties of hydrogenated silicon thin films deposited by pulsed laser ablation of a poly-Si target in a controlled hydrogen atmosphere. The structural properties of the samples were investigated by Raman scattering, X-ray diffraction and infrared absorption. The results were correlated with the dark and photo electrical conductivity measurements.

EXPERIMENTAL

Samples were grown by pulsed laser ablation of high purity c-Si target. A Q-switched Nd:YAG laser beam [532 nm, 150 mJ pulse energy, 10 ns full width at half maximum (FWHM), 20 Hz repetition rate] was focused onto the target, mounted on a rotating holder. The target to substrate distance was 40 mm. Both c-Si and Corning 7049 glass were used as substrates, held at room temperature. Hydrogenation of the samples were achieved by performing the ablation process in a controlled H_2 atmosphere (P_{H_2}). Hydrogen contents up to 14% were estimated by measuring the integrated infrared absorption of the Si-H wagging band. Dark and photo-conductivity measurements were carried out in the temperature range 77-500 K. Raman scattering measurements were obtained using a U1000 ISA double monochromator equipped with an Olympus BX40 microscope. As exciting source the 514.5 nm line of an Ar^+ laser was used, spectra were collected by a LN_2 cooled 1024 elements CCD sensor.

RESULTS AND DISCUSSION

In c-Si only phonons having wavevector $\mathbf{q}$ values equal to zero are Raman active, as a consequence of both energy and momentum conservation. Its Raman spectrum is dominated by a line centered at nearly 521 cm^{-1} and attributed to the transverse optical phonon mode (TO). In systems with no long range order, the $\mathbf{q}$=0 selection rule relaxes and normal vibrational states having a spatial extent smaller than the light wavelength become first order Raman active. In amorphous silicon the Raman spectrum is characterized by a broad band in the 450-500 cm^{-1} region [5]. In Fig.1 are shown the Raman spectra for the samples deposited at 5, 0.5 and 0 mbar of P_{H_2}. The estimated H content of the two hydrogenated samples was 14% and 6%, respectively. Raman spectra are typical of a mixed phase material showing both the crystalline (I_c) and the amorphous (I_a) contributions. Upon deconvolution of the two contributions, an estimate of the crystalline fraction $X_c = I_c/(I_c + yI_a)$ can be obtained. In the previous relation y is the ratio of the Raman cross section for the amorphous and crystalline phases and depends on the grain size L, $y(L) = 0.1 + \exp[-(L/250)]$ [6]. The grain size L can be can be estimated through the shift $\Delta\omega$ of the observed TO phonon mode peak with respect to its position in c-Si, by

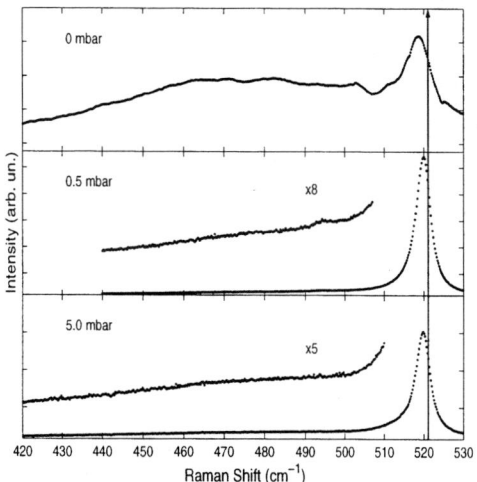

FIGURE 1. Raman spectra for samples deposited at different hydrogen partial pressures. The arrow shows the position of Raman peak in c-Si (520 cm^{-1}).

using the relation $L = 2\pi\sqrt{B/\Delta\omega}$ with $B = 2.0$ cm^{-1}nm^2. As can be seen from Fig. 1, the presence of hydrogen during the deposition drastically influences the structure of the samples. The sample without hydrogen showed the lower X_c value (12%), while samples grown in presence of hydrogen showed much higher values up to 80%. At the same time the crystalline TO line narrows and shifts towards right; L reaches values as high as 12 nm. From X-ray diffraction studies on a wider set of samples [4], we have found that the highest X_c and L values can be obtained for P_{H_2} values around 1 mbar. The above outlined structural differences should strongly influence the electronic properties of the samples. In order to shed light on the structural-electronic properties relationship, we have investigated the dark and photo electrical conductivities by evaluating their temperature dependences. The samples were highly stable, not suffering from light soaking effects and from surface oxidation once exposed to air. The dark conductivity results, reported in Fig. 2, show the presence of a thermally activated region for temperatures above 280 K. The insertion of hydrogen caused a small increase in the dark conductivity room temperature values σ_d^{RT} and a large one (over two orders of magnitude) in σ_{ph}^{RT} values. Hydrogen, therefore, removes defect states which can act as traps for the photo-excited carriers, determining the observed large increase in the conduction process. On the other hand, the activation energy E$_a$ show its minimum value and σ_d^{RT} its maximum for the 0.5 mbar deposited sample, i.e. the more microcrystalline one. Such a behaviour is characteristic of a composite structure consisting of an amorphous tissue with a high crystalline fraction, in which the reduction of the density of grain boundaries, as grains size increases, favors the tunnelling of carriers between grains and lowers the activation energy [7]. It's well known that the role

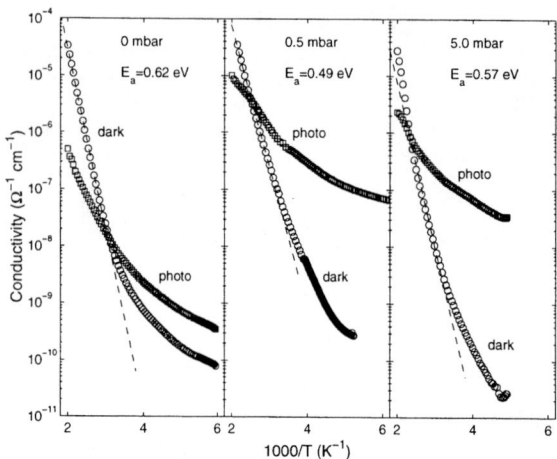

FIGURE 2. Photo and dark electrical conductivity as a function of temperature for samples deposited at different hydrogen partial pressures.

of hydrogen in a-Si:H, through its incorporation in the amorphous matrix, is to passivate defect states lying in the gap. This in turn results in an improvement of the opto-electronic properties of the material. Nevertheless, in PLD deposited samples, hydrogen seems also to influence the growth mechanism itself, playing a crucial role in determining the structure of the material. In particular we have found that, at partial pressures around 1 mbar, its presence during the deposition favors the growth of samples with a high microcrystalline fraction. This finding is similar to the behaviour observed in films prepared by plasma assisted CVD techniques as a function of silane/hydrogen gas mixture dilution [1–3].

Acknowledgements - This work has been partially supported by Consiglio Nazionale delle Ricerche (Progetto Finalizzato MADESS II).

REFERENCES

1. Guozhen, Y.,Lorentzen, J.D., Jing, L., and Daxing, H., *Appl. Phys. Lett.* **75**, 492 (1999).
2. Wang, K.C., Hwang, H.L., Leong, P.T., and Yew, T.R., *J. Appl. Phys.* **77**, 6542 (1995).
3. Hsu, K.C., Chang, H., and Hwang, H.L., *J. Appl. Phys.* **73**, 4841 (1993).
4. Trusso, S., Vasi, C., and Neri, F., *J. Vac. Sci. Tech. A* **17**, 921 (1999).
5. Cardona M., *Light Scattering in Solids II*, edited by M. Cardona and G. Gunterodt, Topics Appl. Phys. Vol. 50 (Springer, Berlin, 1982).
6. Bustarret, E., Hachicha, M.A., and Brunel, M., *Appl. Phys. Lett.* **52**, 1675 (1988).
7. Garrido, B., Perez-Rodriguez, A., Morante, J.R., Achiq, A., Gourbilleau, F., Madelon, R., and Rizk, R., *J. Vac. Sci. Tech. B* **16**, 1851 (1998).

Low Energy Vibrational Excitations In Silver Borate Glasses

A. Bartolotta*, G. Carini, G. D'Angelo, G. Salvato*, G. Tripodo

Dipartimento di Fisica and INFM, Università di Messina, 98166 S. Agata, Messina, Italy
**Istituto di Tecniche Spettroscopiche del C.N.R., 98122 Messina, Italy*

Abstract. Analysis of low temperature specific heat data (1.5 K-25 K) in two glasses of the AgI-Ag$_2$O-B$_2$O$_3$ system reveals that the temperature dependencies of C$_p$ are in qualitative agreement with the predictions of the Soft Potential Model. It has been also found that a decreasing connectivity of the network enhances the deviations of C$_p$ from a T^3 behavior.

I. INTRODUCTION

Anomalies in a wide range of physical properties at low temperatures are a dominant feature of glasses [1]. The most characteristic features are the linear behavior of the specific heat and the T^2-dependence of the thermal conductivity in the temperature region below 1 K, which are well described by the phenomenological theory of two-level systems (TLS's). Above 1 K, however, the thermal properties differ greatly from the predictions of the Debye theory and cannot be accounted for by the TLS model. The existence of additional low energy vibrations besides acoustic phonons must be included in order to explain the excess of specific heat and the "plateau" in the thermal conductivity in the region between 1 and 20 K. Furthermore it has been shown that there is a common origin for the excess specific heat and the "Boson Peak" observed in the low frequency (below 100 cm^{-1}) Raman and inelastic neutron scattering spectra of glasses [2,3]. The distortion of the structures due to the topological disorder in glasses has the main effect to introduce additional excitations in the low energy region of the density of vibrational states g(ω) (DVS), whose nature - localized or delocalized [4], harmonic or anharmonic [5] - is a hot topic at present widely debated. In particular the DVS deduced from inelastic neutron scattering data, when reported as g(ω)/ω^2 vs ω, exhibits a broad maximum indicating again a non-Debye behavior as a universal characteristic of the glassy state. To establish a common microscopic origin for the vibrational anomalies of glasses, studies of a wide range of properties including the particularly instructive thermal effects are needed. The present concern is to report a study of the low energy excitations in AgI-Ag$_2$O-B$_2$O$_3$ glasses using specific heat measurements. Several theoretical models have been proposed to account for the universal form of the excess DVS, the most attractive and general assuming the presence of localized vibrations which arise from soft anharmonic potentials (SPM) and provides a unified description of both the tunneling

CP513, *Nuclear and Condensed Matter Physics,* edited by A. Messina
© 2000 American Institute of Physics 1-56396-929-7/00/$17.00

and low-frequency vibrational states [6]. The present experimental observations have been interpreted in terms of the SPM and found to be consistent with its predictions.

II. RESULTS AND DISCUSSION

Two glasses of the system $(AgI)_x(Ag_2O \cdot 4B_2O_3)_{1-x}$, X being the molar fractions were prepared in the *binary* (X=0) and *ternary* (X=0.4) form using the same procedure elsewhere described [7a]. The specific heat was measured in the range between 1.5 and 25 K using a calorimeter which operated by the thermal relaxation method. The experimental results of specific heat, obtained for silver borate glasses and plotted as $C_p(T)/T^3$, are compared in Fig. 1a. The Debye contributions have been evaluated using the Debye temperatures Θ_D, as determined by the ultrasonic waves velocities at room temperature: Θ_D=335 K for $Ag_2O \cdot 4B_2O_3$ and Θ_D=237 K for the glass with AgI. They are indicated by arrows in Fig.1a and evidence in both the samples an excess specific heat having the shape of a broad peak. The addition of AgI to the silver borate glass enhances the excess specific heat and lowers the maximum temperature: the hump shifts from about 8 K to about 6.5 K. In the SPM model [6], the low-frequency vibrational dynamics in a glass is described in terms of localized

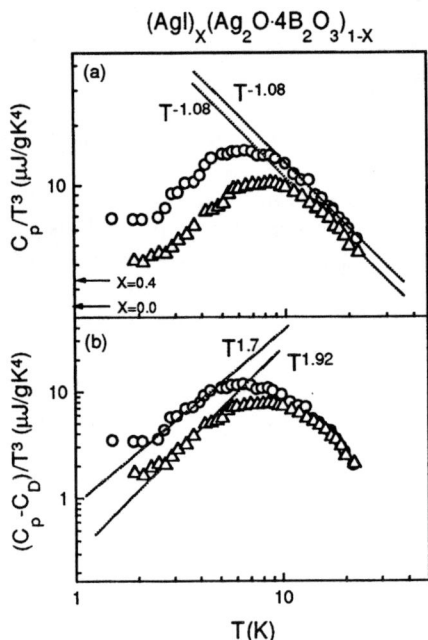

FIGURE 1. (a) Temperature dependencies of C_p/T^3 for $(AgI)_x(Ag_2O \cdot 4B_2O_3)_{1-x}$ glasses: X=0.0, (Δ); X=0.4, (O). (b) Excess specific heat $(C_p-C_D)/T^3$ vs temperature for $(AgI)_x(Ag_2O \cdot 4B_2O_3)_{1-x}$ glasses.

excitations which coexist with ordinary phonons. These excitations are realized in anharmonic or soft atomic potentials, represented by a fourth degree polynomial:

$$V(x) = E_0 \left[\eta \left(\frac{x}{a} \right)^2 + \xi \left(\frac{x}{a} \right)^3 + \left(\frac{x}{a} \right)^4 \right]$$

where x is an appropriate configurational coordinate having the dimension of a length, E_0 is of the order of the atomic binding energy, η and ξ are small dimensionless parameters with a random distribution of their values. The expression of $V(x)$ includes both double- and single-well potentials. The lowest-energy excitations (double wells) correspond to the well known two level systems (TLS) with a nearly constant density of states. At higher energies there are the quasilocal harmonic oscillators (HO) which occur in single-well potentials and have a density of states following a ω^4- dependence. Both the excitations interact with phonons in the same way, the interaction being either resonant or relaxational. At temperatures $(T<<W/k_B)$ small compared with the energy W, which sets the crossover between TLS and HO excitations, a linear T-dependence for the specific heat is determined by the contribution of TLS's. At higher temperatures, when $T>>W/k_B$, excitations with energies larger than W are responsible for a large contribution to the specific heat provided by the HO's whose rapidly increasing density of states leads to a T^5 contribution over that proportional to T^3 of the usual Debye-like phonons:

$$C_{HO}(T) = \frac{8\pi^6 k_B}{63\sqrt{2}} P_0 \eta_L^{5/2} \left(\frac{k_B T}{W} \right)^5$$

The fourth-power HO density of states changes to a "*collective*" or delocalized vibrational excitations density of states for an energy scale higher than a some characteristic energy E_d. This is because the average distance between the HO's with $E \geq E_d$ becomes smaller than the wavelength of a phonon with about the same energy. Consequently an excitation can directly jump from an oscillator to another producing the "*delocalization*" of the HO excitations. When $E \geq E_d$, the excitations are believed to be neither phonons nor localised soft modes, but rather strongly interacting modes resulting from HO's which have lost their quasilocal character. The DVS must be reconstructed and a linear dependence on the energy results [6]. In turn this should correspond to a T^2-dependence for the *total* vibrational specific heat in the high temperature region. Crossover from the low temperature linear dependence to a T^2-behaviour through a T^5-dependence would lead to a minimum at a temperature T_{min} and to a maximum at a temperature of the order of E_d in the temperature dependence of C_p/T^3. From the position T_{min} of this minimum the value of the characteristic energy W can be determined: $W \sim 2k_B T_{min}$ [6]. For the two glasses analysed, it is not possible to examine the linear dependence predicted below T and to determine precisely T_{min} due to the restrictions on the available temperature range. However the experimental data, shown in Fig. 1a, give a clear indication for a minimum ranging between 1.5 K and 2 K, in close agreement with previous observations in a different composition ($Ag_2O \cdot 2B_2O_3$) of the same system and in sodium borate glasses [7b]. It can be seen that, in the region between the expected minimum and the maximum, the experimental data of $(C_p\text{-}C_{Debye})/T^3$ follow a T^{-2} dependence (see Fig. 1b), while the data of C_p/T^3 above the maximum follow a T^{-1} dependence (see Fig. 1a). These T^{-5} and T^{-2} behaviors of $C(T)$ revealed in the experimental data appear to be a consequence of extra vibrational states having a fourth-order frequency dependence which tends to become linear with increasing energy. This can be interpreted as a consequence of the loss of localization in the population of soft vibrations and is close to that predicted for the quasiharmonic density of states in the SPM [6].

An explanation for the relevant increase of C_p/T^3 in the *ternary* glass can be given by considering the role of AgI as modifier of the borate network. In fact the analysis of the vibrational dynamics [7c] led to believe that both the AgI and the borate matrix preserve their local structures. This implies that AgI tetrahedra, weakly bonded to the BO_3 or BO_4 groups of the borate matrix, must be coordinated in order to allow the fast ionic diffusion typical of this kind of glasses: the connectivity of the borate network decreases in consequence of the inclusion of less tied AgI polyhedra. This gives evidence of the fact that the source of the anomalous specific heat lies in additional low-energy vibrational states whose density decreases with increasing connectivity. The revealed features are consistent with the supposed nature of the low energy vibrations, which are ascribed to quasi-local modes in quasiparticles frozen in the glass. In this context the coherence breakdown due to the inclusion of AgI polyhedra corresponds to the formation of less tied clusters of atoms in the network, which could be the source for the additional modes.

REFERENCES

1. For a review see, *Amorphous Solids : Low-Temperature properties*, edited by Phillips W. A., (Springer, Berlin, 1981); *Phil. Mag. B77 (1998), Special Issue: Sixth Int. Workshop on Disordered Systems*, edited by A. Fontana and G. Viliani.

2. Sokolov A. P., Kislink A., Quitmann D., and Duval E., Phys. Rev **B 48**, 7692 (1993).

3. Fontana A., Rossi F., Carini G., D'Angelo G., Tripodo G., and Bartolotta A., Phys. Rev. Lett. **78**, 1078 (1997).

4. Benassi P., Krisch M., Masciovecchio M., Mazzacurati M., Monaco G., Ruocco G., Sette F., Verbeni R., Phys. Rev. Lett. **77**, 3835 (1996); Foret M., Courtens R., Vacher R., Suck J. B., Phys. Rev. Lett. **77**, 3831 (1996)

5. Taraskin S. N. and Elliot S. R., Phys. Rev. **B 59**, 8572 (1999).

6. Karpov V. G., Klinger M. I., and Ignatiev F. N., Zh. Expt. Teor. Fiz. **84**, 760 (1983) [Sov Phys. JEPT **57**, 439 (1983)]; Parshin D. A., Phys. Solid State **36**, 991 (1994).

7. (a) Carini G., Cutroni M., Federico M., Galli G., Tripodo G., Phys. Rev. B **30**, 7219 (1984); (b) Bartolotta A., Carini G., D'Angelo G., Tripodo G., Solid State Ionics **105**, 97 (1998); (c) Carini G., Cutroni M., Fontana A., Mariotto G., Rocca F., Phys. Rev. B **29**, 3567 (1984).

ESR Evaluation Of Stable Free Radicals Produced By Ionizing Radiation In Multifunctional Substances. Application For Absorbed Dose Measurements in Radiotherapy.

Bartolotta A[1], Brai M[2], De Caro V[3], D'Oca C[1], Giannola L[3] Teri G[2]

(1) Dipartimento Farmacochimico, Tossicologico e Biologico, via Archirafi 32 90123 Palermo Italy
(2) Istituto della Biocomunicazione e Unità INFM Palermo
(3) Dipartimento di Chimica e Tecnologie Farmaceutiche
Università di Palermo

Abstract. Electron Spin Resonance dosimetry is a useful system for measuring absorbed dose in radiotherapy. This work describes the results obtained at the University of Palermo regarding an experimental study aimed to optimize the properties of alanine based dosimeters and to analyze other materials, that could be alternatives to alanine.

INTRODUCTION

The use of ionizing radiation in cancer radiotherapy has gained today as much importance as surgery and pharmacological treatments; photon and electron beams of various energy, produced by radioactive sources or linear accelerators, are commonly used worldwide; even neutron, ion and proton beams are also employed in particular cases, when these radiation allow a better conformal treatment. Anyway, whatever is used, the efficacy of the radiotherapy (i.e. the probability of tumor control without healthy tissues complications) can be seriously compromised if the overall uncertainty in the absorbed dose to the target volume is higher than ±5% (1). This strict requirement can be fulfilled only if the dosimetric characteristics of the used radiation beam, such as dose rate and dose distribution in water phantom, are accurately known. To this purpose, it should be used a dosimetric system with adequate properties concerning linear range, fading, size, precision, dependence on radiation quality. The electron spin resonance (ESR) dosimetry with alanine has proved to be useful for radiotherapy applications, included *in vivo* dosimetry (2), provided dose values of few gray or less can be measured with high precision. This paper describes work being carried out aimed to realize ESR alanine dosimeters with optimized properties and to analyze the dosimetric behavior of other substances with various functional groups; the stability of free radicals formed after irradiation was also analyzed.

CP513, *Nuclear and Condensed Matter Physics,* edited by A. Messina
© 2000 American Institute of Physics 1-56396-929-7/00/$17.00

MATERIALS AND METHODS

Solid state dosimeters were obtained by pressing in a stainless steel die a blend of the radiation sensitive substance, polyethylene and magnesium stearate (94%, 5% and 1% in weight respectively, all in powder form). A detailed description of the entire procedure has been described in a previous paper (3).The ESR spectra were recorded with a Brucker ECS106 spectrometer equipped with a TE_{102} rectangular cavity operating at 9.7 GHz. A quartz holder with quartz spacers was used for reproducible dosimeter positioning inside the cavity. For each investigated substance, the recording parameters were properly chosen to obtain the strongest signal with negligible distortion. The ESR spectrum recording was repeated four times, the dosimeter being rotated by 90° inside the cavity after each acquisition, and the peak-to-peak signal height H was measured each time; the mean value H_m was normalized to the dosimeter mass M and to the signal height H_S of a steady sample, to eliminate any interspecimen differences coming from differences in alanine content or from spectrometer sensitivity fluctuations. The resulting value

$$H_R = \frac{H_m}{M H_S}$$

was used as the dose sensitive parameter. Irradiation to ^{60}Co gamma rays and to electrons were performed at the Oncology Hospital of Palermo; proton irradiation at the National Institute of Nuclear Physics in Catania, where an irradiation facility for the treatment of melanoma will start to operate by the end of 2000.

RESULTS

A comparative test was carried out using various amino acids and multifunctional substances as radiation sensitive materials in the dosimeters, concerning the formation of stable free radicals after irradiation, and the ESR signal to noise ratio (S/N). For each one of them, the value of S/N after irradiation at 30 Gy with ^{60}Co was measured; on the basis of this preliminary test, we have focused our attention to alanine and ammonium tartrate dosimeters (4), that gave S/N of about 20, much higher than the other substances (S/N between 1 and 6). A comprehensive comparison of their dosimetric properties was carried out.

Mass Distribution And Zero-Dose Signal

One group of 300 dosimeters was prepared with alanine (AL) and another with ammonium tartrate (AT). Mass and dimensions of 50 randomly chosen samples from each group were measured; results are reported in Table 1. The AT dosimeters have higher mass than AL ones, and a more symmetric distribution around the mean value. Moreover, they showed lower friability and fragility. The ESR signal intensity of twelve non irradiated dosimeters for each type was also measured. The lowest dose that produces a signal significantly different from noise is 2 Gy and 3 Gy for AL and AT, respectively, on the basis of the "zero-dose" signal mean value and fluctuations.

TABLE 1. Dosimeters mass and dimensions.

Parameter	Alanine	Ammonium tartrate
Thickness/mm	0.87 ± 0.04	0.83 ± 0.05
Diameter/mm	4.90 ± 0.05	4.90 ± 0.05
Mass/mg	17.3 ± 1.1	19.8 ± 1.0

Linear Range, Dose Resolution, Accuracy

The value of H_R was studied as a function of dose in the $(0.5 - 40)$ Gy range – of interest in radiotherapy - with the ^{60}Co source. For each dose value three dosimeters of each type were irradiated at the same time in a Perspex phantom. Absorbed dose rate at the mid-plane of dosimeters was previously measured with a calibrated ionization chamber, with an overall uncertainty of $\pm 4\%$ (95% confidence level). To compare the performance of AL and AT dosimeters, the S/N ratio was calculated and analyzed as a function of dose; for both types of dosimeters the linear range was found to be between 5 and 40 Gy (higher doses were non tested). The AT dosimeters showed higher S/N values than AL ones and the corresponding fitting straight line equation had a greater slope; they seems therefore to be more promising for applications in the low dose range. The results in the linear range are shown in Figure 1.

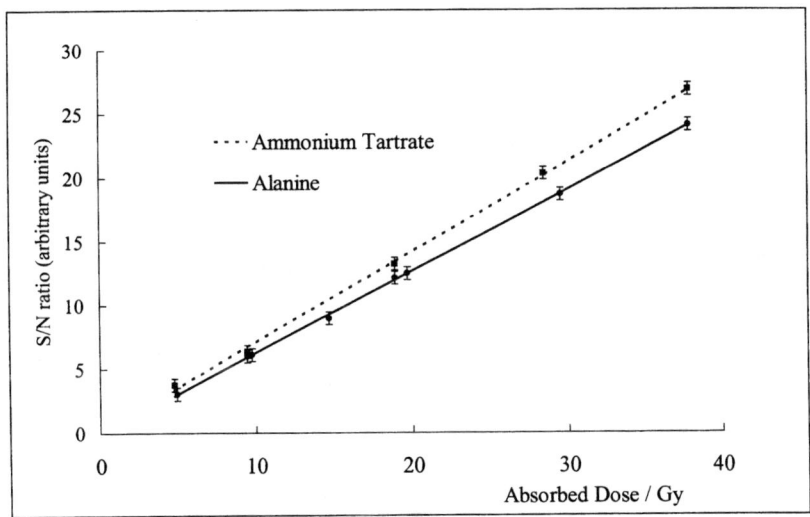

FIGURE 1. Signal to noise ratio vs absorbed dose for alanine and ammonium tartrate dosimeters irradiated with ^{60}Co gamma rays.

Dose resolution, i.e. the least difference between two dose values that can be meaningfully distinguished, was evaluated as the dose increment that causes a variation in H_R of at least one standard deviation. Dose resolution at 5, 15 and 35 Gy is reported in Table 2, according to the linear fit.

TABLE 2. Dose Resolution (Gy).

Absorbed Dose /Gy	Alanine	Ammonium Tartrate
5	0.3	0.6
15	0.4	0.6
35	0.7	0.7

The ESR signal stability was checked measuring periodically H up to six months after irradiation; no significant modification in the ESR signal intensity was detected for both type of dosimeters. Taking into account all the contributions, the overall uncertainty in dose evaluation was estimated to be ±5% (95% confidence level).

Response To Charged Particle Irradiation

The ^{60}Co calibration function was used to calculate absorbed dose even when AL and AT dosimeters were irradiated to 21 MeV electrons and 28 MeV protons, and these values compared with those measured with ionization chambers, to emphasize any dependence of the response on beam quality. Preliminary results didn't show any effect with electrons, whereas under- and over-estimation of dose seems to be present with proton beams.

DISCUSSION

The results obtained on the properties of alanine and ammonium tartrate dosimeters pointed out that both systems are suitable for photon, electron and proton beams used in radiotherapy. More efforts are required to improve dose resolution of ammonium tartrate, that, on the other hand, seems to produce a higher S/N than alanine.

ACKNOWLEDGMENTS

This work was supported with a grant of Ministero Università Ricerca Scientifica e Tecnologica. We sincerely thank Vittorio Caputo for photon and electron irradiation, and Giacomo Cuttone for proton irradiation.

REFERENCES

1. International Committee on Radiation Units and Measurements, "Determination of absorbed dose in patient irradiated by beams of X or gamma rays in radiotherapy procedures", Bethesda, USA, 1976.

2. "ESR Dosimetry and Applications", Proceedings 4th Int. Simp., Munich, *Appl. Radiat. Isot.* , 1996, pp.1151-1687.

3. Bartolotta, A., Brai, M., Caputo, V., De Caro, V., Giannola, L.I., Rap, R., and Teri, G., *Radiat. Prot. Dos.* **84**, 293-96 (1999).

4. Olsson, S.H., Bagherian, S., Carlsson, G.A., and Lund, A., *Appl. Radiat. Isot.* **50**, 5-9 (1999).

Thermoluminescence Curves Of LiF Dosemeters Irradiated With Proton Beams

A. Bartolotta*, S. Basile[§], M. Brai[§], G. Bruno[§], G. Cuttone[^]
R. Rey Zabalza* and G. Teri[§]

[§]Istituto della Biocomunicazione and Unità INFM, Università di Palermo
Via G. Parlavecchio 3, 90127 Palermo,Italy;
*Dipartimento Farmacochimico Tossicologico e Biologico,Università di Palermo.
[^]LNS-INFN, Catania, Italy

Abstract. A glow curve analysis has been carried out to study the response of some LiF dosimeters when irradiated by radiations with different LET. The aim of the study was to have a better understanding of the role of material characteristics in determining their glow curve response when irradiated with ^{60}Co gamma rays and proton beams. Glow curves for TLD-100 and GR-200 LiF dosimeters have been analyzed and deconvolution of the peak structure has been performed within the frame of first order kinetics.

INTRODUCTION

Last years have seen a major development of radiotherapy with proton beams since it allows irradiating small size targets close to critical organs. Within the more general context of dosimetry in radiotherapy, the termoluminescence dosimetry (TLD) has attracted a lot of attention because it allows using dosemeters with large linearity range, small size and good reproducibility. Last but not least, they also have a high sensitivity and can be considered, to a large extent, as tissue equivalent. For the above reasons, the study of thermoluminescent response of different materials to non-standard radiation, i.e. proton and ion beams (carbon and oxygen), becomes important, to perform comparisons with response to radiation widely used in radiotherapy, i.e. photons and electrons.

Two different dopings of lithium fluoride are mostly used: LiF:Mg,Ti and LiF:Mg,Cu,P. Depending on the dopant concentrations, these materials will show different response to the radiation, since the glow curve will be mainly determined by the defect structure that will show up in it by their activation energies and temperatures. The two materials are also characterized by a different response as a function of dose; LiF:Mg,Ti will generally show a supralinear behavior, while LiF:Mg,Cu,P will show a sublinear behavior. The radiation Linear Energy Transfer (LET) is another source of variability in TLD response. In fact it is well known that radiations with different LET produce different ion densities that take place by the transfer of the F center electrons to the impurity ions in LiF dosimeters.

The glow curve of an irradiated thermoluminescent dosimeter is made up by contributions due to the activation of different electron-hole traps of various energies.

CP513, *Nuclear and Condensed Matter Physics*, edited by A. Messina
© 2000 American Institute of Physics 1-56396-929-7/00/$17.00

A good deconvolution of the glow curve is then needed to obtain a reliable estimate of the kinetic parameters (activation energy, peak temperature and FWHM). The trapping levels are within the forbidden gap and can be occupied by carriers (electrons or holes) after irradiation. These trapped carriers can be released if thermal energy is supplied. In LiF:Mg,Ti ten glow peaks are approximately observed in the range between room temperature and 400 °C and each of them is associated with an observed absorption band and, therefore, a specific trap level. In LiF:Mg,Cu,P the glow curve is less structured and six peaks will give a good deconvolution. The temperature range used is between room temperature and 240 °C, to prevent chip damaging.

MATERIALS AND METHODS

Two commercially available thermoluminescent dosimeters have been used in this work, namely TLD-100 (LiF:Mg,Ti, 7.5% ^{6}Li and 92.5 % ^{7}Li, Mg^{2+} at 180 ppm concentration and Ti^{4+} at 10 ppm) and GR-200A (LiF:Mg,Cu,P, 7.5% ^{6}Li and 92.5 % ^{7}Li, 0.05 mol % of Cu^{2+}, 0.2 mol % of Mg^{2+} and 2.3 mol % of P^{5+}).

Three different sources have been used for chip irradiation; namely, a ^{60}Co source to calibrate dosimeters, a 26.7 MeV and a 18 MeV initial energy proton beam. Standard international recommendations have been followed for dose assessment in calibration. For TLD-100 a one hour at 400 °C. For the GR-200 the annealing procedure consists of a single ten minutes at 240 °C step. In both procedures attention must be paid to have a fast cooling rate to avoid sensitivity loss. A Harshaw 3500 was used as TLD reader. For both kinds of dosemeters a heating rate of 1 °C s^{-1} has been used. The operating temperature ranges were 50–320°C for the TLD-100 and 50–240°C for the GR-200.

A Randall and Wilkins first order kinetics has been used to fit data and to accomplish the deconvolution of glow curves. This model assumes the presence of a single defect within the forbidden gap of the phospor and that the thermally stimulated electron has a negligible chance of re-trapping. The trapped electrons have a Maxwellian thermal energy distribution, moreover, the Podgorsak-Moran-Cameron approximation has been used to get analytic expressions for the single peak shape.

RESULTS AND DISCUSSION

In figure 1 glow curves of TLD 100 and GR 200 detectors are compared after irradiations with relative low doses of ^{60}Co gamma rays (8 Gy and 4 Gy respectively). The glow curve structure of the LiF detectors depends strongly on dopant type and concentration. TLD 100 were deconvoluted in eigth peaks, while GR 200 glow curves were well fitted with six peaks. The dots are the experimental data and the continuous line is the sum function of first order kinetic curves. The TLD100 chips were calibrated in the (0.5-5.0) Gy dose range (five chips for each selected dose) and the GR-200 in the (0.5–3) Gy dose range.

For doses higher than 10 Gy a supralinearity behaviour appears for TLD100, while over 8 Gy GR-200 dosemeters show a sublinearity. Differences between nominal and calculated doses were not higher than 3%.

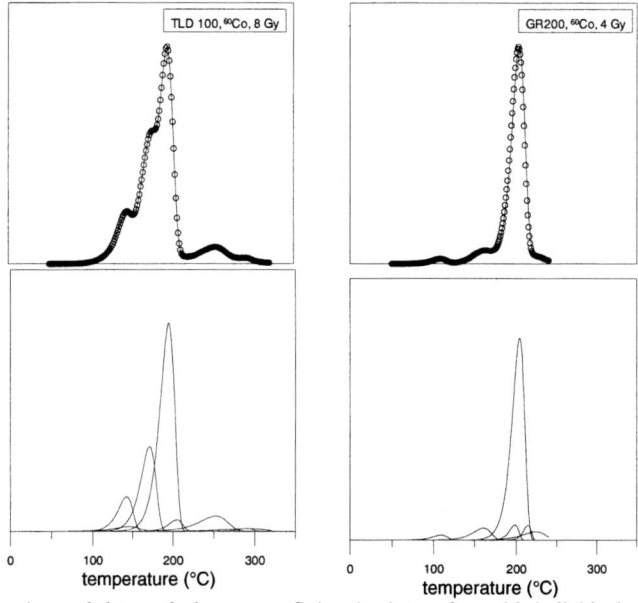

Figure 1. Experimental data and glow curve fitting (top) together with individual peak deconvolution (bottom) are shown for a TLD100 (left) dosimeter irradiated at 3 Gy and a GR200A (right) dosimeter irradiated at 2 Gy. A ^{60}Co gamma ray source has been used in both cases. The different peak structure of the glow curve is clearly visible.

TABLE 1. Activation energies (E), as obtained from fitting of glow curves, are shown for TLD100 (irradiation dose 8 Gy) and of GR200 dosimeters (at a dose of 4 Gy). The data, averaged on five chips, are relative to 3-8 peaks for TLD100 and 1-5 for GR200.

TLD-100 E (eV)				GR-200 E (eV)			
T (K)	[60]Co	Protons 26.5 MeV	Protons 17.4 MeV	T(K)	[60]Co	Protons 26.5 MeV	Protons 17.4 MeV
420	1.39	1.08	1.14	382	1.47	0.91	0.79
444	1.86	2.02	2.06	433	1.51	1.49	1.48
467	2.20	2.39	2.18	472	3.46	3.40	3.23
477	2.56	3.37	1.92	478	2.66	2.55	2.52
519	1.00	0.78	1.47	487	4.12	4.17	3.90
542	2.33	1.51	1.71	496	1.41	1.86	1.78

In table 1 the activation energies are shown for the predominant peaks in TLD100 type dosimeter and the GR200. The peak temperatures are the same for all three irradiation fields, so indicating that the structure of the defect is unchanged. It is worth to point out that the activation energies at low and high temperatures for the lowest energy proton beam decrease in TLD100 and .remain constant in GR200.

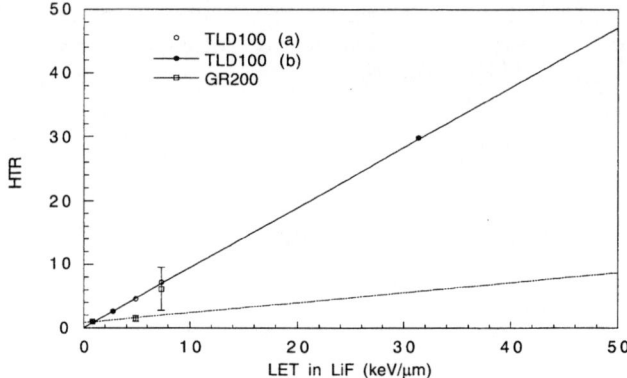

Figure 2. High temperature ratio HTR plotted against LET of absorbed radiation for TLD100, (a) our experiments, (b) data of ref.2, and GR200.

In figure 2 the ratios of the area of the high temperature peaks to the main dosimetry peak are reported. After absorption of radiation with higher LET, peak ratios change significantly for the TLD100, while the values for GR200 vary slowly. The large vertical error bar over the point of HTR of GR200 calculated for protons of 17.4 MeV, LET=7.34 keV μm-1, indicate that GR200 was easily damaged.

Data of HTR and activation energy indicate that TLD100 is suitable for mixed field of radiation to separate low and high LET component, while at very low doses GR200 can be a better candidate for monoenergetic field due to its independence on LET.

ACKNOWLEDGMENTS

The authors gratefully acknowledge the collaboration of Dr. Vittorio Caputo from the Servizio di Fisica Sanitaria, Azienda di Rilievo Nazionale ad Alta Specializzazione "Civico, Giuseppe Di Cristina e Maurizio Ascoli" in Palermo. This work has been partially supported by Istituto Nazionale di Fisica Nucleare, TLIP project.

REFERENCES

1. Bartolotta A., Barone Tonghi L., Brai M., Cuttone G., Fattibene, P., Onori S., Raffaele L., Rovelli A., Sabini M. G., Teri G. Response characteristic of thermoluminescence and alanine-based dosemeters to 16 and 25 MeV proton beams. Rad Prot Dos, 85, 353-356 (1999)

2. Schoner W., Vana N., Fugger M. The LET dependence of LiF:MG,TI dosemeters and its application for LET measurements in mixed radiation fields. Rad Prot Dos, 85, 263-266 (1999)

PHASE-DIFFERENCE EFFECT IN TWO-COLOR DETACHMENT OF H⁻

S. Bivona, R. Burlon and C. Leone

Istituto Nazionale di Fisica della Materia and Dipartimento di Energetica ed Applicazioni di Fisica
Viale delle Scienze 90128 Palermo, Italy

Abstract. We show photoelectron energy spectra of a negative ion irradiated simultaneously by a weak, photodetaching, electromagnetic field of frequency ω_{H1} and an intense low-frequency bichromatic field (LF) when the frequencies ω_{L1} and ω_{L2} of its components are commensurate.

INTRODUCTION

In the last years different schemes have been proposed to study and observe quantum control of elementary atomic processes; in particular the so-called phase control is based on the interference between different excitation pathways followed by the atomic system to go from a given initial state to the same excited state [1]. The control of the process is achieved by opportunely changing the relative phases of two or more electromagnetic radiation sources. Recently, we have proposed a scheme of excitation for observing interference in two-color experiments involving states of negative ion lying in the continuum. In this schemes, a single photon of frequency excites into the continuum a negative ion in the presence of a bichromatic low-frequency (LF) field whose components have commensurate frequencies ω_L and $2\omega_L$, and relative phase δ [2]. Calculations of photodetachment cross section of a model negative ion simulating have shown that, by varying the relative phase δ between the components of the bichromatic LF field, considerable modifications occur in the angular distributions of the electrons ejected into various photodetachment channels. These modifications come from the interference of the different pathways the electron follows to reach the same final state. As far as the photoelectron spectra are concerned, it has been found that when $\delta = \pi/2$ they show a pronounced, narrow maximum when the photoelectron absorbs by the bichromatic field such an amount of energy $Nh\omega_L$ almost equal to the ponderomotive shift. It is the aim of the present communication to discuss more deeply this last feature. It will be shown that when the ratio of the frequencies of the bichromatic field components is changed, by opportunely choosing their relative phase δ, a sharp peak appears in the electron energy spectra when the photoelectron absorbs by the LF field an energy equal to the ponderomotive shift.

CP513, *Nuclear and Condensed Matter Physics*, edited by A. Messina
© 2000 American Institute of Physics 1-56396-929-7/00/$17.00

THE PHOTOELECTRON ENERGY SPECTRA

Photoelectron energy spectra when the bichromatic and the photodetaching electric fields are respectively taken as $E_{LF}=E_0$ [$\sin(\omega_L t) + \sin(2\omega_L t + \delta)$] z and $E_H(t)=E_{0H}\cos(\omega_H t)$ z have been obtained in Ref.2 under simplifying assumption that will be recapitulated below. Fig.1 shows typical electron energy spectra obtained for two different values of δ , and ω_H far from the field-free photodetachment threshold. The electromagnetic fields were treated in dipole approximation and were assumed to be linearly polarized along the z-direction . Effects connected with the rescattering of the ejected electron by the residual atom were ignored, and the electron-atom interaction was approximated by a zero range potential that supports only a bound state with energy I_0 taken equal to -0.75 eV in our calculations. The main steps leading to the photodetachment cross section have been outlined in [2], therefore we quote the basic formulas giving the probability amplitude T_N of the event in which the electron is detached in a single step after absorbing a single photon of energy $\hbar\omega_H$ and exchanging all the possible numbers of photons n_1 and n_2 of frequency, respectively, ω_L and $2\omega_L$, such that $n_1+2n_2=N$. So, the probability amplitude may be expressed as (hereafter atomic units will be used)

$$T_N = \frac{1}{2\pi} \int_{-\pi}^{+\pi} \exp\left\{-iN\alpha + i\frac{\zeta(\alpha,\delta)}{\omega_L}\right\} M(q_N,\alpha,\delta)d\alpha \tag{1}$$

$$\zeta(\alpha,\delta) = -(\frac{q_N^2}{2} + \Delta)\alpha + \frac{1}{2}\int_\alpha [q_N + K_L(\alpha',\delta)]^2 d\alpha' \tag{2}$$

$$M(q_N) = -iB(\frac{b}{\pi^2})^{1/2} \frac{E_{0H}[q_N + K_L(\alpha,\delta)] \cdot \hat{z}}{\left\{b^2 + [q_N + K_L(\alpha,\delta)]^2\right\}^2} \tag{3}$$

with B the empirical constant equal to $(2.65)^{1/2}$, $b = (-2mI_0)^{1/2}$, $K_L(\alpha,\delta)$ the oscillating momentum imparted by the LF to the electron, $q_N^2/2$ the drift energy of the electron and $\Delta=(5/16)E_0^2/\omega_L^2$ the ponderomotive potential due to the LF field. To arrive at Eq.(1) the weak e.m. field has been treated perturbatively and the final continuum states describing the ejected electron have been approximated by the nonrelativistic Volkov wave function. Proceeding in the usual way the total cross section are arrived at. As it is easily seen from Fig.1, for $\delta=\pi/2$ the electron spectrum exhibits a peak when the energy given to the electron by the LF field is almost equal to the ponderomotive shift. In order to illustrate this feature, we evaluate the integral of Eq.(1) by means of the stationary phase method. The main contribute to the integral comes from the points of stationary phase satisfying

$$\frac{1}{2}\left[\mathbf{q}_N + \mathbf{K}_L(\alpha_s,\delta)\right]^2 = \omega_H - I_0 \tag{4}$$

from which it follows that

$$q_N \cos\vartheta + K_L(\alpha_s,\delta) = \pm\sqrt{2\left[(\omega_H - I_0)\cos^2\vartheta - (N\omega_L - \Delta)sin^2\vartheta\right]} \tag{5}$$

where ϑ denotes the angle between q_N and z, and the sign on the right-hand side of Eq.(5) are associated with the values the projection of the kinetic momentum of the photoelectron along the z-axis takes at $\omega_L t_s = \alpha_s$ According to the stationary phase method, the Eq.(1) may be approximated as follows

$$T_N \approx \sum_l \left(\frac{2\pi\omega_L}{\zeta''(\alpha_l)}\right)^{1/2} \exp(-i\arg\zeta'')M(q_N,\alpha_l)\exp\left\{-iN\alpha_l + i\frac{\zeta(\alpha_l)}{\omega_L}\right\} \tag{6}$$

where the index l numbers the stationary phase point in the interval $-\pi\leq\alpha\leq\pi$, $\zeta''(\alpha_l)$ is the second derivative of. When $N\omega_L$ approaches Δ, for $q_N^2 \cos^2\vartheta \gg N\omega_L - \Delta$, from Eq.(5) it is found

$$K_{L+} \approx \frac{(\Delta - N\omega_L)}{q_N|\cos\vartheta|} \quad ; \quad K_{L-} \approx 2q_N|\cos\vartheta| \tag{7}$$

For $\delta=\pi/2$ the real solutions of K_{L+} are respectively located near $\pi/2$ and $3\pi/2$, while the solutions of K_{L-} may be real or complex. When $N\omega_L$ is close to Δ, $\zeta''(\alpha_1)$ decreases very rapidly for decreasing value of $|\Delta - N\omega_L|$, while $\zeta''(\alpha_l)$ with $l=2,3,4$, do not vary appreciably. Then, when the integer part of $|\Delta/N\omega_L - N|$ is a small integer, the sum in Eq.(6) is dominated by the $l = 1$ term: moreover, as the absolute value of Eq.(6) turns to be, on the average, an increasing quantity when of $|\Delta/N\omega_L - N|$ decreases , the electron energy spectrum assumes the characteristic form shown in Fig.1. The above analysis shows that, for $N = \Delta/\omega_L$, the condition to have a peak in the photoelectron energy distribution is that along with $K_L(\alpha_l)$ also $K'_L(\alpha_l)$ become a very small quantity. We remark that, a part from a constant, $K_L(\alpha_l)$ and $K'_L(\alpha_l)$ represent respectively the low-frequency vector potential and electric field. Therefore, for arbitrary ratios of the frequencies of the LF fields, provided the frequencies are commensurate, by choosing the relative phase of the components of the bichromatic field in such a way that the vector potential and the electric field are simultaneously zero, photoelectron energy distribution should be expected to peak at $N=\Delta/\omega_L$. By taking, for instance, LF fields of the form $E_{LF}(t)=E_0[\sin(\omega_{L1} t) + \sin(2\omega_{L2}t + \delta)]$ z the condition $K(\alpha)=K'(\alpha)=0$ is satisfied for $\alpha=\pi/2$ when $\omega_{L2}=4\omega_1$ and $\delta=-\pi/2$, and for $\alpha=3\pi/2$ in the case in which $\omega_{L2}=3\omega_1$ and $\delta=0$. Results of numerical calculations of the cross section for these cases are shown in Fig.2.

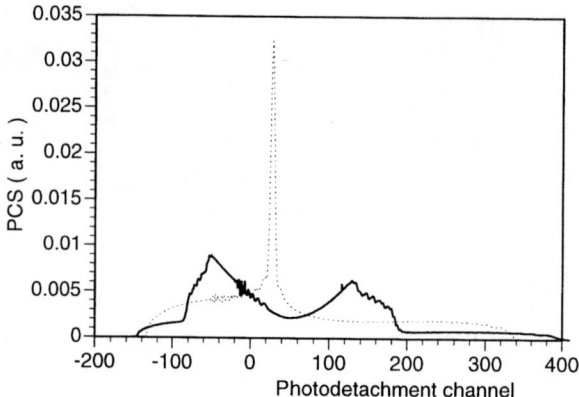

Fig.1 Photodetachment cross sections (PCS) versus the photodetachment channel N. $\hbar\omega_H$=1.5 eV , $\hbar\omega_{L1}$=0.004 eV and I_L=10^7 W/cm^2. Continuous curve : δ = 0; dotted curve: δ = $\pi/2$.

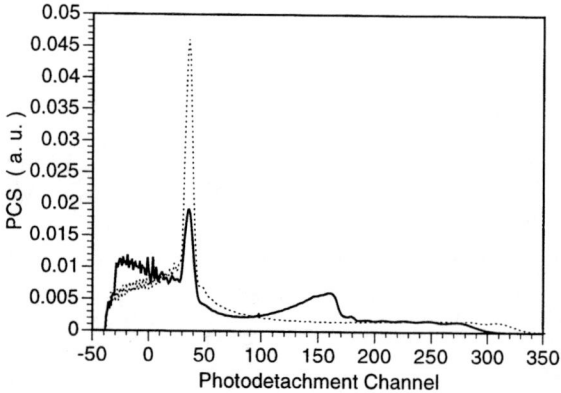

Fig.2 As in Fig.1 Dotted curve: energies of the LF photons $\hbar\omega_{L1}$=0.01eV and $\hbar\omega_{L2}$=3 $\hbar\omega_{L1}$. Continuos curve: $\hbar\omega_{L1}$=0.01eV and $\hbar\omega_{L2}$=4 $\hbar\omega_{L1}$. The intensities of the LF fields are taken equal and in such a way that $\Delta/N\omega_L$=36.

REFERENCES

1. Veniard,.V, Taieb,R. and Maquet, A., *Phys.Rev.Lett.* **74**, 4161 (1995).

2. Bivona, S., Burlon, R. and Leone, C., *Laser Physics* **8**, 78 (1998).

Backscattering from model atmospheric ice microcrystals

Ferdinando Borghese[1,2,3] Paolo Denti[1,2,3] M. Antonia Iatì[1] and
Rosalba Saija[1,2]

[1] *Università di Messina, Dipartimento di Fisica della Materia e Tecnologie Fisiche Avanzate*
[2] *Istituto Nazionale di Fisica della Materia, Unità di Messina*
[3] *Centro Siciliano per le Ricerche atmosferiche e di Fisica dell'Ambiente, Messina*

INTRODUCTION

Recent studies on the physics of the atmosphere agree on the conclusion that the cirruses give a relevant contribution to the greenhouse effect. In fact, the cirruses are mainly composed of ice microcrystals that backscatter the infrared radiation emitted by earth. [1] Some aspects of this phenomenon are addressed by the present paper which discusses the preliminary results of our calculation of the backscattered intensity from dispersions of model ice microcrystals.

In order to build a reliable model, we recall that the most common shapes of ice microcrystals are sticks, crosses and hexagons, with an overall size of $\approx 50\,\mu$m. [1] All these highly anisotropic shapes can be conveniently modelled as aggregates of spherical scatterers of appropriate radii and refractive indexes so that the scattering properties of such aggregates can be calculated in the framework of the transiton matrix approach. [2] Since the transition matrix has well defined transformation properties under rotation of the coordinate frame, when dealing with a dispersion of identical anisotropic particles, it is an easy matter to perform weighted averages over the orientations of the latter. [3] In this respect, we notice that, in several instances of the orientational weighting function, the average can be performed analitically.

The procedure for calculating the transition matrix of aggregates of spheres has been published elsewhere in full detail, [4] so that, in the following sections the main stress will be put on the orientational averaging procedure for physically significant choices of the weight function. The resulting procedure will then be applied to the calculation of the backscattered intensity from a dispersion of model ice microcrystals.

CP513, *Nuclear and Condensed Matter Physics*, edited by A. Messina
© 2000 American Institute of Physics 1-56396-929-7/00/$17.00

ORIENTATIONAL AVERAGES

Let us recall that the transition matrix relates the (known) multipole amplitudes W of the incident plane wave to the multipole amplitudes A of the field scattered by a particle through the equation

$$A_{\eta lm}^{(p)} = - \sum_{p'l'm'} S_{l'm',lm}^{(p,p')} W_{l'm'}^{(p')}(\hat{\mathbf{u}}_{I\eta}, \hat{\mathbf{k}}_I).$$

Once the transition matrix has been calculated the components of the scattering amplitude of the particle are given by

$$f_{\eta\eta'}(\hat{\mathbf{k}}_S, \hat{\mathbf{k}}_I) = -\frac{i}{4\pi k} \sum_{plm} \sum_{p'l'm'} W_{lm}^{(p)*}(\hat{\mathbf{u}}_{S\eta'}, \hat{\mathbf{k}}_S) S_{lm,l'm'}^{(p,p')} W_{l'm'}^{(p)}(\hat{\mathbf{u}}_{I\eta}, \hat{\mathbf{k}}_I).$$

Since both the transition matrix and the scattering amplitude depend on the orientation of the (anisotropic) particle with respect to the incident field, we attach to the particle itself a local set of axes, $\overline{\Sigma}$, whose orientation with respect to the laboratory axes, Σ, is individuated by the Euler angles $\Theta \equiv (\alpha, \beta, \gamma)$. Then the transformation rule holds

$$S_{lm,l'm'}^{(p,p')} = \sum_{\mu\mu'} D_{\mu m}^{(l)*}(\Theta) \overline{S}_{l'm',lm}^{(p,p')} D_{\mu'm'}^{(l')}(\Theta),$$

so that, when the transition matrix is known in the local set of axes it is also known in the laboratory for any choice of the orientation Θ af the scattering particle. As a consequence the components of the scattering amplitude transform according to the equation

$$f_{\eta\eta'}(\Theta) = -\frac{i}{4\pi k} \sum_{plm} \sum_{p'l'm'} \sum_{\mu\mu'} W_{lm}^{(p)*}(\hat{\mathbf{u}}_{S\eta'}, \hat{\mathbf{k}}_S) D_{\mu m}^{(l)*}(\Theta) \overline{S}_{l'\mu',lm}^{(p,p')} D_{m'\mu'}^{(l')}(\Theta) W_{l'm'}^{(p')}(\hat{\mathbf{u}}_{I\eta}, \hat{\mathbf{k}}_I).$$

When dealing with the macroscopic optical properties of a dispersion of identical particles the distribution of their orientations can be taken into account by taking the appropriate ensemble averages. In our preceding papers [4] we considered in full detail the average

$$\langle f_{\eta\eta'} \rangle = \int N(\Theta) f_{\eta\eta'}(\Theta) \, d\Theta, \tag{1}$$

with weighting functions

$$N(\Theta) = \delta(\Theta - \Theta_0), \qquad N(\Theta) = 1/8\pi^2,$$

i. e., when all the particles are oriented alike or are randomly oriented, respectively. For the stick-shaped microcrystals a random distribution is likely to occur, whereas, for hexagons and cross-shaped crystals, elementary aerodynamical considerations lead to the conclusion that none of the above distributions is appropriate.

Nevertheless, we can choose the local frame $\overline{\Sigma}$ with the $\overline{z}$-axis coincident with the high-symmetry axis. Then, a reliable orientational distribution could be the one characterized by random values of α and γ while β is weighted by a suitable function $n(\beta)$. In this case we get

$$\langle f_{\eta\eta'} \rangle = -\frac{i}{4\pi k} \sum_\ell F_{\eta\eta'\ell 00} \int_0^\pi n(\beta) P_\ell(\cos\beta) \sin\beta \, d\beta,$$

where $P_\ell(\cos\beta)$ are Legendre polinomials and

$$F_{\eta\eta'\ell,\overline{m}-\overline{m}',m-m'} = \sum_{pp'} \sum_{ll'} \sum_{\overline{m}\overline{m}'} (-)^{m'-\overline{m}'} C(l,l',\ell;\overline{m},-\overline{m}') C(l,l',\ell;m,-m')$$
$$\times W_{lm}^{(p)\star}(\hat{\mathbf{u}}_{S\eta'}, \hat{\mathbf{k}}_S) S_{l\overline{m},l'\overline{m}'}^{(p,p')} W_{l'm'}^{(p')}(\hat{\mathbf{u}}_{I\eta'}, \hat{\mathbf{k}}_I).$$

RESULTS

The procedure described in the preceding section has been applied to the calculation of the backscattered intensity from a dispersion of hexagon-shaped ice microcrystals modelled as a cluster of 6 identical spheres sitting at the vertices of a regular hexagon. The radius of the spheres is $a = 2.5\,\mu$m and the neighbouring spheres are in mutual contact. Accordingly, the whole model has a radius of $\approx 10\,\mu$m. The refractive index of the spheres has been taken from the paper of

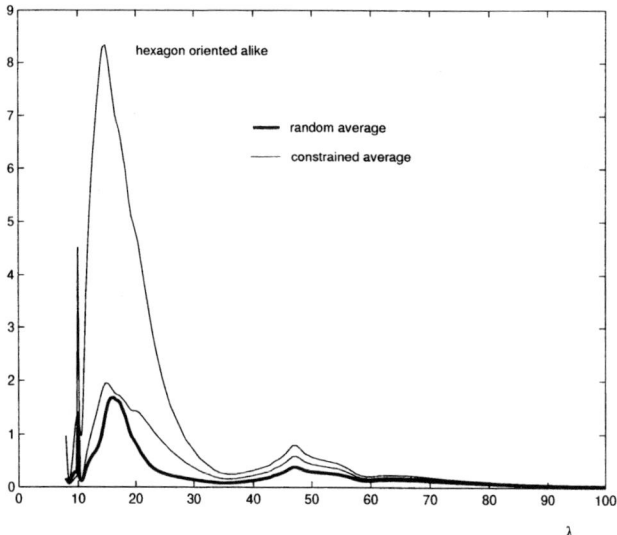

FIGURE 1. Backscattered intensity in μm^2 from a dispersion of hexagon-shaped microcrystals.

45

Warren [5] whose tabulation covers the range of interest. The orientational weight function has been chosen in the form $n(\beta) \propto \cos^4 \beta$ that is peaked for $\beta = 0°$ and ensures that $\approx 1/2$ of the particles have β in the range $0° \leq \beta \leq 30°$.

In Fig. 1 we report the intensity for hexagons all oriented alike with $\beta = 0°$, the intensity for random orientational distribution and the intensity for orientational distribution weighted by $n(\beta) \propto \cos^4 \beta$. Note that all the results are given for unit incident intensity, so that the percentage of backscattered radiation can be read off the scale on the vertical axis. The sharp peak that occurs at $\lambda = 10\,\mu$m is an optical resonance that corresponds to the overall size of the microcrystal. We notice that the backscattered intensity is higher when the hexagons are orientationally distributed according to $n(\beta) \propto \cos^4 \beta$ than when they are randomly oriented. Anyway, the curves for the latter two cases show that a noticeable amount of radiation ($\approx 20\,\%$) at $\lambda \approx 15\,\mu$m is backscattered.

REFERENCES

1. H. R. Pruppacher and J. D. Klett, *Microphysics of clouds and precipitation*, Kluver Academic Pub. (1997).
2. P. C. Waterman, Phys. Rev. D **3**, 227–235 (1971).
3. F. Borghese, P. Denti, R. Saija, G. Toscano and O. I. Sindoni, Nuovo Cim. B **81**, 29–50 (1984).
4. F. Borghese, P. Denti, R. Saija, G. Toscano and O. I. Sindoni, J. Opt. Soc. Am. A **4**, 1984 (1987).
5. S. G. Warren, Appl. Opt. **23**, 120 (1984).

Solute-Solvent Interaction Strength Of Disaccharide Aqueous Solutions: Trehalose Primate

C. Branca[a], A. Faraone[a], S. Magazù[a], G. Maisano[a], P. Migliardo[a], P. Mineo[b], V. Villari[a]

[a]Dipartimento di Fisica and INFM dell'Università di Messina, C.da Papardo, S.ta Sperone n°31, 98166 Messina, Italy
[b]Istituto per la Chimica e la Tecnologia dei Materiali Polimerici, CNR-Catania, Viale A. Doria, 6, 95125 Catania, Italy

Abstract. Results of density, ultrasonic velocity, and DSC measurements performed on aqueous solutions of the homologous disaccharides trehalose, maltose and sucrose, are reported. To get some insight into the mechanisms of cryopreservation that characterize these systems, and to clarify the reasons that make trehalose the most effective bioprotector, we investigate the volumetric and thermic properties of trehalose, maltose and sucrose aqueous solutions as a function of concentration and temperature. What conclusively emerges is the presence of a more collapsed conformation for trehalose in respect with the other disaccharides, indicative of a much more marked solute-solvent interaction strength. Moreover, DSC findings indicate a greater effectiveness of trehalose in destroying the tetrahedral network of water compatible with the formation of ice and support the hypothesis of a higher "fragile" thermodynamic character of the trehalose-water system at high dilution.

INTRODUCTION

During the last years, there has been a rapid growth in biochemical and biomedical studies involving disaccharides. Trehalose, for example, has been found to be an effective bioprotectant, a unique property observed in some soil dwelling organisms, cysts, yeasts and so on, whose common feature is the synthesis of the disaccharide[1]. Thanking to this mechanism, these organisms survive for decades, and in some cases centuries, in a state of suspended animation under conditions of water deficiency, resuming perfectly their metabolism when re-hydrated. Even if other disaccharides, such as maltose and sucrose, have shown similar properties, particularly at high concentration[2], it clearly appears that trehalose is, with a great extent, the most effective. The extraordinary properties of trehalose have determined a growing technology addressed to stabilize biomaterials of various composition during air-drying or freeze-drying. It has been used widely as excipients during freeze-drying of a variety of materials including products in the pharmaceutical industry. It has also been predicted that trehalose can be used as an ingredient for dried and processed

CP513, *Nuclear and Condensed Matter Physics*, edited by A. Messina
© 2000 American Institute of Physics 1-56396-929-7/00/$17.00

food, as well as a non-toxic cryoprotectant of vaccines and organs for surgical transplants[3]. Notwithstanding the extraordinary cryoprotective and osmoregulative properties of trehalose are well established empirically, the underlying molecular mechanisms remain cryptic. The main purpose of this work is to gain a better understanding of the freezing processes in disaccharides solutions to unravel the mechanisms of cryopreservation that characterize these systems. To carry out this study we have performed density, ultrasonic velocity and DSC measurements on trehalose, maltose and sucrose aqueous solutions as a function of concentration and temperature.

EXPERIMENTAL SET-UP

The solutions were prepared by weight using double distilled deionized water. The solutions were filtered with 0.45 μm Amicon filters and stored in the dark to minimize biological and photo-chemical degradation. Ultrasonic velocity measurements were performed by the pulse echo technique using a home-made acoustic interferometer working at the frequency of 3 MHz. The corresponding density measurements, necessary to evaluate the adiabatic compressibility of the solutions, were performed using a standard picnometer technique. The temperature range investigated was 20 °C÷80 °C with a temperature control better than 0.1°C. Differential Scanning Calorimetry (DSC) were performed by using a Mettler DSC 30 calorimeter in the range from -100°C to 50°C, equipped with a sub-ambient system cooled with liquid nitrogen, and a Mettler DSC 20S in the range from 30°C to 450 °C. All DSC data were collected in heating mode. The heating rate was of 10°C/min, under a dynamic nitrogen anhydrous atmosphere (50 ml/min). All data was collected and processed with a Mettler TA processor.

RESULTS AND DISCUSSION

Volumetric properties. We assume, following a well established molecular model [4], that the volume of the solution can be partitioned into two contributions: the hydration volume, where significant interactions between the disaccharide and water occur, and the bulk water volume. Following this model, under the hypothesis of *negligible compressibility for hydrated units*, we obtain the solute-solvent interaction strength parameter ISP:

$$ISP = n_w / n_d \left(1 - \beta / \beta_w \right) \tag{1}$$

being β is the adiabatic compressibility of the solution, n_d and n_w the mole numbers of disaccharide and of water respectively. In Fig. 1, as an example, the behaviour of the *ISP* parameter as a function of concentration is reported for trehalose, maltose and sucrose at T=25°C. The continuous lines represent the results of a polynomial fitting procedure. It clearly emerges that, in respect to the other disaccharides, the trehalose-water system is characterized, in all the investigated concentration range, by both the highest value of the interaction strength parameter, and of hydration number obtained by the extrapolation at infinite dilution of eq.1. In spite of the simplicity of the model the obtained n_H values are in good agreement with those calculated, at room temperature, by Grigera et al. [5]by means of MD simulation. The higher value of the hydration number for trehalose solution in respect with sucrose and maltose solution, suggests the existence of a more "collapsed" conformation, indicative of a much more marked solute-solvent interaction strength.

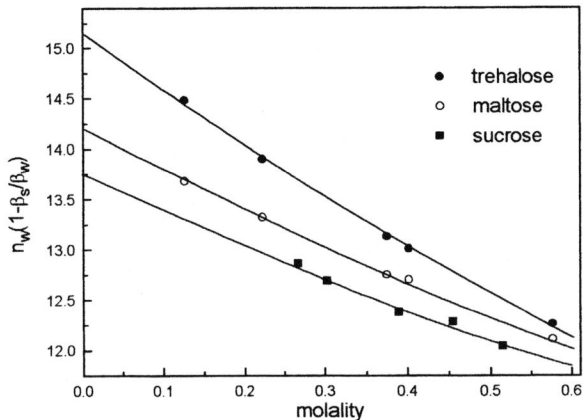

FIGURE 1. Behaviour of the solute-solvent interaction strength parameter of the three disaccharides as a function of concentration. The continuous lines represent the results of a polynomial fitting procedure.

DSC data. It is well known that DSC yealds information about the heat associated with thermal events. In Fig.2, as an example, a comparison of the DSC runs obtained by heating from −100 °C to 40 °C trehalose, maltose and sucrose aqueous solution at a molality, m, of 2.92, is reported. In the heat flow vs temperature profile, it is possible to distinguish three different regions: the glass transition region within the temperature range −50 °C and -42 °C, the ice dissolution region, in the range from -42 °C to −28 °C, and the melting region in the temperature range −28 °C÷5 °C. First of all, for what concerns the glass transition temperatures, the obtained data confirm that in the concentrated region, trehalose is characterized, among the three disaccharides, by higher values of the glass transition temperature[6]. Moreover, in comparison with maltose and sucrose, trehalose shows both the lowest values of the Onset temperature, -25.7 °C, and of heat of melting, ΔH_{melt}=-84.7 (J/g). The obtained values indicate that during heating the amount of ice melted in the trehalose solution is less than that in the sucrose solution, ΔH_{melt}=-95.7 (J/g), and far less than that in maltose solution $\Delta H_{melt}=-103.8$ (J/g). Consequently it can be argued that a less water amount was frozen in the case of the trehalose solution. This could be ascribed to the disaccharide destructuring effect, which results much more marked for trehalose than for sucrose and maltose[7], on the tetrahedral hydrogen bonded network of water (to which liquid water tends as it is supercooled).

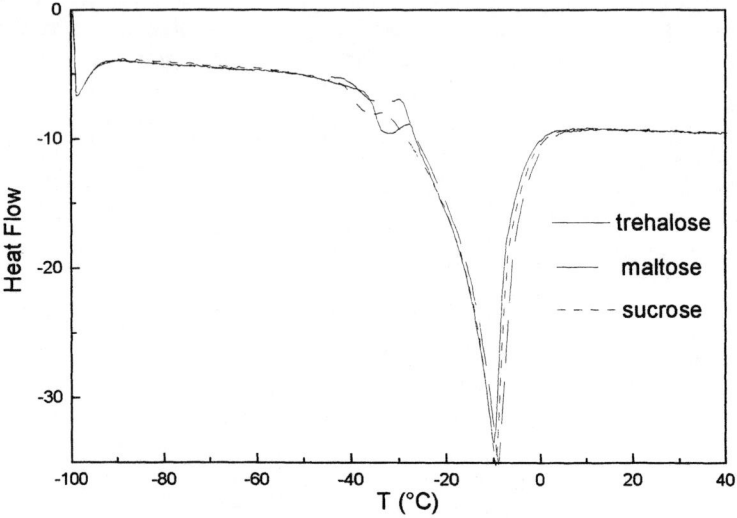

FIGURE 2. DSC curves for trehalose, maltose and sucrose aqueous solution at m=2.92.

REFERENCES

1. Crowe, J. H., and Crowe, L. M., *Biological Membranes*, edited by D. Chapman, New York: Academic Press, 1984, Vol 5, pp 57; Fox, K. C., *Science* **267**, 1922 (1995).

2. Chan, R. K., Pathmanathan, K., and Johari, G. P., *J. Phys. Chem. B* **90**, 6358 (1986).

3. Paiva, C. L., and Panek, A. D., *Biotechnol. Annu. Rev.* **2**, 293 (1996).

4. Magazù, S., Migliardo, P., Musolino, A. M., and Sciortino, M. T., *J. Phys. Chem B* **101**, 2348 (1997).

5. Donnamaria, M. C., Howard, E. I., and Grigera, J. R., *J. Chem. Soc. Faraday Trans.* **90**, 2731 (1994).

6. Green, J. L., and Angell, C. A., *J. Phys. Chem.* **93**, 2880 (1989).

7. Branca, C., Magazù, S., Maisano, G., and Migliardo, P., *J. Chem. Phys.* **111**, 281 (1999); Branca, C., Magazù, S., Maisano, G., and Migliardo, P., *J. Phys. Chem.B* **103**, 1347 (1999).

Theoretical investigations of simple model protein solutions

C. Caccamo, G. Pellicane, and D. Costa

Istituto Nazionale per la Fisica della Materia (INFM) and
Dipartimento di Fisica, Università di Messina
Contrada Papardo, Salita Sperone 31, 98166 Messina, Italy

Abstract.
Thermodynamically self-consistent integral equation theories and computer simulations are applied to the investigation of thermodynamic and structural properties as well as to the determination of the phase diagram of the hard-core Yukawa fluid. We consider different Yukawa-tail screening lengths λ, going up to $\lambda = 9$, when the potential becomes so short-ranged that the interaction can be considered fairly similar to that present between macroparticles in colloidal suspensions and protein solutions.

Theoretical approaches are found to give a reasonably accurate description of the physical properties of the system at high λ's. The relative position of the sublimation *vs.* the liquid-vapor binodal line, known from computer simulations to play a crucial role in the onset of crystallization in protein solutions, seems qualitatively reproducible.

We suggest on the basis of our results the possibility to extend such investigations to more realistic models of protein solutions, so to take into account the true multicomponent nature of these fluids, a physical situation whose description still challenges the currently available computer simulation capabilities.

Several authors have recently investigated model protein solutions by employing to such an aim both statistical mechanical theories and computer simulations (see [1–3] and references therein). The reason of such an interest essentially stems from the problem of protein crystallization, a phenomenon whose predictability and control seem still far from being achieved [4].

Although the modelization of such systems is far from being trivial, it is commonly accepted that, at least for globular proteins, the effective interaction can be roughly approximated in terms of a strongly repulsive potential at short range plus a rapidly decaying and attractive tail [1,3]. In this context, it would be highly desiderable the availability of an accurate theoretical approach, able to predict, even roughly, the phase behavior of the system at issue. For such a reason, we have recently undertaken an extensive investigation of the performances of several fluid state theories for a system of hard-sphere particles interacting through an attractive Yukawa tail (HCYF, [5]); we have implemented, in particular, a numerical solution procedure of the Generalized Mean Spherical Approximation (GMSA, [6,7]). We also apply the Modified Hypernetted Chain approximation (MHNC, [8]), as well as simple versions of the Self-Consistent Ornstein Zernike Approximation (SCOZA, [9,10]) and the Hierarchical Reference Theory (HRT, [11]); the solution schemes of these theories have been developed elsewhere by other authors.

We consider different values of the Yukawa-tail screening length λ, ranging from $\lambda \simeq 2$ up to $\lambda \simeq 9$, the latter being a realistic screening length for HCYF modelizations of colloidal suspensions and protein solutions [1,12,13]. We show results for the liquid-vapor binodal and for the freezing line of the system as obtained through the adoption of a one-phase freezing criterion due to other authors [14].

TABLE 1. Theoretical and simulation critical point parameters. †: finite-size scaling MC simulation of Ref. [10]; §: Gibbs Ensemble MC simulations of Ref. [15]; ◦: Gibbs Ensemble MC simulations of Ref. [12]; ‡: MHNC calculations with Verlet-Weis bridge functions of Ref. [16].

	$\lambda = 1.8$		$\lambda = 4.0$		$\lambda = 7.0$	
	T_{cr}	ρ_{cr}	T_{cr}	ρ_{cr}	T_{cr}	ρ_{cr}
MC	1.212 (2)†	0.312 (2)†	0.576 (6)§	0.377 (21)§	0.411 (2)◦	0.50 (2)◦
GMSA	1.199	0.312	0.576	0.324		
MHNC	1.193	0.326	0.581	0.412		
MHNC‡	1.21	0.28				
HRT	1.214	0.312	0.599	0.394	0.435	0.424
SCOZA	1.219	0.314	0.591	0.3895	0.419	0.4575

We find that thermodynamic and structural properties are well predicted by the MHNC, in comparison with Monte Carlo (MC) simulation data, especially for the internal energy. The same theory is slightly less accurate in predicting the pressure, the compressibility and $g(\sigma)$, the contact values of the radial distribution function. The SCOZA is the best of the theories for the equation of state. The GMSA is reasonably accurate for the thermodynamic quantities and for $g(\sigma)$ only at low λ's. The HRT appears to be on a comparable level of accuracy. The MHNC also appears to predict in a fairly quantitative manner the overall pattern of the radial distribution function. The GMSA, SCOZA and HRT $g(r)$ are reasonably good at low λ's. The SCOZA and the HRT $g(\sigma)$'s are too low for all states sampled.

As far as the investigation of the phase diagram is concerned, the binodal results confirm that the SCOZA predicts with good accuracy the equation of state over the whole interval of the Yukawa screening parameter studied (see Table I). At low λ's ($\lambda = 1.8$) all the theories reproduce fairly well the simulation phase diagram. At higher λ's ($\lambda \geq 7$) only the SCOZA is able to maintain good agreement with the computer simulation binodal (see Fig. 1). The freezing line, as determined according to a criterion based on the vanishing of the residual multiparticle entropy Δs [14], is not as satisfactorily predicted as the binodal line (see Fig. 1). In fact, the SCOZA and the MHNC yield a freezing density which is not significantly sensitive to the variation of the potential range as is known, instead, to be the case from simulation studies [12]. The latter show that at $\lambda > 7$ the solid-liquid coexistence line actually becomes a sublimation line running above the (metastable) binodal line, with the liquid-vapor critical point falling just beneath the vapor-solid transition line. The GMSA $\Delta s = 0$ turns out to be able to follow in a qualitative manner the modification with λ of the freezing line. In particular, at $\lambda = 9$, the locus of vanishing residual multiparticle entropy shifts well above the liquid-vapor coexistence line in a manner which fairly mimics the relative location of the freezing and binodal line in this λ regime.

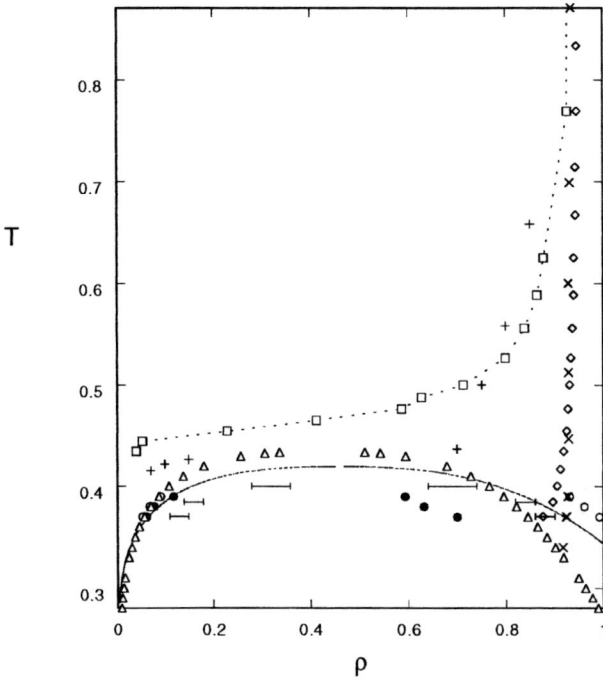

FIGURE 1. Phase diagram for the HCYF with $\lambda = 7$. Liquid-gas coexistence: horizontal bars: Gibbs Ensemble MC results of Ref. [12]; dotted line: SCOZA; triangles: HRT; full circles: GMSA; open circles: MHNC; $\Delta s = 0$ *locus*: diamonds: SCOZA; pluses: GMSA; crosses: MHNC; dashed line with squares: sublimation line of Ref. [12].

In view of the good performances of the MHNC, it may be worth to further improve its predictions through the use of accurate bridge functionals [17]. The GMSA turns out to be reasonably accurate only at low λ's. This suggests to impose the thermodynamic consistency constraints in a more sophisticated version than that here adopted. The binodal liquid-vapor line appears rather satisfactorily predicted by the SCOZA all over the λ range investigated, while the HRT turns out to be on a comparable level of accuracy of the other theories only at low λ; an expression for $c(r)$ more suited to deal with very short-ranged potentials might be a possible way to improve its results.

The sublimation-freezing line is not as satisfactorily predicted as the binodal at large λ; the only theory that turns out to follow in a qualitative manner the evolution of the vapor-solid stability boundary with λ is the GMSA. It appears from these results that the combination of information from the GMSA and the SCOZA might be able to provide a description of the relative position of phase coexistence lines in the phase diagram of the HCYF in the very short-ranged potential regime. This circumstance might prove useful in order to extend the present investigation to two-component models of protein solutions, a situation in which the theory should not face the difficulties typically encountered in the simulation of fluids containing highly diluted and strongly asymmetric particle species.

REFERENCES

1. M. Malfois, F. Bonneté, L. Belloni, and A. Tardieu, J. Chem. Phys. **105**, 3290 (1996).
2. A. Lomakin, N. Asherie, and G.B. Benedek, J. Chem. Phys. **104**, 1646 (1996).
3. P.R. ten Wolde and D. Frenkel, Science **277**, 1975 (1997).
4. A. McPherson, *Preparation and Analysis of Protein Crystals* (Krieger, Malabar, FL, 1982).
5. C. Caccamo, G. Pellicane, D. Costa, D. Pini and G. Stell, Phys. Rev. E **60** (Nov. 1999).
6. E. Waisman, Mol. Phys. **25**, 45 (1973).
7. J.S. Høye, J.L. Lebowitz, and G. Stell, J. Chem. Phys. **61**, 3253 (1974); J.S. Høye and L. Blum, J. Stat. Phys. **19**, 317 (1978); L. Blum, J. Stat. Phys. **22**, 661 (1980).
8. Y. Rosenfeld and N.W. Ashcroft, Phys. Rev. A **20**, 1208 (1979).
9. J.S. Høye and G. Stell, J. Chem. Phys. **67**, 439 (1977) and Mol. Phys. **52**, 1071 (1984).
10. D. Pini, G. Stell, and N.B. Wilding, Mol. Phys. **95**, 483 (1998).
11. A. Parola and L. Reatto, Phys. Rev. Lett. **53**, 2417 (1984); A. Parola and L. Reatto, Phys. Rev. A **31**, 3309 (1985); A. Meroni, A. Parola, and L. Reatto, Phys. Rev. A **42**, 6104 (1990); M. Tau, A. Parola, D. Pini, and L. Reatto, Phys. Rev. E **52**; 2644 (1995); A. Parola and L. Reatto, Adv. in Physics **44**, 211 (1995).
12. M.H.J. Hagen and D. Frenkel, J. Chem. Phys. **101**, 4093 (1994).
13. D. Rosenbaum, P.C. Zamora, C.F. Zukoski, J. Cryst. Growth **169**, 752 (1996).
14. P.V. Giaquinta and G. Giunta, Physica A **187**, 145 (1992); P.V. Giaquinta, G. Giunta, and S. Prestipino Giarritta, Phys. Rev. A **45**, 6966 (1992).
15. E. Lomba and N.E. Almarza, J. Chem. Phys. **100**, 8367 (1994).
16. C. Caccamo, G. Giunta, and G. Malescio, Mol. Phys. **84**, 125 (1995).
17. Y. Rosenfeld, Proceed. of NATO-ASI *New Approaches to Old and New Problems in Liquid State Theory*, Patti (Messina, Italy), C. Caccamo, J. P. Hansen and G. Stell eds. (Kluwer, Dordrecht, 1999).

Generalized Mean Spherical Approximation with internal thermodynamic consistency constraints: an application to hard sphere mixtures

C. Caccamo[1], G. Pellicane and R. Ricciari

Istituto Nazionale di Fisica della Materia (INFM) and Dipartimento di Fisica
Sez. Teorica, C.P. 50, Contrada Papardo, 98166 Messina (Italy)

Abstract. We apply the Generalized Mean Spherical Approximation (GMSA) to hard sphere mixtures by imposing the thermodynamic consistency of the theory by means of both internal and external constraints. Structural results are obtained for a moderate to high size-asymmetry regime and compared with those we also obtain in the Rogers-Young theory, with previous Modified-Hypernetted Chain theory results and with Monte-Carlo simulation data. The overall good performances of the thermodynamically self-consistent GMSA and the semianaliticity of its solution scheme, thanks to which one could determine the phase diagram of the system in a fairly rapid manner, strongly suggest the opportunity to generalize our theoretical procedure to hard-core Yukawa mixtures, a model that is of relevance for the description of phase equilibria in globular protein solutions.

In perturbation theories of one-component fluids, the thermodynamic properties of a given liquid can be related to those of a hard sphere reference system, the attractive part of the potential being treated as a perturbation [1]; however, while this approach is made feasible by the complete knowledge of the structural properties of the hard sphere fluid, its extension to multi-component systems is not trivial since the role of reference system should be played, in this case, by hard sphere mixtures (HSMs), a system whose properties are not easy determined over the full range of the defining physical parameters. During the last few years, there have been significant advancements in our knowledge of binary HSMs both from the theoretical [2–7] and the simulation [8,9] point of view; these studies, however, have not yet provided a reliable structural information in the very asymmetric and dilute regimes of the component species. Such regimes are in fact quite demanding to be described in terms of the computer simulation approach; on the other hand, they are of great interest for the scientific community because of the experimentally

---

[1] E-mail: caccamo@vulcano.unime.it

CP513, *Nuclear and Condensed Matter Physics*, edited by A. Messina
© 2000 American Institute of Physics 1-56396-929-7/00/$17.00

observed phase separation in colloidal mixtures with size ratio between 0.069 and 0.294 [10].

These two circustamces prompt to the study of such multicomponent fluids by means of refined microscopic theoretical approaches, as integral equation theories (IETs, [11]) of the fluid state. IETs are usually written by starting from the Ornstein-Zernike (OZ) equation

$$h_{ij}(r) = c_{ij}(r) + \sum_{k=1}^{2} \rho_k \int c_{ik}(|\mathbf{r} - \mathbf{r}'|) h_{kj}(\mathbf{r}') d\mathbf{r}', \tag{1}$$

with $h_{ij}(r) = g_{ij}(r) - 1$ and $c_{ij}(r)$ the pair and the direct correlation function, respectively, and providing a closure to eq. (1) in terms of an 'ansatz' functional form of $c_{ij}(r)$ that may contain adjustable parameters. The latter can then be used in order to impose the thermodynamic consistency of the theory. Thermodynamically consistent integral-equation theories (TC-IETs) constitute a valid approach to the study of the HSMs [6]; in particular, the phase diagram can be determined through the calculation of the Gibbs free energy, a task, however, that unless the theory is solvable in a semianalytic manner turns out to be very cumbersome. Actually, when the solution has to be obtained through iterative procedures, one usually meets convergence problems that are not always clearly attributed to a thermodynamic instability of the system, as it may happen in the neighborhood of a phase coexistence line. We here consider two different theories with a parametrized closure to the OZ equation:

(a) the Rogers-Young (RY, [12]) approximation, in which the closure is written as

$$g_{ij}(r) = \exp[-\beta v_{ij}(r)] \left\{ 1 + \frac{\exp\{f_{ij}(r)[h_{ij}(r) - c_{ij}(r)]\} - 1}{f_{ij}(r)} \right\}, \tag{2}$$

where $f_{ij}(r) = 1 - \exp[\xi_{ij} r]$ and the ξ_{ij} are adjusted so to satisfy the thermodynamic consistency of the theory. In the present implementation of the theory we assume all $\xi_{ij} = \xi$ and we fix this parameter by requesting the equality of the fluctuation and virial isothermal compressibility. The solution algorithm in this case is entirely numerical.

(b) The Generalized Mean Spherical Approximation (GMSA, [13]) in which the closure to the OZ equation is

$$c_{ij}(r) = K_{ij} \frac{\exp[-z(r - \sigma_{ij})]}{r/\sigma_{ij}} \qquad r > \sigma_{ij}. \tag{3}$$

In this case we do not experience convergence problems because, for HSMs, the GMSA solution is achieved through algebraic means; moreover, explicit analytical expressions for a number of thermodynamic quantities [14] are also obtained. Therefore, the GMSA solution scheme takes, in general, much less time than is

FIGURE 1. Left-column panels: rdfs at particle-sizes fraction $\Lambda = 0.3$, concentration of the bigger particle $\chi_2 = 0.0625$, total packing fraction $\eta = 0.4$. Modified-Hypernetted Chain (MHNC) theory and MC rdfs from ref. [18]. Right-column panels: rdfs at $\Lambda = 0.166$, $\chi_2 = 0.064$, $\eta = 0.3$.

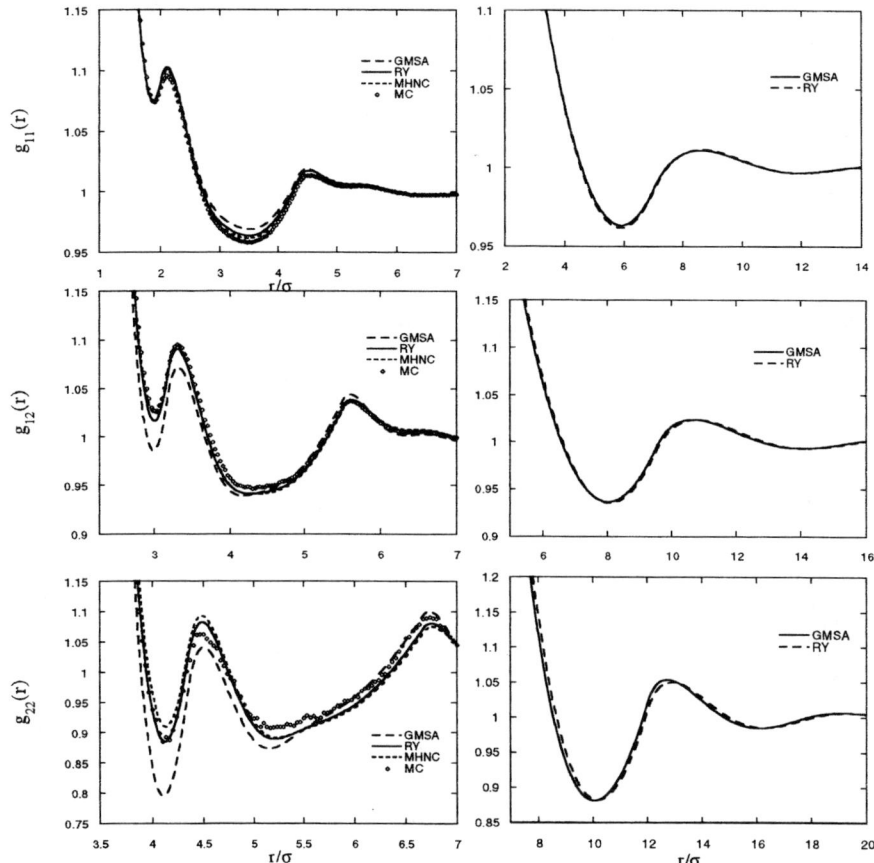

usually requested in iteratively solved TC-IETs; this quality makes the GMSA suitable for extensive investigations of the structure, and eventually of the phase diagram of HSMs.

In our implementation of the GMSA, we fix the four adjustable parameters $K_{ij}(i, j = 1, 2)$ and z, appearing in eq. (3), by fitting the virial pressure to the Mansoori-Carnahan-Starling-Leland equation of state [15], by imposing the equality of two osmotic compressibilities as evaluated from the virial and fluctuation routes [11] and by requiring $K_{12} = \sqrt{K_{11}K_{22}}$. Radial distribution functions (rdfs), for a moderate asymmetry, are shown on the left column panels in figure 1. The comparison clearly shows the quantitative agreement generally obtained between the RY and the Monte Carlo (MC) rdfs, while the GMSA seems able to predict well only the main features of the simulation patterns. A better agreement of the theories with the simulation emerges, in general, when considering the contact values of the rdfs, (not shown here) except for some discrepancy in the very dilute regime of the bigger particles, that we interpret as due to the inadequacy of the simulation to give a correct statistical sampling of the phase space.

For high asymmetries, where no simulation data are available in literature, we compare the RY and the GMSA rdfs with each other in the right column panels of figure 1. The very good agreement found between the two theories for such asymmetries, supplemented by the quantitative agreement between the RY and MC rdfs for lower asymmetries, is indicative of the GMSA ability to successfully deal with systems composed of quite different sized particles. This result seems remarkable since it suggests the possibility to investigate the phase diagram of highly asymmetric hard-core Yukawa mixtures (a model relevant to the determination of phase equilibria in globular protein solutions [17]), in the framework of a little time-consuming solution scheme [16] as the one proper of the GMSA.

Acknowledgments. *Work performed in the framework of the Advanced Project of the INFM (Section G) "Theoretical and Simulation Prediction of Phase Coexistence in Complex Fluids"*

REFERENCES

1. J. P. Hansen, I. R. McDonald (1986) *Theory of Simple Liquids* 2nd Ed. (Academic, London)
2. M. Dijkstra, R. van Roij *Phys. Rev. E* **56**, 5594 (1997)
3. M. Dijkstra, R. van Roij, R. Evans *Phys. Rev. Lett.* **81**, 2268 (1998)
4. T. Biben and J. P. Hansen *Phys. Rev. Lett.* **66**, 2215 (1991)
5. Y. Rosenfeld *Phys. Rev. Lett.* **72**, 3831 (1994)
6. C. Caccamo, G. Pellicane (1997) *Physica A* **235**, 149
7. T. Coussaert, M. Baus *J. Chem. Phys.* **109**, 6012 (1998)
8. A. Malijevsky, M. Barosova and W. R. Smith *Mol. Phys.* **91**, 65 (1997)
9. A. Buhot and W. Krauth *Phys. Rev. Lett.* **80**, 3787 (1998)
10. P. D. Kaplan, J. L. Rouke, A. G. Yodh and D. J. Pine *Phys. Rev. Lett.* **72**, 582 (1994); A. Imhof and J. K. G. Dhont *Phys. Rev. Lett.* **75**, 1662 (1995)
11. C. Caccamo *Phys. Rep.* **274**, 1 (1996)
12. F. J. Rogers and D. A. Young *Phys. Rev. A* **30**, 999 (1984)
13. J. S. Høye and G. Stell *J. Chem. Phys.* **67**, 524 (1977)
14. E. Arrieta, C. Jedrzejek, K. N. Marsh *J. Chem. Phys.* **86**, 3607 (1987)
15. G. A. Mansoori, N. F. Carnahan, K. E. Starling, J. Leland *J. Chem. Phys.* **54**, 1523 (1971)
16. E. Arrieta, C. Jedrzejek, K. N. Marsh *J. Chem. Phys.* **95**, 6806 (1991)
17. C. Caccamo, D. Costa and G. Pellicane *Proc. of NATO-ASI on New Approaches to Old and New Problems in Liquid State Theory* (Patti (Messina, Italy)), vol 1 (Kluwer, Dordrecht) p 421 (1999)
18. C. Caccamo, G. Pellicane and E. Enciso *Phys. Rev. E* **56**, 6954 (1997)

The Electronic Properties of Some Transition Metal Chalcogenophosphates

C. Calareso, V. Grasso, L. Silipigni

Dipartimento di Fisica della Materia e Tecnologie Fisiche Avanzate - Università di Messina - Salita Sperone 31 .- I 98166 Messina - Istituto Nazionale per la Fisica della Materia - Unità di Messina

Abstract. We have investigated the core level and photoinduced Pb Auger transition regions of $Pb_2P_2S_6$, a member of the $M_2P_2X_6$ chalcogenophosphate family using the x-ray photoemission spectroscopy. The XPS core level spectra show single peak structures for the Pb, P and S core levels in agreement with crystallographic data. The combined analysis of the Pb $4f_{7/2}$ core level and the Pb $N_6O_{45}O_{45}$ Auger transition has allowed us to deduce the modified Auger parameter for Pb in $Pb_2P_2S_6$ obtaining in this way information on the Pb-S bond in this compound.

1. INTRODUCTION

The transition metal or post-transition metal chalcogenophosphates $M_2P_2X_6$ (X= S or Se) constitute a very interesting compound family in the field of the academic and applied scientific research. In fact these compounds, because of their layered structure and of the presence of available empty interstitial sites, can accommodate, via intercalation, different guest species and acquire physical properties deriving from the combination of those of the guest and of the host. In this way they find application as cathode in rechargeable lithium solid state batteries[1]or as second harmonic generators[2] or as solid polymeric electrolites[3]. Here we have paid our attention on a member of this compound family, the lead thiophosphate ($Pb_2P_2S_6$) which is isotypic to the $Sn_2P_2S_6$ monoclinic modification II, even it crystallizes in a different space group (the centrosymmetric $P2_1/c$ space group[4]). Rather poor data are reported in the literature on the physical properties of this compound, which is classified as a ferroelectric photoconductor with a band gap of 2.57 eV at room temperature[5]. In order to obtain more information on this compound we have analyzed its core levels by using XPS technique. We have also investigated the x-ray excited Pb $N_{67}O_{45}O_{45}$ Auger series, whose analysis, combined with that of the Pb $4f$ core levels, has shed light on the Pb-S bond nature.

2. EXPERIMENTAL RESULTS AND DISCUSSION

The obtained room temperature XPS and XAES spectra were collected by a VG Scientific Spectrometer with a 105° concentric hemispherical analyzer CLAM 100 operating in the constant pass-energy mode (20 eV). The Al $K\alpha$ radiation (1486.6 eV)

CP513, *Nuclear and Condensed Matter Physics,* edited by A. Messina
© 2000 American Institute of Physics 1-56396-929-7/00/$17.00

of a twin-anode Al/Mg $K\alpha$ x-ray source was used as excitation source. During each measurement, the analysis chamber pressure was in the 10^{-9} mbar range. The $Pb_2P_2S_6$ crystals, grown in the Laboratoire de Physique des Matériaux Électroniques (LPME) of the Institut de Physique Appliquée of the École Polytechnique Fédérale de Lausanne, were mounted onto sample holders by means of a double-sided conductive adhesive tape. All the investigated photoelectron and Auger line binding energies are referred to the C $1s$ line at 285.0 eV. Figs. 1-2 show the XPS spectra of the Pb $4f$ (fig.1a) and $5d$ (fig.1b), S $2p$ (fig.2a) and P $2p$ (fig.2b) core levels in $Pb_2P_2S_6$. All the spectra, to which a Shirley-type inelastic background[6] was subtracted, have been fitted with Gaussian-Lorentzian cross-product functions as lineshapes. The resulting best fit is shown with solid line, the Gaussian-Lorentzian subbands with dashed line and the experimental data with open circles. The deduced binding energy positions of the above-cited core levels are summarized in Table I together with those of other observed Pb and S core levels. Apart from the splitting due to the spin-orbit coupling, all the analyzed core levels exhibit single-peak structures, pointing out that each constitutional atom sees the same environment. This means that in $Pb_2P_2S_6$ non-equivalent sites are not present for the Pb, P and S atom. All the investigated Pb XPS peaks are located towards higher binding energies with respect to the metallic lead[7] and the binding energy position of the Pb $4f_{7/2}$ core level is similar to that observed for the same level in PbF_2, where Pb is in a 2+ formal oxidation state[7]. This suggests that in the analyzed $Pb_2P_2S_6$ a bivalent lead cation is present in good agreement with the literature[1]. Moreover, the Pb $4f_{7/2}$ and $4f_{5/2}$ components show a spin-orbit separation of 4.8 eV in agreement with that observed in PbS, where Pb plays the role of a bivalent cation[8]. The Pb $5d$ levels, shown in fig.1b, constitute the most prominent structures of the $Pb_2P_2S_6$ valence band and appear as a resolved doublet with an energy separation of 2.6 eV in agreement with that observed for the same levels in PbS[8]. As regards the P and S $2p$ core levels, the former appears as a single peak (fig.2b), while the latter as a poorly resolved doublet (fig.2a) with a spin-orbit splitting of 1.2 eV. Their binding energy positions agree with those observed in the $Sn_2P_2S_6$ monoclinic phase II[9], confirming the existence of the $(P_2S_6)^{4-}$ structural unit in $Pb_2P_2S_6$. Fig.1c shows the lead $N_{67}O_{45}O_{45}$ Auger series spectrum induced by the Al $K\alpha$ x-ray source. In non-conducting compounds the $N_6O_{45}O_{45}$ component results sharper than $N_7O_{45}O_{45}$ one and its kinetic energy position is used, together with the Pb $4f_{7/2}$ core level binding energy one, for calculating the Pb modified Auger parameter α' in $Pb_2P_2S_6$. This parameter, defined[10] as follows:

$$\alpha'\left(Pb_{Pb_2P_2S_6}\right) = E_K\left(N_6O_{45}O_{45}\right) + E_B\left(4f_{7/2}\right) \qquad (1)$$

is particularly useful when the chemical state spectral interpretation of a given element is difficult due to static charging effects and/or very small chemical shifts. In fact, it is not affected by charging corrections, it does not depend on the chosen reference level and it has a unique value for each chemical state[11]. As shown in table II we have deduced a value of 231.0 eV for the Pb α' in $Pb_2P_2S_6$. This value is intermediate between those observed for Pb in PbS and PbF_2, even if it is closer to the PbS one (see table II). Since in PbS the bond is partially ionic with a percentage of covalence[8], the noted analogy suggests a similar nature for the Pb-S link in $Pb_2P_2S_6$. We have also calculated the change in the modified Auger parameter, $\Delta\alpha'$, defined[12] as:

$$\Delta\alpha' = \alpha'(Pb) - \alpha'\left(Pb_{Pb_2P_2S_6}\right) \tag{2}$$

where $\alpha'(Pb)$ is the modified Auger parameter of the metallic lead whose value is listed in table II. We have obtained a value of 2.0 eV for the Pb $\Delta\alpha'$ in $Pb_2P_2S_6$, as shown in table II, in rough agreement with the 2.57 eV $Pb_2P_2S_6$ E_g energy gap value[5]. This accordance means that for $Pb_2P_2S_6$ the following equation[13] is valid:

$$\Delta\alpha' = 2\delta R^{ea}(1hole) \approx E_g \tag{3}$$

where R^{ea} is the extra-atomic relaxation energy for the metal with the core hole. The observed relation between δR^{ea} and E_g suggests that, in analogy to some lead chalcogenides[13], in this compound a non-local screening mechanism is present excluding in this way that the electronic transition associated to its optical gap is of the *d-d* or *charge-transfer* type.

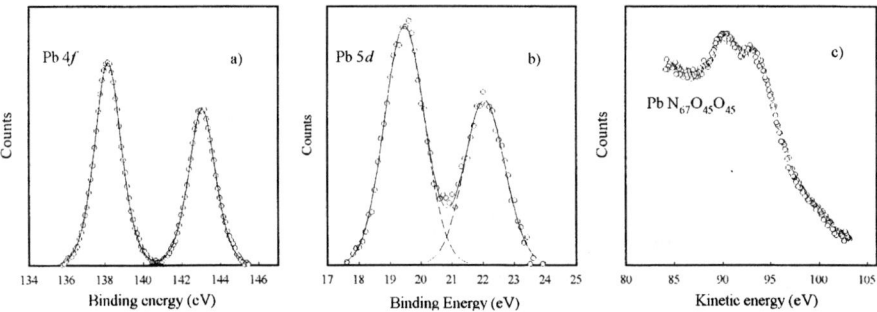

FIGURE 1. Pb $4f$ (*a*) and $5d$ (*b*) core level XPS and x-ray induced Pb $N_{67}O_{45}O_{45}$ Auger series (*c*) spectra in $Pb_2P_2S_6$.

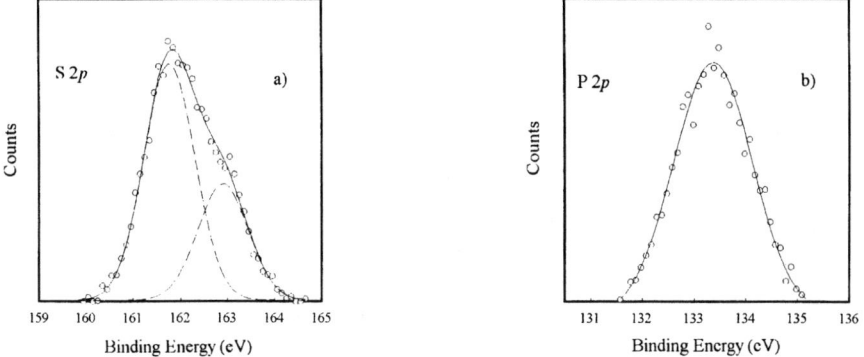

FIGURE 2. S $2p$ (*a*) and P $2p$ (*b*) core level XPS spectra in $Pb_2P_2S_6$

TABLE I. $Pb_2P_2S_6$ core level binding energies referred to the $C1s$ line at 285.0 eV

Core levels (eV)	Pb	P	S	
$5d_{5/2}$	19.5			
$5d_{3/2}$	22.1			
$4f_{7/2}$	138.2			
$4f_{5/2}$	143.0			
$4d_{5/2}$	413.3			
$4d_{3/2}$	435.4			
$2p$		133.4	161.8	163.0
$2s$			226.1	

TABLE II. Pb $4f_{7/2}$ core level binding energies, $N_6O_{45}O_{45}$ Auger line kinetic energies, α', $\Delta\alpha'$ and E_g in $Pb_2P_2S_6$ and in other Pb compounds. All values are in eV

	$4f_{7/2}$	$N_6O_{45}O_{45}$	α'	$\Delta\alpha'$	E_g
Pb [a]	136.2	96.8	233.0		
PbS [a]	136.9	95.1	232.0	1.0 [c]	0.37 [c]
$Pb_2P_2S_6$	138.2	92.8	231.0	2.0	2.57 [b]
PbF_2 [a]	137.9	91.2	229.1	3.9	

[a] Ref. [7]: 0.4 eV subtracted (B.E.) or added (K.E.) being data referred to $C1s$=284.6 eV, [b]Ref.[5], [c]Ref.[13]

REFERENCES

1. Brec, R., Ouvrard, G., Louisy, A., and Rouxel, J., *Solid State Ionics* **6**, 185 (1982)

2. Lagadic, I., Lacroix, P.G., and Clement, R., *Chem. Mater.* **9**,2004 (1997)

3. Jeevanandam, P., and Vasudevan, S., *Chem. Mater.* **10**, 1276 (1998).

4. Carpentier, C.D., and Nitsche, R., *Mat. Res. Bull.* **9**, 401 (1974)

5. Agekyan, V.F., and Muzyka, L.N., *Sov.Phys. Solid State* **28**, 1815 (1986)

6. Shirley, D.A., *Phys. Rev. B***5**, 4709 (1972)

7. Pederson, L. R., *J. Electron. Spectrosc. Relat. Phenom.* **28**, 203 (1982)

8. Shalvoy, R.B., Fisher, G.B., and Stiles, P.J., *Phys. Rev. B* **15**, 1680 (1977)

9. Currò, G.M., Grasso,V., Neri, F., and Silipigni,L., *Il Nuovo Cimento D* **20**,1163 (1998)

10. Wagner, C.D., Gale, L. H., Raymond, R. H., *Anal. Chem.* **51**, 466 (1979)

11. Waddington S.D., "Auger Chemical Shifts and the Auger Parameter", in *Pracitcal Surface Analysis*, 2nd ed., edited by D. Briggs et al., Wiley, Chichester, England, 1990, vol.I, App.4,p.587

12. Wagner, C.D., *Faraday Discuss. Chem. Soc.* **60**, 291 (1975)

13. Moretti, G., *J. Electron. Spectrosc. Relat. Phenom.* **50**, 289 (1980)

Contributions to the Photoluminescence Activity in the UV Range in Amorphous-SiO$_2$

M. Cannas, S. Agnello, R. Boscaino, F.M. Gelardi and M. Leone

INFM and Department of Physical and Astronomical Sciences, University of Palermo,
via Archirafi 36, I-90123 Palermo, Italy

Abstract. We report a detailed experimental study on the photoluminescence activity in the UV range (4.0-4.5 eV) performed in natural and synthetic silica types both as grown and after γ exposure. Our results allow us to add new insight on the well-known emissions at 4.2 and 4.4 eV as regard their vacuum-UV excitation spectra and their kinetics behaviors. Moreover, a new contribution to the photoluminescence at 4.4 eV, excited within the absorption E band at 7.6 eV and exhibiting a strong temperature dependence, is identified and discussed in the light of the structural models reported in literature.

1. INTRODUCTION

Photoluminescence (PL) spectroscopy has been to be proven a powerful technique to reveal the presence of point defects in silica glasses. Since the first experimental observation by Carino Guanina [1], large attention has been devoted to the centers responsible for the luminescence in the UV spectral region where two close PL emissions centered at ~4.2 eV (α_E band) and ~4.4 eV (α_I band) have been distinguished [2]. On the basis of PL excitation (PLE) spectra, α_E and α_I were associated with the inverse transition of the ones giving rise the optical absorption (OA) bands B$_{2\beta}$ (5.15 eV) and B$_{2\alpha}$ (5.02 eV), respectively [3].

In spite of the fact that the α_E and α_I emissions have been well characterized under UV excitation, few data are till now available for their excitation with vacuum UV radiation [4-6]. We report here experimental results on the spectral and kinetics properties of the PL activity as excited in the vacuum UV range by using synchrotron radiation (SR). The aim of the present work is to complete the description of the excitation processes of the emissions at 4.2 and 4.4 eV occurring via higher excited states and clarify the correlation of these emissions with the vacuum UV absorption transitions.

CP513, *Nuclear and Condensed Matter Physics,* edited by A. Messina
© 2000 American Institute of Physics 1-56396-929-7/00/$17.00

2. SAMPLES AND METHODS

Our measurements were carried out in different types of as grown and γ irradiated silica of commercial origin where the different contribution to the PL activity in the UV range could be singled out. We have examined a natural sample Herasil 1 (H1) [7] where α_E is revealed, an oxygen-deficient type (QC) [8] showing α_I and a synthetic Suprasil 300 (S300) [7], which, as manufactured, has no PL emission under excitation at 5 eV and shows an OA centered at 7.6 eV (E band). The γ irradiation was performed in a ^{60}Co source at room temperature.

The PL activities of our samples were investigated in the excitation energy range 6.0-8.5 eV by using the SR at the superlumi experimental station on the I-beamline of HASYLAB at DESY (Hamburg, Germany). PL and PLE spectra were obtained under multi-bunch operation, with 0.25 and 5.0 nm as excitation and emission bandwidth, respectively. The PL lifetimes were measured under single-bunch operation by scanning 192 ns between adjacent SR pulses, 0.5 ns wide. The experiments were carried out both at 300 and 10 K, by using a sample chamber with a helium-flow cryostat.

3. RESULTS

In Figure 1, PL spectra obtained in the as grown samples H1 (a) and QC (b) are reported together with the corresponding PLE profiles. In H1 the PL band is centered at 4.26 ± 0.03 eV and has a PLE spectrum centered at 7.5 eV. In QC the emission is peaked at 4.42 ± 0.02 eV and its PLE spectrum has a maximum at 6.9 eV. These two PL bands are clearly recognizable as the α_E and α_I emissions when observed under excitation at 5 eV. The PLE bands at 7.5 and 6.9 eV identify the transition energies from the ground to the second excited state in the centers responsible for α_E and α_I, respectively [9, 10]. As no OA band centered at 7.5 eV in H1 and at 6.9 eV in QC was detected, we rule out any correlation between these transitions and the E band at 7.6 eV which is observed in some silica samples [2, 4, 5].

A further experimental analysis regarding their kinetic behavior evidenced that both PL emissions at 4.2 and 4.4 eV have a lifetime of few nanoseconds which is independent on the excitation energy. More quantitatively, at T=300 K, α_I has a lifetime $\tau \approx 4.1$ ns whereas the decay of α_E is fitted by a single exponential law with $\tau \approx 5.5$ ns only for t ≥ 2 ns. At T=10 K, a lengthening of the lifetimes for both bands occurs and we measured $\tau \approx 4.6$ ns and $\tau \approx 7.3$ ns for α_I and α_E, respectively.

The PL measurements carried out on the S300 sample, irradiated with a γ dose of 100 Mrad, allowed us to single out other contributions to the 4.4 eV emission. In Figure 2 we report the vacuum-UV excitation profiles of the 4.4 eV PL as detected at T=300 K (a) and at T=10 K (b). The room temperature spectrum, peaked at ~6.9 eV,

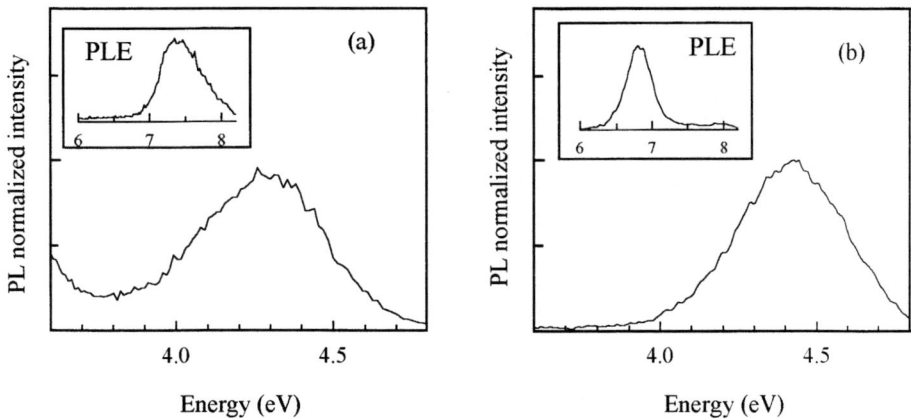

FIGURE 1. PL and PLE spectra as detected in H1 (a) and QC (b) samples under vacuum-UV excitation.

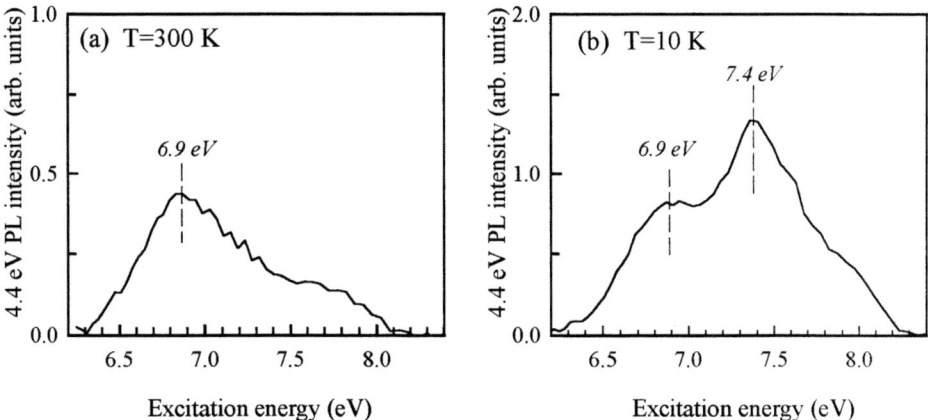

FIGURE 2. Vacuum-UV excitation spectra as detected in the 100 Mrad γ irradiated S300 sample at T=300 K (a) and T=10 K (b).

evidences the γ induced growth of the α_R band that exhibits close similarities to α_I as regards its peak energy, the PLE spectrum and the decay time [9]. At variance, the low temperature PLE profile appears quite different for the presence of an additional structure centered at 7.4 eV. This result provides evidence for a further contribution to the PL activity at 4.4 eV, strongly dependent on the temperature and named α_T [11]. The features of α_T were also investigated in the as grown S300 sample, which is characterized by the presence of the OA band E (7.6 eV) whereas the $B_{2\alpha}$ band and the related 4.4 eV emission are absent. The α_T band could be clearly distinguished from the isoenergetic α_I and α_R on the basis of its peculiar temperature dependence, excitation profile (7.4 eV) and lifetime ($\tau \approx 2.3$ ns). In particular, the PLE spectrum suggests that we should relate α_T to the 7.6 eV OA. As regards the excitation pathway of α_T involving the absorption transition within the E band, a two-configuration (TC) model, as proposed by Nishikawa et al. [5], can account for our results. In this scheme, the vacuum-UV excitation causes the conversion of the absorbing center from its stable configuration (C_S) to an unstable one (C_U), via the excited states. The relaxation from the excited state of C_U can take place either by emission at 4.4 eV toward the ground state (α_T band) or by a non-radiative back-conversion process to C_S whose occurrence affects the lifetime and the temperature dependence of the PL band. We note that in this model, C_U, from which the α_T emission originates, coincides with the defect responsible for the isoenergetic bands α_I and α_R, and all of the observed peculiarities of α_T are due to the different excitation process.

ACKNOWLEDGMENTS

The authors express their gratitude to Professor E. Calderaro (Department of Nuclear Engineering, University of Palermo) to taking care of the γ irradiation in the irradiator IGS-2. SR experiments were carried out at Hasylab (DESY, Hamburg) (Project no. I-96-08 EC, Project Leader Professor G. Spinolo). This work is a part of a National research project supported by Ministero Italiano della Ricerca Scientifica e Tecnologica, Roma, Italy

REFERENCES

1. Garino-Canina, V. K., *Comptes Rendus Acad. Sci.* **238**, 875-879 (1954).
2. Tohomon, R., Mizuno, H., Ohki, Y., Sasagane, K., Nagasawa, K., and Hama, Y., *Phys. Rev. B* **39**, 1337-1345 (1989).
3. Skuja, L. N., *J. Non-Cryst. Solids* **149**, 77-95 (1992).
4. Trukhin, A. N., Skuja, L. N., Boganov, and Rudenoko, *J. Non-Cryst. Solids* **149**, 96-101 (1992).
5. Nishikawa, H., Watanabe, E., Ito, D., and Ohki, Y., *Phys. Rev. Lett.* **72**, 2101-2105 (1994).
6. Anedda, A., Congiu, F., Raga, F., Corazza, A., Martini, M., Spinolo, G., and Vedda, A., *Nucl. Instrum. Meth. B* **91**, 405-409 (1994)
7. Heraeus Quartzglas, Hanau, Germany, catalogue POL-0/102/E.
8. Starna Ltd., Romford, England.
9. Boscaino, R., Cannas, M., Gelardi, F. M., and Leone, M., *Phys. Rev. B* **54**, 6194-6199 (1996).
10. Cannas, M., Barbera, M., Boscaino, R., Collura, A., Gelardi, F. M., and Varisco, S., *J. Non-Cryst. Solids* **245**, 190-195 (1999).
11. Boscaino, R., Cannas, M., Gelardi, F. M., and Leone, M., *J. Phys.: Condens. Matter* **11**, 721-731 (1999)

Influence of Thermal Irreversibilities on Thermoacoustic Engines

G. Cannistraro[1], C. Giaconia[2] and A. Piccolo[1]

Dipartimento di Fisica, Università degli Studi di Messina
Contrada Papardo, Salita Sperone 31, 98166 (S. Agata) Messina
Dipartimento di Energetica ed Applicazioni di Fisica, Università defli Studi di Palermo
Viale delle Scienze, 90128 Palermo

Abstract. We present an analysis of the thermal irreversibilities which dominate thermoacoustic refrigerators and assess, qualitatively, their impact on the engine performance.

INTRODUCTION

Performance optimization of real machines necessary requires an analysis of the sources of irreversibility which affect machine operation. As it is well known, irreversibility is a necessary cost to pay for operating with non-zero power output engines. The classical maximum-efficiency Carnot engine, in fact, requires reversible processes and this implies a zero power output since an infinite time is required to generate a finite amount of work. Irreversible finite-time processes, therefore, are essential for the operation of practical machines and their analysis is critical to establish limits on efficiency at maximum power. In view of the growing interest in thermoacoustic technology, in the present work we apply such analysis to thermoacoustic refrigerators.

ANALYSIS

A thermoacoustic refrigerator[1] may be idealized as composed of an acoustic resonator that houses a stack of parallel plates whose ends are thermally connected to the "cold" (C) and the "hot" (H) reservoir through heat exchangers. When such a system is excited at its fundamental resonance frequency by an acoustic energy source, such as an electroacousic transducer, a temperature gradient arises along the stack plates as a consequence of an hydrodynamic heat flow from the cold to the hot side of the stack within a the fluid thermal penetration depth.

As in all practical heat engines, we can recognize the first source of irreversibility in the heat transfer processes through finite temperature differences representing the driving forces for the heat fluxes Q_c and Q_H respectively from (C) and to (H). These unavoidable so called "external" irreversibilities, although contribute to system inefficiency, allow the engine to pump a finite amount of heat in a finite time, as

CP513, *Nuclear and Condensed Matter Physics*, edited by A. Messina
© 2000 American Institute of Physics 1-56396-929-7/00/$17.00

pointed out about two decade ago by Curzon and Ahlborn[2] who emphasized the maximization of power rather than thermodynamic efficiency.

The coefficient of performance (COP_R) of a refrigerator is defined as the ratio of the heat flux (Q_c) removed from the "cold" reservoir to the total power input (W_e):

$$COP_R = \frac{Q_C}{W_e} \qquad (1)$$

As we are dealing with an acoustically driven machine, we can individuate a source of "internal" irreversibility in the conversion process from electric (W_e) to acoustic (W_{ac}) energy performed by the electroacoustic transducer at an efficiency

$$\xi_{ea} = \frac{W_{ac}}{W_e} \qquad (2)$$

Nevertheless, only a fraction, W_2, of W_{ac}, is effective for engine operation. This power represents the acoustic energy dissipated through (a) irreversible thermal conduction between the gas and the stack-plates and (b) friction within the fluid viscous penetration depth. The source of irreversibility (a) is "productive" and favorable since it furnishes the "natural" time phasing for the elementary heat pumping cycle performed by a single fluid particle. Contribute (b) has, obviously, a deleterious effect and can be minimized properly choosing the working fluid and operating conditions. Fluids with low Prandtl number are strongly recommended. The Prandtl number, in fact, determines the ratio of viscous to thermal boundary layer thickness, so, a lower Prandtl number results in lower viscous losses for the same thermal diffusion. The rest of W_{ac} is dissipated by viscous shear and by thermal relaxation on the surface of the resonator and of the heat exchangers. So, we can write:

$$W_{ac} = W_2 + W_{res} + W_{exc} \qquad (3)$$

This argument allows us to define an "acoustic power efficiency" as

$$\xi_{ac} = \frac{W_2}{W_{ac}} \qquad . \qquad (4)$$

which accounts for the overall acoustic power dissipation within the resonator. On the other hand, the useful thermoacoustic effect takes place only in the stack whose intrinsic performance is usually defined as:

$$COP_{stack} = \frac{Q_2}{W_2} \qquad (5)$$

Q_2 being the pumped heat flow including in addition to a convective component a conductive one. The mean temperature gradient in the fluid and in the solid, in fact, causes adverse heat conduction, effectively reducing the amount of heat pumped by the fluid. COP_{stack} depends on a moltitude of thermophysical, geometrical and configurational parameters and its optimization is generally carried out defining design algorithm in multidimensional parameter spaces[3].

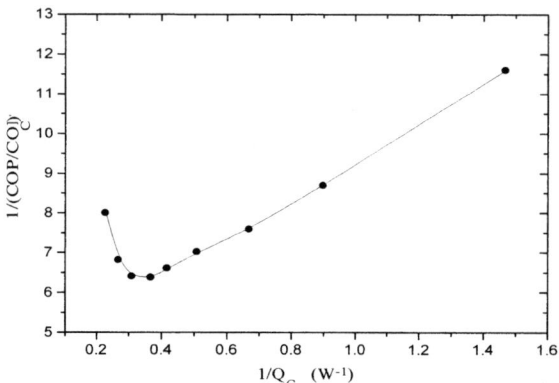

FIGURE 1. Inverse COP_R Relative to the Ideal Carnot COP_C as a Function of the Inverse Heat Load (Data Extracted from Ref. 6)

It should be noted that Q_2 differs from Q_c ($Q_2 > Q_c$) since the cold heat exchanger is also loaded by heating contributes arising from dissipative effects; namely, the acoustic power losses on the surface of the cold heat exchanger, $Q_{cexc} = W_{cexc}$ and of the cold portion of the resonance tube, $Q_{cres} = W_{cres}$:

$$Q_2 = Q_c + Q_{cexc} + Q_{cres} \qquad (6)$$

Analogously, the hot heat exchanger has to reject to the environment the heat

$$Q_H = Q_2 + W_2 + Q_{Hexc} + Q_{Hres} + Q_{vc} \qquad (7)$$

where $Q_{vc} = (W_e - W_{ac})$ is the heat generated by the resistive losses in the acoustic driver voice coil. Equation (6) shows that Q_2 would be the maximum amount of heat pumped from the cold reservoir if dissipation of acoustic power would not take place. In this way, it is possible to define an effectiveness ξ_{cexc} for the cold heat exchanger as the ratio of the heat transferred to the maximum heat transferable as:

$$\xi_{cexc} = \frac{Q_c}{Q_2} \qquad (8)$$

Obviously, ξ_{cexc} maximization entails minimization of the acoustic power losses W_{cres} and W_{cexc}; this, in turn, has a positive effect on ξ_{ac}. Swift estimated the term W_{exc} to be of the order of 25% of the produced acoustic power[4]. Substituting equation (2) (4) (5) (8) into equation (1) it follows:

$$COP_R = \xi_{ea}\xi_{cexc}\,\xi_{ac}COP_{stack} \qquad (9)$$

that is the same result obtained in Ref. 3. Eq. (9) shows that COP_{stack} represents the upper limit of COP_R; this limit is attainable only if the dissipative effects embodied by parameters ξ_{ea}, ξ_{ac} and ξ_{cexc} are minimized. The relative influence of the irreversibilities previously discussed can be seen qualitatively in the curve $1/COP_R$ vs $1/Q_c^5$ (Fig 1). COP_R increases at low cooling rates due to the internal irreversibilities such as thermoviscous dissipation on the resonator surface; these losses are responsible for the slope of the linear region of the curve. Losses deriving from finite-rate heat transfer, on the other hand, dominate at high cooling rates so that COP_R decreases as cooling rate increases. Irreversibilities oppose to very fast or very slow cooling rates so COP_R vanish in both these limits exhibiting a maximum in the intermediate region of moderate cooling rates.

CONCLUSIONS

The overall COP_R of thermoacoustic refrigerators has been expressed as the product of four partial efficiencies, each one deriving from a different source of irreversibility These irreversibilities account for the fact that, at the present time, efficiencies of thermoacoustic refrigerators are at 20% to 30% of Carnot efficiency, still below those of conventional technologies. Part of this lower efficiency is due to the intrinsic irreversibility associated with the thermoacoustic effect. This intrinsic irreversibility, on the other hand, constitutes the favorable aspect of the cycle since it makes for mechanical simplicity, with few or no moving parts. Improvements in COP_R are expected mainly from optimization of the heat exchangers. It has been also argued that efficiency should increase due only to the fact that thermoacoustic refrigerators are well suited for proportional control as opposed to conventional vapor compression chillers which have constant displacement compressors and are therefore only capable of binary (on/off) control. Proportional control reduces the inefficiencies in the heat exchangers, since the proportional system can operate over smaller temperature differences between the coolant fluid and the heat load.

REFERENCES

1. Swift, G. W., *J. Acoust. Soc. Am.* **92**, 1145-1180 (1988).

2. Curzon, F. L., and Ahlborn, B., *Am. J. Phys.* **43**, 22-24 (1975)

3. Wetzel, M., and Herman, C., *Int. J. Refrig.* 20, 3-21 (1997).

4. Swift, G. W., *J. Acoust. Soc. Am.* **92**, 1551-1563 (1992).

5. Gordon, J. M., and Ng, K C., *Int. J. Heat Mass Transfer* **38**, 807-818 (1995).

6. Garret, S. L., Adeff, J. A., and Hofler, T. J., *J. Therm .Heat Transfer* 7, 595-599 (1993).

Analogue Memories for High Speed Detectors Signals

Luigi M. Caponetto*

*Physics Dept. and I.N.F.N. Catania Corso Italia 57, Catania, Italy I-95129
email: Luigi.Caponetto@ct.infn.it

Abstract. An analogue memory chip family has been developed as a switched capacitor cell array. The family is an ongoing prototyping project within the readout system of the Silicon Drift Detectors for the Inner Tracker System of the ALICE experiment. All chips consist of several capturing channels whose cells number is varying within the family. Chips design choices and performances are briefly discussed here.

INTRODUCTION

The problem of capturing fast transient waveforms of modest duration with acquisition rates up to 100 MHz, is common to many High Energy Physics (HEP) experiments [2]: samples are acquired in a storing unit until a trigger decision is made. Depending on this decision, samples are either retrieved at a lower rate for digitization and post processing, or discarded. Applying this process to huge quantities of analogue signals - hundred thousands in such cases - could easily affect important system costrains such are density, cost and power dissipation. Furthermore, in HEP applications, only parts of the analog input waveforms are of interest, doesn't requiring a continous digitization [3]. Solutions often relies on GaAs CCDs as well as on fast analog to digital converters followed by digital memories, but the feasibility of using switched capacitor techniques to address these drawbacks was showed in many different applications [2].

A Switched Capacitor Array (SCA) analog memory family has been developed to address the requirements of the readout system of the ALICE SDDs.

FIGURE 1. The ADeLine1 chip microphotography.

CP513, *Nuclear and Condensed Matter Physics,* edited by A. Messina
© 2000 American Institute of Physics 1-56396-929-7/00/$17.00

SYSTEM DESCRIPTION

A massively parallel data acquisition system as the one described in [5], both uses time delaying for trigger latency and time stretching of the stored waveforms to match the lower bandwidth of the analogue-to-digital converters processing the data. These functions are provided by the ADeLine (Analog Delay Line) family with only slight differences between the various versions. ADeLine1 was the first attempt to study feasibility of the basic memory cell when built in silicon. It was (Figure 1) a single acquisition channel with 8 memory cells. While completely functional, its readout rate was quite lower (100 KHz) in respect to the 1 MHz target. With the ADeLine3 and ADeLine4 chips several improvements were made both at the physical silicon layout and at the system level. A new memory cell capacitor was designed and the internal address circuitry was redesigned to meet the target 256 cells/channel depth costrain. The channels count for these chips was respectively 3 and 4. Both were able to drive 15 pF loads at 1 MHz readout rate. ADeLine5 and ADeLine6 chips eventually meet the 16 channels requirement offering a totally renewed digital control internal interface (ADeLine6). Table1 resumes target values for the final chip.

All the chips within the family are realized in the AMS CYE (0.8 μm double metal double poly CMOS) process. Memory operation consists of a continous 40 MHz parallel sampling phase of all channels inputs, interruptible by issuing an external trigger. During this phase new samples simply rewrite old cells everytime the single channel depth has been reached. Cells addressing is linear within a channel: a new cell is addressed at any positive edge of the system clock. When the write phase is paused, memory begins its lower frequency read phase. Then

 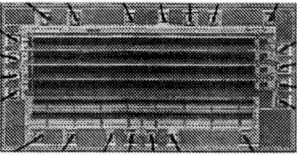

FIGURE 2. ADeLine3 (left) and ADeLine4 (right) chip microphotographies.

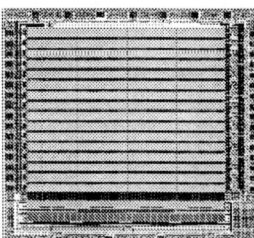 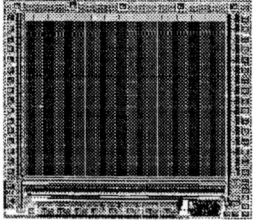

FIGURE 3. ADeLine5 (left) and ADeLine6 (right) design layout views.

TABLE 1. Main charactheristics.

	Value	Unit
Write/Read frequency	MHz	40/1
Cells×channels		256×16
Power/channel	mW/ch	≤ 4
Linear range	V	1
Dynamic range	V	3
Pedestal var.	mV rms	≤ 1
Resolution	bit	11
DC gain		1

each output presents the values stored inside the cells. Each channel uses its own output buffer for the readout of the stored samples, thus allowing parallel readout. Power dissipation requirements bound the maximum rate of this phase to 1 MHz over a 15pF capacitive load. The same addressing control circuit is used for both phases.

The principle of the sampling technique adopted is showed in Figure 4(a). Here the write mechanism consists of a sampling of the input voltage for the whole active phase of a 50% duty-cycle 40 MHz external clock (Voltage Sampling technique). The same figure shows the sampling MOS switch bound at a fixed external voltage at its source. This particular topology allows a better reconstruction of the input signal shapes and makes the pedestal error V_{ped} and the MOS induced noise independent from the input voltage [1]. The voltage stored in the sampling capacitor is presented to the output trough a transcoductance amplifier with a voltage transfer mechanism (see Figure 4(b)), allowing a cell gain insensitive to the capacitance spread along the channel [1].

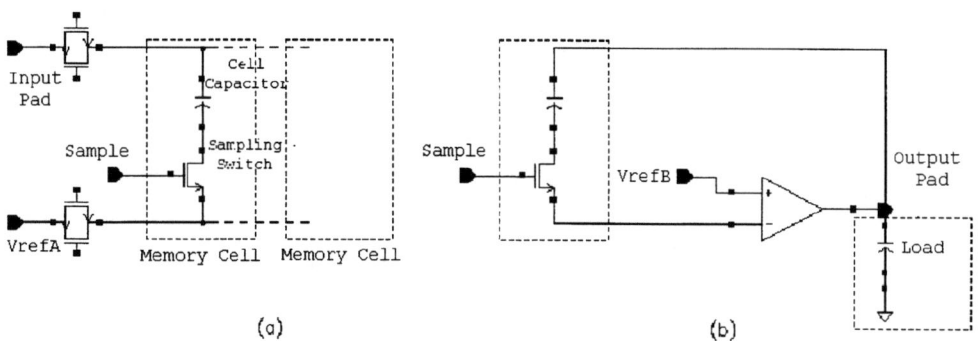

FIGURE 4. ADeLine sampling (a) and readout (b) schematics.

DESIGN AND PERFORMANCES

The main design costrains for our purposes are: (a) low power (less than 4 mW/channel) due to overall thermal costrains; (b) small silicon area to reduce multiple scattering of particles inside the ITS; (c) high output resolution (less than 1 mV) for a good reconstruction of the input shape. Some design optimizations has been made in order to meet all target requirements.

Clock feedthrough. This is a common effect for all SC-based circuits. It can be shown (see [1]) that it depends on the source voltage of any MOS transistor switching at one end of a capacitor. As already mentioned, the circuit topology in Figure 4(a) allows to make the charge injected in the sampling cell independent from the input voltage . The use of CMOS switches has been avoided within the memory cell: the compensation achieved by using complementary devices would be hampered by the increased fluctuation in the rms value of the V_{ped} voltage. CMOS switches are used only for connection to the rail busses.

Capacitors layout. Polysilicon-oxide-polysilicon capacitors has been designed for the memory cell. Because of the topology choosen, the cell-gain variation along the channel depends substantially on the mismatch of the sampling capacitors. In order to minimize the relative capacitance error due to edge variations, a square or circular shape should be adopted. To avoid the undercut effect in the realization of the capacitor, identically sized smaller units are connected in parallel to construct a larger one. Tests on ADeLine4 and ADeLine5 chips shows a resolution of 10 bit with a V_{ped} variation of $0.821mV$ in a voltage range of $3.5V$.

CONCLUSIONS AND FUTURE DEVELOPMENTS

All of the main design challenges were met with the ADeLine6 chip though it will not be available for test until the end of November. Test measurements on previous chips anyway, show some minor adjustment is required to fit final costrains. Next step will be a new prototype featuring an integrated preamplifier inside the memory chip [4].

REFERENCES

1. Panebianco S. *et al. NIM A* **434**, (1999) 424–434.
2. Brönnimann. *et al. NIM A* **420**, (1999) 264–269.
3. Haller, G.M. *et al. IEEE Trans. on Nuclear Science* **41**, 4 (1994) 1203–1207.
4. Randazzo, N. *et al. Proc. of* 5th *Workshop on Electronics for LHC Experiments (LEB99)* to be published.
5. ALICE Collaboration *ALICE - A Large Ion Collider Experiment Technical Design Report 1999* http://www.cern.ch/Alice/TDR/

Anharmonicity And Fragility In Linear And Reticulated Polymers

G. D'Angelo, G. Carini, G. Tripodo, A. Bartolotta* and V. P. Privalko[+]

Dipartimento di Fisica and INFM, Università di Messina, 98166 S. Agata, Messina, Italy
*Istituto di Tecniche Spettroscopiche del C.N.R., 98122 Messina, Italy
[+]Institute of Macromolecular Chemistry, National Academy of Sciences of Ukraine, 253160 Kyiv, Ukraine

Abstract. The mechanical characteristics in the region of the primary relaxation have been measured in linear (bisphenol-A-polycarbonate and polyurethane) and reticulated (a fully crosslinked heterocyclic network) amorphous polymers. It is found a correlation between the fragility and anharmonicity of the systems, the most fragile polymer being characterized by the largest anharmonicity.

I. INTRODUCTION

Polymer systems exhibit a complex dynamic behavior, whose study is make very hard mainly by their complicated inhomogeneous microstructures and different binding forces within and between chains. The influence of molecular structure and local environment of polymer chains on the physical properties is more relevant at higher temperatures and it is believed that it characterizes the extent of deviation from linearity and from exponentiality of dynamics of segmental relaxation in the glass transition range. At low temperatures polymeric chains are frozen and the thermal vibrations and elastic properties are controlled mainly by the weak Van der Waals interchain forces. In this paper we will present a study of the extent to which interchain binding and morphology of amorphous polymers regulate the dynamics of segmental relaxation in the glass transition region. Measurements of the temperature behavior (120K-600K) of the complex dynamic modulus $E^* = E' + iE''$ have been performed on various polymers including two linear polymers (bisphenol A-polycarbonate (BPA-PC) and polyurethane (LPU) and a fully crosslinked heterocyclic polymer network (HPN). The present analysis shows that the tightly packed structure of reticulated network hinders the thermal degradation of the amorphous chains, enhancing their ability to preserve the short and medium range order, as compared to the two linear polymers. In the latter systems the presence of weaker Van der Waals interchain forces makes easier the mechanism driving the collective phase changes of several chain segments giving rise the glass transition. The experimental findings are analyzed in a general outlook assuming that the anharmonicity of the inter- and intra-chain vibrations, connected to Van der Waals

CP513, *Nuclear and Condensed Matter Physics,* edited by A. Messina
© 2000 American Institute of Physics 1-56396-929-7/00/$17.00

and covalent potential respectively, regulate the relaxation dynamics in amorphous polymers.

II. RESULTS AND DISCUSSION

The temperature behaviors of E' of studied polymers are characterized by a sharp drop associated to the glass-rubber (α_a-) transition (see Fig. 1). In the region below T_g, E' in all samples shows a roughly linear decrease with temperature, which is independent of frequency in 3-30Hz range. The shift of the drop of E' to higher temperatures going from LPU to HPN indicates increasing values of T_g.

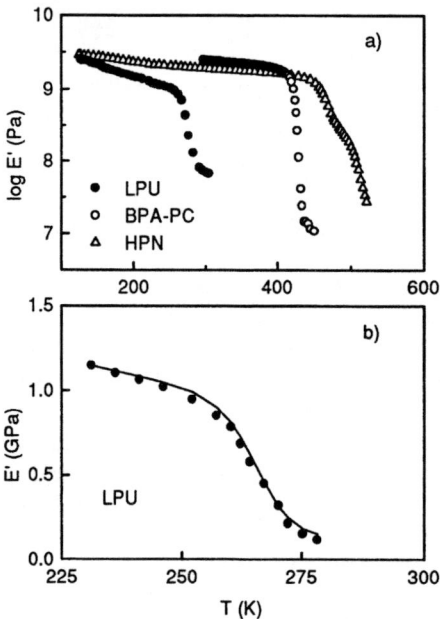

Figure 1. (a) E' (T) at a driving frequency of 3 Hz in BPA-PC, LPU and HPN; (b) theoretical fit (solid line) by Eq. (1) of E'(T) in the region of the α_a-relaxation for LPU; the mechanical frequency is 0.3 Hz.

As discussed elsewhere [1], the temperature dependence of the elastic behavior in the region of T_g, when not affected by dissipative secondary processes induced by configurational transitions of molecular segments or side groups, can be described by two overlapping different mechanisms. The first one is the anharmonic vibrational contribution which causes the nearly linear temperature variation of the elastic constant and the second one is the cooperative primary relaxation which produces the sharp drop in the modulus. In this approach, the modulus E' (ω,T) in the region of the α_a-relaxation is given by the relation:

$$E'(\omega,T) = [E_o - b'(T-T'_o)]_{anh} + \left[E'_\infty + \delta E' \int \frac{d\phi(t)}{dt}\cos(\omega t)dt\right]_{rel} \quad (1)$$

where E_0 is the value of E' at the lowest temperature T'_0 in the experiment and the parameter b' is mainly determined by the Grüneisen coefficient, which account for the anharmonic interactions between the vibrational modes. The relaxation term is represented by the Fourier transform in the frequency domain of the stress relaxation function E'(t) [2], where E' (t) = E_0' + (E_∞' - E_0') ϕ(t) , E_∞' and E_0' are the high-frequency (unrelaxed) and low-frequency (relaxed) storage moduli, respectively. In this expression, ϕ(t) = exp [-(t/τ)$^\beta$], is the Kohlrausch-Williams-Watt (KWW) "stretched exponential function" and $0 < \beta < 1$ is a phenomenological measure of the non-exponentiality (the width of the relaxation times spectrum).

Finally the relevant relaxation time τ is expressed by the Vogel-Fulcher-Tamman (VFT) equation, $\tau = \tau_0$ exp [B/(T - T_0] , where τ_0 is the characteristic time and B and T_0 are the empirical fitting constants. The values of the parameters of Eq. (1), except for b' which has been directly evaluated trough the slope of E' in the glassy region, were determined by a minimum search program that provides the simultaneous best fit of the experimental data at all the available frequencies for each sample.

TABLE 1. Values of typical parameters for the studied samples. β, B, To, are the coefficients obtained by fitting the mechanical primary relaxation, b' is the anharmonicity coefficient, m is the fragility parameter and γ_{th} is the Grüneisen parameter. The values of density ρ refer to room temperature.

Samples	ρ (gr cm^{-3})	T_g (K)	M	β	B (K)	T_0 (K)	b' (10^{-3}GPaK^{-1})	γ_{th}
PVC	1.27	320	171				27.3	1.0
BPA-PC	1.20	418	119	0.41	1512	370	11.0	0.6
LPU	1.24	255	89	0.32	1612	213	6.9	
HPN	1.21	404	45	0.35	3314	351	5.5	

The analysis of the data reported in Table 1 reveals an interesting feature concerning the values of b'. It decreases going from BPA-PC to HPN and follows the behavior of the "fragility" parameter m of the systems, which describes the inability of glass-former systems to preserve the short and medium range order against the thermal degradation [3]. The "fragility" has been introduced by Angell to classify the glass-former liquids in terms of the nature of bonds present in the system. High values of m (~100) set weak molecular and simple liquids (*fragile*) and low values (~20-30) the covalently bonded liquids (*strong*).

The fragility decreases markedly by going from linear BPA-PC, through LPU, to cross-linked HPN, as a consequence of chemical crosslinks which make the HPN more resistant to thermal degradation. Concerning the linear polymers, the smallest m in LPU is believed to arise from additional forces between neighboring chains due to the sharing of hydrogen atoms by neighboring segments, which contain the strongly electronegative nitrogen atoms. These forces cause an additional binding (besides the Van der Waals interactions) between molecular chains whose strength is intermediate between the covalent and Van der Waals binding forces, reducing the cooperative effects typical of *fragile* systems. Moreover, as before discussed, the anharmonicity factor b' depends on Grüneisen coefficients, which can be quantified by the mean thermal Grüneisen parameter $\gamma_{th} = 3\alpha B^s / \rho C_p$. In this expression α is thermal expansion coefficient, ρ the density, B^s the isentropic bulk modulus and C_p the specific heat measured at constant pressure. The Grüneisen parameter is a measure of the asymmetry of the binding potentials involved in the molecular structure or of the anharmonicity of oscillations. It is believed that, in polymers investigated, the b' values describe the decrease of the interchain potential asymmetry, which is correlated to a growing strength of the interchain bonds, going from nearly pure Van der Waals forces in BPA-PC, to covalent bonds in crosslinked HPN. To establish that the revealed correlation between b', γ_{th} and *m* represents a general feature of amorphous polymers, we have extended the analysis to another linear polymer, the PVC (poly(vinyl chloride)). The high value of *m* (=171) sets PVC such as the most fragile among the polymers investigated. It has been observed (see Table 1) that also b', obtained by the temperature dependence of E' [4], follows the same trend. More importantly the values of γ_{th}, also reported in Table 1 for PVC and BPA-PC [5], allow an useful check of the existence of the claimed relation between b' and γ_{th}. It results that γ_{th} increases with increasing m, the most fragile glass-former showing the highest anharmonicity and, even if the comparison is restricted to two polymers only (BPA-PC and PVC), the behavior of b' reflects that of γ_{th}. Thus it can be concluded that the class of amorphous polymers appears to be regulated by a correlation, which relates the "fragile" or "strong" character of these glass formers to a larger or smaller anharmonicity.

REFERENCES

1. Bartolotta, A., Di Marco, G., Lanza, M., Carini, G., 1993, Phys. Rev. **B48**, 10137.

2. Nowick, A. S., and Berry, B. S., 1972, *Anelastic Relaxation in Crystalline Solids*, (Ac. Press, New York).

3. Böhmer, R., Ngai, K. L., Angell, C. A., Plazek, D. J., 1993, J. Chem. Phys. **99**, 4201.

4. D'Angelo G, Tripodo G., Carini G., Bartolotta A., Di Marco G., Salvato G., J. Chem. Phys. **109**, 7625 (1998).

5. Hartwig G., *Polymer Properties at Room and Cryogenics Temperatures*, (*Plenum Press, New York 1994*).

Tensor-Product States and Local Indistinguishability: an Optical Linear Implementation

A.Carollo[1], G.M.Palma[1], C.Simon[2], A.Zeilinger[2]

[1]Dipartimento di Scienze Fisiche ed Astronomiche, Universita' di Palermo & Unita' INFM
Via Archirafi 36,I - 90123 Palermo, Italy
[2]Institut fuer Experimentalphysik, University of Vienna
Boltzmanngasse 5, 1090 Wien, Austria

Abstract. In this paper we investigate the properties of distinguishability of an orthogonal set of product states of two three level particle system by a simple class of joint measures. Here we confine ourselves to a system of analysis built up of linear elements, such as beam splitters and phase shifters, delay lines, electronically switched linear devices and auxiliary photons. We present here the impossibility of realization of a perfect never falling analyzer with this tools.

THE "SAUSAGE" STATES

The most celebrated phenomena of quantum non-locality are often considered as manifestations of the entangled nature of multipartite systems: i.e. compound systems that admit no description in terms of the states of their parts. Indeed entangled states exhibits a very immediate non local property: they cannot reliably be distinguished by locally measuring their separate parts. Due to this properties this states have been used to provide strong evidences for the validity of Quantum Mechanics as their behavior cannot be described in classical terms. On the other hand, no non local properties are expected to manifest themselves for a set of orthogonal product states of a bipartite system. This however is not generally true. Bennett et al. [1] have shown the existence of a set of orthogonal unentangled states of a bipartite system that are fully distinguishable only by a global joint measurements of the separate parts. More specifically they have constructed a product state basis for a system of two three level system whose elements cannot be reliably discriminated by two observers that are allowed any sequence of local measurement and classical communication with each other. In other word: let's consider a couple of systems, say A and B, each of which described by a three dimensional Hilbert space. In the nine dimensional Hilbert space of the compound system A⊗B take the following orthogonal product state basis:

CP513, *Nuclear and Condensed Matter Physics*, edited by A. Messina
© 2000 American Institute of Physics 1-56396-929-7/00/$17.00

$$|\psi_0\rangle = |2\rangle_A \otimes |2\rangle_B$$

$$|\psi_{1,2}\rangle = 1/\sqrt{2}\left(|1\rangle_A \pm |2\rangle_A\right)\otimes |1\rangle_B$$

$$|\psi_{3,4}\rangle = 1/\sqrt{2}\,|1\rangle_A \otimes \left(|2\rangle_B \pm |3\rangle_B\right)$$

$$|\psi_{5,6}\rangle = 1/\sqrt{2}\left(|2\rangle_A \pm |3\rangle_A\right)\otimes |3\rangle_B$$

$$|\psi_{7,8}\rangle = 1/\sqrt{2}\,|3\rangle_A \otimes \left(|1\rangle_B \pm |2\rangle_B\right)$$

suppose now that the compound state of A⊗B is prepared in one of the above states and that system A and B are given to two separate observer allowed to do only repeated local measurements and classical communication . It can be shown that in spite of the fact that such states are unentagled there is no way by which they can be identified unless a global joint measurement on A⊗B is performed.

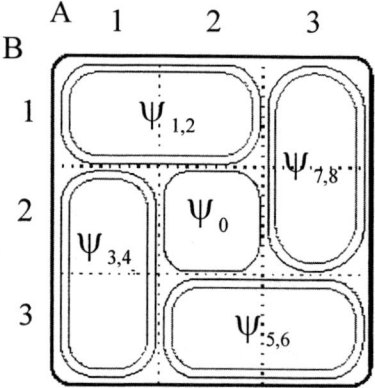

Figure 1 Graphical representation of the 'sausage states' as a system of dominos. The fact that even if these states are globally orthogonal their parts are not, is evident in the picture, where the measures are represented as cut along dashed lines.

LINEAR OPTICAL IMPLEMENTATION

The problem of an experimental realization of such states an of their joint measurement remains still an open problem. In this short communication we will confine ourselves to an experimental implementation involving only linear quantum optical elements.

In our scenario systems A and B are two photons, their Hilbert space being spanned by three different modes of the electromagnetic field, i.e. each photon can follow one of three different possible path and in this way we obtain two three level systems using six independent modes. Although we assumed to work with spatial mode, it should be noted that this scheme can be easily generalized, considering any possible degrees of freedom of photons, for example we could include polarization of photons to characterize particles A and B, or decide to span the Hilbert space of a particle using modes distinguishable by frequency, time or anything else allows us to define independent modes. As for the measuring apparatus we restrict ourselves to linear elements only, this means that we couple the input modes only with linear devices, like beam splitter and phase shifters and generalizations[3,4], in order to map the input modes with some configuration of output states . In this way placing detectors on the output mode we can perform a broad class of measurement on this sausage states. We will also suppose the detectors to be ideal, i.e. 100% efficient. Another important assumption we make in our analysis, is that this detectors are able to distinguish the number of photons by which they are triggered. Although this assumption is not realistic indeed, if we show that it is impossible to discriminate these states with ideal devices a fortiori it will not be possible to discriminate them with real devices. With these assumptions we can interpret the detection as a projection of the input states into a particular output mode Fock states.

To improve the capability of the measuring apparatus there could be good strategies which make use of auxiliary modes coupled to the six original input modes, as to perform some measurement with the help, for example, of some 'ancillary states'. In our implementation we keep in mind also this possibility. Another tool that can be useful to introduce is the ability to perform conditional measurements. In our scheme we could place a detector on a selected mode and interpret it as a control device, whose outcomes influence the successive transformation acting on the remaining modes.

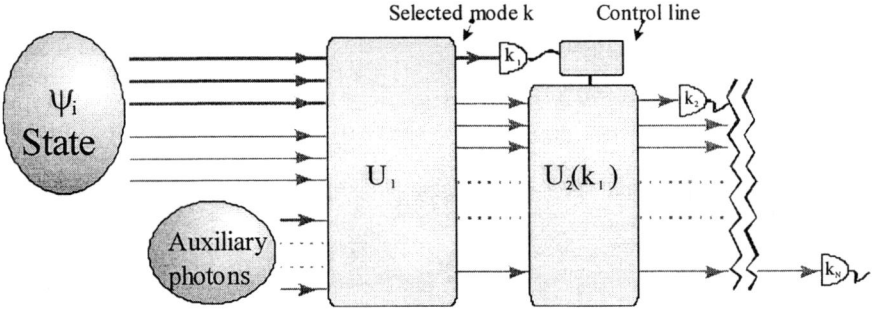

Figure 2 General scheme of our concatenated measurement

Finally, the general scheme of the type of measurement we can perform can be summarized in a succession of elementary blocks of transformation, as sketched in the figure 2. Each block is composed by a unitary linear transformation followed by a detector monitoring a selected mode This influences the subsequent transformation on the remaining modes.

We have found, following similar arguments used by Luetkenhaus et al.[2] in the case of a Bell discriminator, that this class of instrument are not able enough to discriminate completely the 'sausage states'[5]. It is possible to demonstrate that even after a single step of analysis represented by the first block shown in figure 2, the output states are in general non-orthogonal and this cause the states to became undistinguishable. One might expect that mutually indistiguishability of orthogonal states by linear devices were a peculiar feature of entangled states. We have shown that this is not the case.

On the other hand we have found a class of measurements involving simple non linear devices that are able to discriminate completely the sausage states[6].

ACKNOWLEDGMENTS

This work was supported in part by the European TMR Research Network under contracts FMRX-CT-97-0143 and by the ERASMUS student mobility scheme.

REFERENCES

1. C. Bennett, D. DiVincenzo, C.A. Fuchs, T. Mor, E. Rain, P.W.Shor, J.A. Smolin, W.K. Wootters , *Phys. Rev. A* **59**, 1070 (1999)

2. N. Luetkenhaus, J. Casamiglia, K.-A. Suominen, *Phys. Rev.* A **59**, 3295 (1999)

3. M.Zukowski, A. Zeilinger, M.A.Horne, *Phys.Rev.* A **55**, 2564 (1997)

4. M. Reck, A. Zeilinger, H.J.Bernstein, and P.Bertani, *Phys.Rev.Lett.* **73**, 58 (1994)

5. A.Carollo, G.M.Palma, C.Simon and A. Zeilinger, in preparation

6. A.Carollo, G.M.Palma, B.Sanders and A. Zeilinger, in preparation

Thermal Broadening Of Lb Band Of "Trehalose Coated" Tyrosine And Phenylalanine

Rita Carrotta, Vincenzo Sanfratello, Maurizio Leone
and Lorenzo Cordone

Instituto Nazionale di Fisica della Materia and Department of Physical and Astronomical Sciences, University of Palermo, 90123 - Palermo, Italy.

Abstract. We studied the thermal broadening of Lb band of tyrosine and phenylalanine embedded in a trehalose matrix. Aim of this work is to obtain information on the effects of "trehalose coating" on the coupling of the electronic transition to low frequency modes in the surrounding of the chromophore. The results obtained for the two molecular complexes put in evidence that O-H groups are involved in blocking these structures within the solid trehalose matrix and shed light on the role played by hydrogen bonds on the interactions that keep "trehalose coated" proteins rigid and solid-like.

INTRODUCTION

Several organisms can afford stresses due to extreme drought and/or high temperature in a state of "suspended animation" called anhydrobiosis; they can remain in this state for long time and resume their lifecycle following rehydration. These organisms contain, in the dry state, large amounts of trehalose that has been found to be the most active saccaride for preservation of biostructures.

In a study aimed to understanding the effects of "trehalose coating" on protein dynamics, Cordone et al. (1,2) reported that embedding carbonmonoxy myoglobin (MbCO) in a trehalose glass hinders the large amplitude *protein specific motions*, i.e. the non harmonic motions arising from thermal fluctuations of a protein among conformational substates. It was found a dramatic reduction of quasi diffusive modes of large scale, as detected by Mossbauer spectroscopy and by neutron diffusion, whereas the effect was found to be much smaller for the highly localized motions of the iron with respect to the heme plane, as detected by optical absorption spectroscopy. In agreement with the above results, recently Librizzi et al. (3,4) showed a sizeable effect of trehalose coating on the distribution of the A substrates of the CO bound and on their interconversion, related with the amount of water present in the solid matrix. With the aim of explaining the mechanisms through which biostructures are blocked in solid trehalose matrices, Crowe (see e.g. 5) first proposed the so-called water replacement hypotesis, developed for trehalose coated membranes. According to this hypotesis, trehalose hydrogen-bonds with the polar headgroups of lipids that constitute

CP513, *Nuclear and Condensed Matter Physics,* edited by A. Messina
© 2000 American Institute of Physics 1-56396-929-7/00/$17.00

the membranes, thus replacing the water of hydration at the membrane-fluid interface and maintaining the membrane headgroups at their hydrated position.

In this paper we report the thermal broadening of the near UV band (arising from the Lb $\pi-\pi^*$ electronic transition) of tyrosine and phenylalanine embedded in a trehalose matrix; this, in turn, give information on the low frequency motions of the matrix and/or of the chromophore itself, coupled to the electronic transition. Since tyrosine and phenylalanine differ only for the presence of an O-H group at the end of their residual, the comparison between the two amino acids shed light on the role of O-H groups at the protein surface as *"pinning points"* that keep the proteins rigid within the solid saccharide matrix.

MATERIALS AND METHODS

Sample solutions contained 0.2 M trehalose, 10^{-1} M phosphate buffer pH 7, and tyrosine (to saturation at room temperature) or phenylalanine ($\sim 1,5 \cdot 10^{-2}$M) respectively. Vitreous samples were obtained by drying 1 ml of the above solutions layered on a suitable quartz surface (1 cm × 4 cm). The drying procedure was performed by first putting the sample for ~ 5 h in a silica gel dessicator and then for ~ 1 h in an oven at 80EC. Drying within the oven was necessary in order to obtain non crystalline, perfectly transparent, vitreous samples.

Spectra were obtained by using a Cary 2300 spectrophotometer with a spectral resolution of 0.5 nm. The spectral band profiles were analyzed by a deconvolution procedure that explicitly takes into account the vibronic coupling of normal modes of the aromatic ring with the $\pi-\pi^*$ electronic transition, responsible for the observed Lb band; due to its electronically forbidden nature, besides the usual Franck-Condon progression of vibronic replica, we must introduce an expansion, to the first order, of electronic dipole moment in the normal vibrational modes, following an Herzberg-Teller approach (6,7). The fine structure of the bands consists in a series of Lorentians, each modulated in width by the coupling of the electronic transition with a bath of soft vibrational modes (Gaussian broadening).

Within the Einstein approximation, the low frequency bath can be considered as a set of N harmonic oscillators of average frequency $<v>$ and average coupling constant S. The temperature dependence of the Gaussian half-width can be expressed as:

$$\sigma_{harm}^2(T) = N S <v>^2 \coth\left[\frac{h<v>}{2K_BT}\right] + \sigma_{in}^2 \qquad (1)$$

The subscript "harm" indicates that the above expression is valid only in the harmonic regime; the term σ_{in}^2 represents the inhomogeneous broadening of the band that is considered as a Gaussian distribution of the 0-0 transition frequency.

84

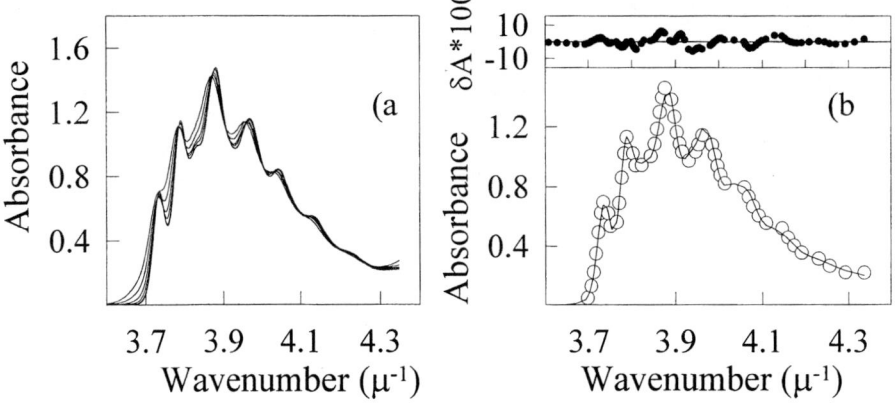

FIGURE 1. (a) The spectra of "trehalose coated" phenylalanine as a function of temperature in the range 300-20 K. (b) Deconvolution of the 20 K spectra; circles are the experimental points, continuous line represents the overall synthesized profile; the residuals are also reported, in an expanded scale.

RESULTS AND DISCUSSION

Fig. 1a shows the thermal behavior of spectral band profile for "trehalose coated" phenylalanine. The fitting of 20 K spectrum according to the above reported procedure is shown in Fig. 1b; analogous fitting quality was obtained in the whole temperature range, both for tyrosine and phenylalanine.

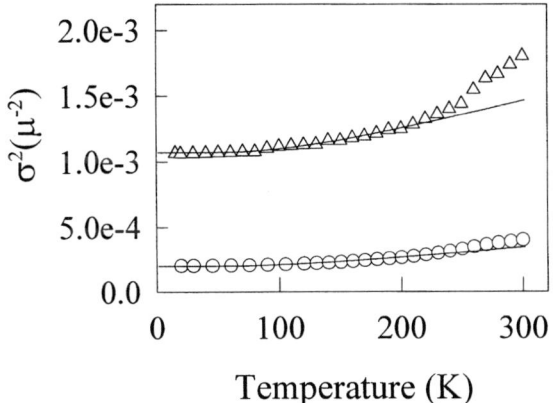

FIGURE 2. Temperature dependence of σ^2 for phenylalanine (circles) and tyrosine (triangles). Continuous lines represent the best-fit of Eq. 1 to the experimental points.

85

The temperature dependence of the parameter σ^2 both for the phenylalanine and tyrosine is reported in Fig. 2. Data in Fig. 2 can be fitted in terms of Eq. 1 only in a limited temperature interval; deviation from the harmonic behavior (continuous line in Fig. 2) are evident at high temperatures (T>230 K); these deviations are much larger for tyrosine, wherein an O-H group at the end of the residual is present. This finding stems for hydrogen bonding of tyrosine with the surrounding solid matrix. In fact, in view of the coupling of the Lb electronic transition with the vibrational normal mode of the O-H groups, one expects an harmonic behavior of the band linewidth up to temperatures at which the proton starts oscillating between the acceptor and the donor group, thus bringing about the non harmonic σ^2 behavior evident in Fig. 2. Accordingly this effect is absent in phenylalanine due to the lack of the OH. Moreover, the quite larger σ_{in} value obtained for tyrosine with respect to phenylalanine, suggests that the presence of hydrogen bond largely influences the structure/energy landscape in the surrounding of this amino-acid.

REFERENCES

1. Cordone, L., Galajda, P., Vitrano, E., Gassmann, A., Ostermann, A., and Parak, F., *Eur Biophys. J.* **27**, 173-176 (1998).
2. Cordone, L., Ferrand, M., Vitrano, E., and Zaccai, G., *Biophys. J.*, **76**, 1043-1047 (1999).
3. Librizzi, F., Vitrano, E., and Cordone, L., *Biophys. J.*, **76**, 2727-2734 (1999).
4. Librizzi, F., Vitrano, E., and Cordone, L., in *Biological Physics*, edited by H. Frauenfelder et al., AIP Conference Proceedings 487, New York: American Institute of Physics, 1999, pp. 132-138.
5. Crowe, J. H. and Crowe, L.M, *Science*, **223**, pp. 701-703 (1984).
6. Sanfratello, V., *Ruolo delle distorsioni indotte da leganti di grosso ingombro sterico sulle proprietà stereodinamiche del sito attivo di emoproteine,* Thesis, 1998.
7. Sanfratello, V., Boffi, A., Cupane, A., and Leone, M., (1999), submitted.

Measurement Of Electrical Field Amplitude By A Needle-Like Detector

G. Compagno, F. Persico

Istituto Nazionale di Fisica della Materia and Dipartimento di Scienze Fisiche ed Astronomiche dell'Università di Palermo, via Archirafi 36, 90123 Palermo, Italy

Abstract. The Bohr-Rosenfeld measurement theory of the electromagnetic field amplitude is applied to a needle-like detector. The expression of the time average of the self-force due to the measurement process is explicitly obtained. It is shown that the appearance of divergences does not allow the approximations required by the Bohr-Rosenfeld theory.

In QM field strengths are measured by measuring the moment transferred to massive charged bodies. It was shown by Landau and Peierls, for pointlike charged bodies, that fundamental quantum uncertainties made impossible to distinguish the momentum imparted by the external electromagnetic field to be measured from the self-field created by the same body during the process of measurement[1]. The resulting uncertainty in the measured field amplitude results in the impossibility of revealing the quantum features of the same field. This state of affairs appeared to devoid of physical significance the concept of field amplitude in QM. Subsequently Bohr and Rosenfeld (B-R)[2] modified the measurement procedure by introducing a rigid test body of large mass M with finite linear extension $a \sim V^{1/3}$ and with constant charge density ρ. The test body, here called pointer, is constrained to move along the direction of the field component to be measured and is neutralized by a fixed opposite charge $-\rho$. The measurement lasts a finite time and a measurement protocol is followed where two momentum measurements are performed in short time intervals at the beginning and at the end of the time interval. This procedure is assumed to induce the pointer to follow a gatelike trajectory $Q(t) = Q\Theta(t''-t)\Theta(t-t')$ (with $t''-t'=\tau$). By solving Maxwell equations under the above conditions they find that the time averaged self-force is proportional to Q. Although the limits on the uncertainty obtained for the field amplitude is much lower than the one obtained with the pointlike pointer, because of its dependence on the properties of the free field only it is of fundamental nature. (B-R) taking advantage of the linear dependence on Q of the average self force, by introducing an elastic force of non electromagnetic nature to compensate the electromagnetic self force, are able to render arbitrarily small the uncertainty of the electrical field amplitude. Recently an *ab initio* QED calculation of the pointer-field system has been performed using standard quantum optical methods [3]. It has been found a QM expression for the self force on the pointer which differs

CP513, *Nuclear and Condensed Matter Physics,* edited by A. Messina
© 2000 American Institute of Physics 1-56396-929-7/00/$17.00

from the (B-R) expression, in particular its time average doesn't show a linear dependence on Q, but is in agreement with the usual textbook expression[4]. This QM expression for the self-force leads to an uncertainty in the field amplitude different from the (B-R) result. The origin of this uncertainty may be traced to the finite duration of the field measurement. In fact the measurement produces, because of the time-energy uncertainty principle, an uncertainty on the field energy density of the order of $\sim \hbar / \tau V$ and this in turn produces a corresponding uncertainty in the average field amplitude. To resolve the discrepancy between the (B-R) and QED results the classical measurement process has been reanalyzed following the (B-R) method[3]. The expression for the average self force in the presence of a neutralized body is

$$
\bar{F}_D = \frac{1}{\tau} \int_\tau dt_2 F_D(t_2)
$$

$$
= -\frac{\rho^2}{\tau} \int_\tau dt_2 \int_V d^3x_1 \int_V d^3x_2 \int_\tau dt_1 Q(t_1) \left(\frac{\partial^2}{\partial x_1 \partial x_2} - \frac{1}{c^2} \frac{\partial^2}{\partial t_1 \partial t_2} \right) \frac{1}{r} \delta\left(t_2 - t_1 - \frac{r}{c} \right)
$$

1)

The above expression, after expansion of the displacement $Q(t_1)$ in series of r/c and performing the time integration over t_1 and t_2, gives the standard average radiation reaction force being thus in agreement with the QED result. It however coincides with the average of the (B-R) expression for the self-force if we could take $Q(t_1)$ out of all integrals in F_D. This is not however a trivial step due to the singular nature of the multiplying factor of $Q(t_1)$ in the in integrand in eq.(1) Here we shall consider a simplified model of pointer that will allow to perform explicitly the integrals over the space variables and over t_2 and to obtain an explicit expression for the factor ,$f(t_1)$, of $Q(t_1)$ in the integrand of eq. (1). In particular we shall consider a needle-like pointer of length L, negligible transverse dimensions and with its major axis lying along the x direction. We take the charge density distribution of the form $\rho(\mathbf{x})=\rho\delta(z)\delta(y)\Theta(L-x)$ Substituting this expression in eq. (1) , performing the integration on the transverse variables and introducing the new variables $l=(x_2-x_1)$, $X=(x_2-x_1)/2$ we obtain for the time averaged self-force due to an electric field E pointing along the x direction

$$
\bar{F}_D = -\frac{\rho^2}{\tau} \int_\tau dt_1 Q(t_1) \int_{-L}^{+L} dl \left\{ \frac{1}{l} \Theta(t''-t_1 - \frac{|l|}{c}) + \frac{1}{c^2} \frac{1}{|l|} \partial_t \delta(t''-t_1 - \frac{|l|}{c}) \right\}
$$

2)

It is easy to see that for measurements of time short enough to satisfy the condition $\tau \ll L/c$ it must be $\Theta(t''-t_1-L/c)=0$ while $t''-t_1-l/c$ is always outside of the integration interval. Thus the Θ dependent term in eq. (2) does not contribute to the integral. The integration on l can the be explicitly performed and we finally obtain

88

$$\overline{F}_D = \int_\tau dt_1 Q(t_1) \left\{ 2\rho^2 L \frac{1}{c\tau} \frac{1}{(t''-t_1)^2} \right\} \qquad 3)$$

Where the term inside the curly brackets represents the function $f(t_1)$ for our needlelike pointer. The explicit form obtained for our chosen simplified model shows that $f(t_1)$ is unbounded inside the integration interval. This indicates that at least in the case considered, the approximations required to obtain from eq. (1) the correspondent (B-R) expression for the self force are not allowed. Thus the self force obtained by QED [3] is substantially different from the (B-R) form.

ACKNOWLEDGMENTS

The authors acknowledge partial financial support by Ministero dell' Università e della Ricerca Scientifica e Tecnologica, Cofinanziamento MURST, Istituto Nazionale di Fisica della Materia, and Assessorato BB.CC.AA. Regine Siciliana.

REFERENCES

1. Landau L. and R. Peierls, Z. Phys. **69**, 56 (1931)

2. Bohr N. and L. Rosenfeld, Mat. Fys. Medd. K. Dan. Vidensk. Selsk. **12**, No 8 (1933)

3. Compagno G. and F. Persico, Phys. Rev. A 57, 1595 (1998)

4. Jackson J. D. *Classical Electrodynamics*, (Wiley New York,1962)

Relaxation and phase transformations in simple atomic systems with short range interactions

D. Costa, P. Ballone and C. Caccamo

Istituto Nazionale per la Fisica della Materia
Università degli Studi di Messina
Dipartimento di Fisica, Contrada Papardo, C.P. 50, 98166 Messina, Italy

Abstract. We simulate by molecular dynamics the long time evolution of metastable phases for particles interacting via a spherical, short range pair potential. We analyse the dynamics of the first order phase transitions taking place in the system, as well as the relaxation processes by which the system anneals the extended defects generated during the first stages of the phase transformations.

Introduction. The relaxation kinetics of metastable structures, and the dynamics of first order phase transitions are fundamental problems underlying a variety of different phenomena, ranging from the formation and the aging of glasses, to the crystallization of proteins. The difficulty of these problems is due to the wide range of time and length scales encompassed by these phenomena, and by their co-evolution with faster degrees of freedom, collectively identified as noise.

Molecular dynamics simulation, with its detailed description of the motion of each atom, could in principle provide the full characterisation of the time evolution in metastable phases. However, practical and conceptual limitations have prevented an extensive application of this method to the study of metastable systems and of the dynamics of phase transitions. First of all, MD is usually restricted to microscopic length (~ 100 Å) and time ($\sim 10^{-8}$s) scales, while the phenomena of interest often involve mesoscopic systems. Even more challenging is the identification, in the over-abundant information provided by MD, of the relevant degrees of freedom that characterize the collective transformations under study.

We explore the ability of molecular dynamics (MD) simulation to provide a direct view of the crystallization kinetics in a system of spherical particles interacting via a pair potential. We focus on a short range potential, that has been used in the past to model solutions of globular proteins. For reasons of space, we report here a preliminary account of an extensive study described in detail elsewhere [1].

The model and the equilibrium phase diagram. We simulate a system of spherical particles interacting with a pair potential given by a generalization of the Lennard-Jones model [2], which has been used extensively to study colloids and solutions of globular proteins:

$$V(R) = 4\frac{\epsilon}{\alpha^2} \left\{ \left[\left(\frac{R}{\sigma}\right)^2 - 1 \right]^{-6} - \alpha \left[\left(\frac{R}{\sigma}\right)^2 - 1 \right]^{-3} \right\} \qquad (1)$$

The parameter α determines the range of the attractive interaction with respect to the range σ of the repulsive part: large values of α correspond to deep and narrow attractive wells, while low values of α bring the potential closer to the Lennard-Jones prototype. In turn, the range of the attractive potential influences the relative position of the liquid-vapor and solid-gas coexistence curves: for very short range potentials there is no stable liquid phase, and the corresponding liquid-vapor region appears as a metastable feature in the phase diagram underlying the solid-fluid coexistence line.

In the simulations described below, we adopted $\alpha = 50$, corresponding to a short range attractive potential. The phase diagram of this model is reported in Fig. 1.

The non-equilibrium molecular dynamics simulation. Our simulations consist of standard molecular dynamics runs at constant number N of particles, total energy E and volume V. Most of our computations have been performed for systems of 2592 particles, at a density $\rho\sigma^3 = 0.5$. The simulation box is tetragonal, with $c/a = 3$. This choice of the simulation geometry makes it favorable for the system to create an interface perpendicular to the z axis, thus facilitating the identification of the onset of phase segregation.

Starting from a well equilibrated sample in the homogeneous fluid region of the phase diagram ($\rho\sigma^3 = 0.5$, $T/\epsilon = 1$), we induce metastability by progressively quenching the system below the fluid-solid and the (metastable) liquid-vapor coexistence lines.

The quench consists of 10 stages, each of them 10^6 MD steps long. In a few cases, the simulation has been extended beyond 10 10^6 MD steps. For protein solutions, these long runs correspond to a time span of microseconds, i.e., well within the mesoscopic range.

Results Snapshots of MD configurations are collected in Fig. 2 and Fig. 3.

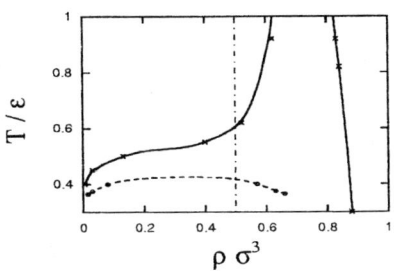

FIGURE 1. Phase diagram of the simulated model

As shown in Fig. 2, the system remains in the homogeneous fluid state well below the solid-fluid coexistence line ($T = 0.43$ in Fig. 2). On approaching the metastable liquid-vapor line, however, the first phase separation readily takes place, initiated by a vapor region nucleating in the fluid phase. ($T = 0.41$ in Fig. 2). The segregation of the high density fluid opens the way for the liquid-solid transformation, giving origin to a defective crystal.

The long time evolution of the defective crystal structure is a second focal point of our simulation. As illustrated in Fig. 3, the solid slowly releases energy by annealing extended defects like grain boundaries and stacking faults. This release of energy is characterised by the almost discontinuous opening of relaxation channels, associated to a broad distribution of relaxation times. The global effect of these relaxations could give rise to a stretched exponential relaxation in extended systems.

Conclusions The large system sizes and long simulation times required to ob-

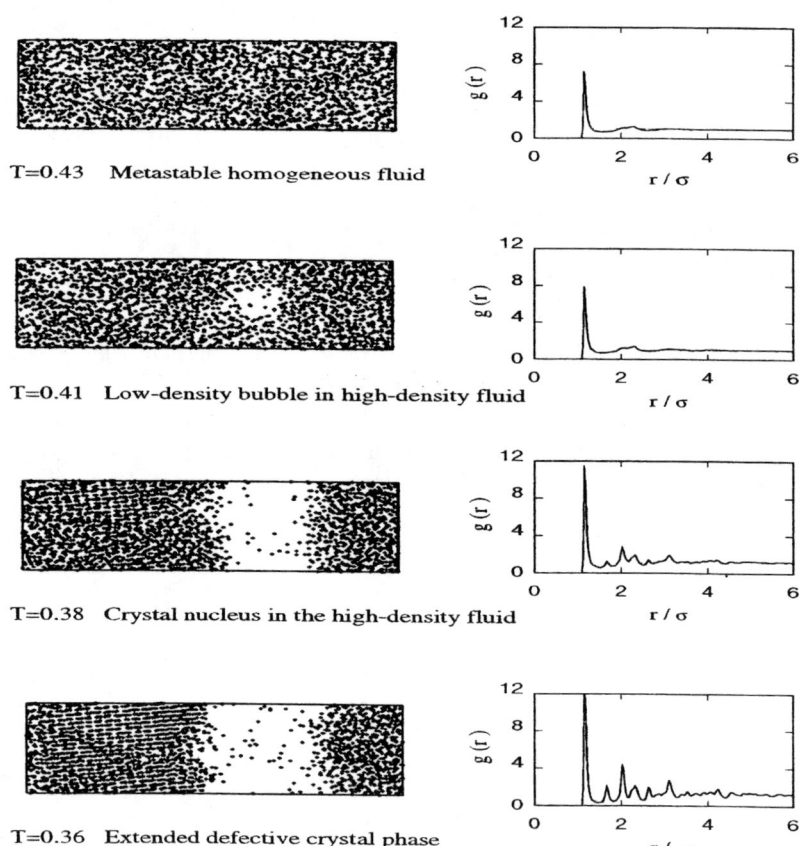

T=0.43   Metastable homogeneous fluid

T=0.41   Low-density bubble in high-density fluid

T=0.38   Crystal nucleus in the high-density fluid

T=0.36   Extended defective crystal phase

FIGURE 2. Selected configurations during the quench of the system from the homogeneous fluid to the solid-fluid coexistence state

tain an unbiased description of the kinetics of phase transformations are starting to be accessible to ordinary computer facilities and simulation methods, at least for systems interacting via simple and short range potentials. Then, molecular dynamics provides a direct and intuitive description of phase changes, and, in general, of relaxation in metastable systems. The slow dynamical variables associated to the collective behavior of the particles in the system can be easily identified on the basis of standard statistical mechanics approaches, like the projection technique and the mode coupling method [3]. The long term goal of our project is to combine these statistical mechanics approaches with a time dependent density functional scheme, in order to bridge the gap between the atomistic and macroscopic description of phase transformations.

REFERENCES

1. D. Costa, P. Ballone and C. Caccamo, to be published.
2. P. R. ten Wolde, and D. Frenkel, Science **277**, 1975 (1997).
3. J. P. Hansen and I. McDonald, *Theory of Simple Liquids*, 2nd edition, Academic Press, London, (1986).

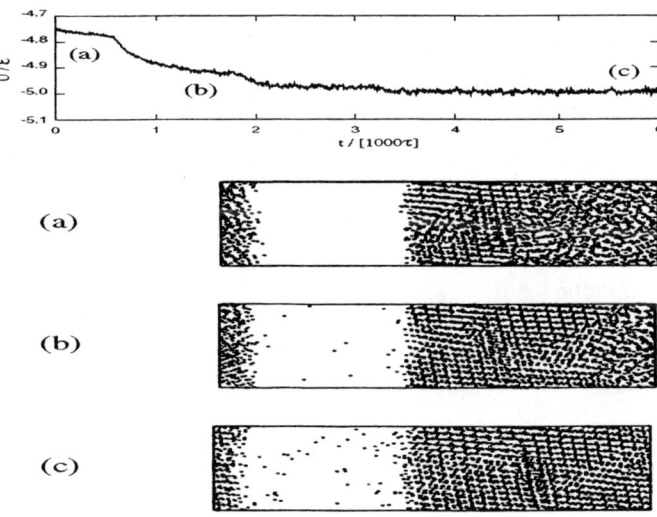

FIGURE 3. Long time relaxation of the defective crystal generated by the fluid-solid phase transition.

Dynamical Properties and Confinement Effects in Complex Liquids

D. Majolino, V. Crupi, G. Maisano, P. Migliardo and V. Venuti

Dipartimento di Fisica & INFM, Università di Messina, P.O.Box 55, 98166 Messina (Italy)

Abstract. We studied the reorientational and vibrational dynamics of ethylene glycol and its homologous systems, namely ethylene glycol monomethyl ether and ethylene glycol dimethyl ether, having a different number of OH groups per molecule. The above systems were studied in bulk and in confined state within a porous glass matrix with 25Å diameter pores by light scattering (Raman and Rayleigh). By comparing the obtained results the evident influence of the chemical and physical traps on the molecular mobility of the liquids forced to diffuse within the nanopores was shown.

INTRODUCTION

Recently the study of the confinement effects on the molecular mobility of liquids has been object of great scientific interest either from a theoretical or an experimental point of view due to the wide range of applicability in the technological and biological fields.[1] In order to understand the dynamical behavior of liquids that diffuse and reorient within porous matrix, one has to take into account essentially two competitive processes related to the surface interactions and to the geometric restrictions, namely called as chemical and physical traps respectively.[2] Porous silica glasses represent one of the most frequently used confining matrix for their chemical and mechanical stability, transparency and the highly pore interconnection. Furthermore the presence of a high density of Si-OH groups on the inner surface of the porous glasses that constitute active sites for the interaction with hydrogen bonded systems, makes these matrixes particularly suitable to study the confinement influence on the reorientational dynamics of this class of liquids. A large variety of different spectroscopic techniques has been applied to probe the dynamical properties of molecular liquids. In particular the experimental data showed an enhancement of correlation times, a shift of the freezing point below the usual solidification temperature of bulk liquids as well as an increased microviscosity that causes a more hindered diffusion in the imbibed liquids.

In this paper we report a detailed analysis of the vibrational and reorientational dynamics of Ethylene Glycol (EG, H-[O-CH$_2$-CH$_2$]-OH) and its homologous, EG monomethyl ether (EGmE, CH$_3$-[O-CH$_2$-CH$_2$]-OH) and EG dimethyl ether (EGdE, CH$_3$-[O-CH$_2$-CH$_2$]-CH$_3$) in bulk state and confined in a sol-gel silica glass with 25Å pores, by means of Rayleigh-wing and Raman scattering. The investigated systems differ from each other in the number of hydroxyl groups, namely two for EG, one for EGmE, zero for EGdE.

CP513, *Nuclear and Condensed Matter Physics,* edited by A. Messina
© 2000 American Institute of Physics 1-56396-929-7/00/$17.00

EXPERIMENTAL SET-UP

Raman and Rayleigh-wing data were collected by a high-resolution fully computerized triple monochromator. As exciting source the 6471 Å line of a Ar^+-Kr^+ Ion Laser was chosen in order to minimize the fluorescence contribution from the Gelsil matrix. The samples were studied in bulk and confined into a sol-gel matrix with treated and untreated internal surfaces, cylindrical in shape (10 mm diameter, 5 mm thick) purchased from GelTech Co., with nominal pore diameter of 26 Å. The matrix contains a great number of Si-OH groups that constitute strong active sites for H-bond with the liquid samples. The hydrogen atom of the Si-OH groups was replaced with the non-active methyl groups CH_3 for minimizing the interactions with the fluid.

RESULTS AND DISCUSSION

The analysis, by IR and Raman spectroscopies, of the O-H stretching vibration gives information about the environments of this molecular group. Our liquids can be in fact considered as constituted by self-associated systems in which the H-bond promotes a set of inter- and intramolecular *transient* structures. The different degrees of association, with the relative populations, generate a different dynamical response. According to Laubereau,[3] we define ω_α the O-H stretching mode of free (monomers) and/or end groups, ω_β (ω_γ) the O-H vibration of proton-acceptor (donor) end groups (dimers), ω_δ the O-H vibration of fully-bonded groups (trimers), and, finally, ω_ε the O-H stretching intramolecular vibration.

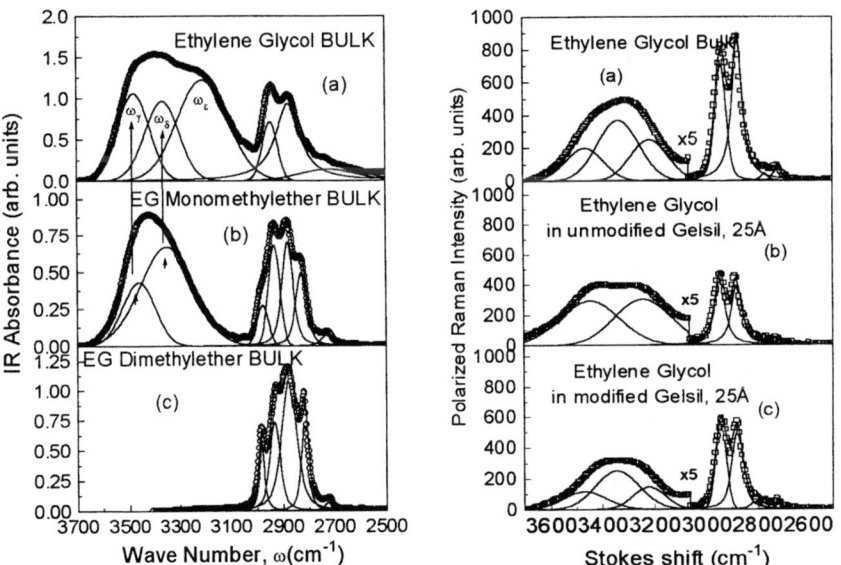

FIGURE 1. IR spectra of bulk systems and Raman data of EG bulk and confined.

In Fig.1 we report the IR spectra of bulk systems, together with the deconvolution components (Voigt profiles) and, as an example, the deconvolution of O-H stretching region in the case of EG bulk and confined in unmodified and modified Gelsil. No end and/or open monomers are present, as evidenced by the absence ω_α of and/or ω_β bands. As far as EG is concerned, the presence of a large variety of aggregates can be explained in terms of the chain dimension and the presence, in its chemical structure, of two OH end groups. The band ω_ε disappears in EGmE not having this system, that exhibits only one OH end group, possibility of organizing intramolecular H-bond. O-H stretching peaks corresponding to dimers and trimers have been observed also in this case, as a confirm of the fact that the increased steric hindrance is not so great to avoid polymeric aggregates of co-ordination number three or four existing in liquid phase. A comparison between experimental Raman spectra of these samples in bulk and confined allowed to clarify the different role played by geometrical restrictions and surface interactions. In particular, in the reported case of EG, the aggregates in bulk and confined in modified Gelsil are the same, and it means that the confinement effects depend essentially on the *physical* traps, that don't affect the structural arrangements present in the bulk. When EG is confined in modified Gelsil, trimers (ω_δ) disappears, probably because of the presence of *chemical* traps, which hinder the sub-bands linked to more extended species present in bulk.

On the basis of the interpretation of IR and Raman measurements, that showed the existence of different structural aggregates, we can justify the application of particular shapes (for example, the Havriliak-Negami profile) for the scattering laws of our Rayleigh wing data. The distribution of relaxation times in the ω-domain clearly revealed in the bulk EG has been well taken into account through the Havriliak-Negami H-N(ω) relaxation function:

$$H - N(\omega) = -\left(\frac{1}{\omega}\right) \mathrm{Im}\left[\frac{1}{1 + (i\omega\tau_{HN})^\alpha}\right]^\gamma \tag{1}$$

where τ_{HN} is a characteristic relaxation time, α and γ are shape parameters ranging between 0 and 1 and related, respectively, to the symmetric and asymmetric line widths. In particular, if $\alpha=\gamma=1$, the H-N(ω) restores the Lorentzian profile. Such a function plays, in the ω-domain, the same role of the Kolrausch-Williams-Watt KWW(t) profile

$$\phi(t) = \exp\left[-\left(\frac{t}{\tau_{KWW}}\right)^\beta\right] \tag{2}$$

in the time domain, characterized by the relaxation time τ_{KWW} and the shape parameter β, which is the function that slow relaxation in complex condensed systems obey. An experimentally verified mathematical relation between is expressed in Ref. 4. The mean relaxation times is obtained from the experimental data by the relation:

$$\langle \tau \rangle = \left(\frac{\tau_{KWW}}{\beta} \right) \Gamma \left(\frac{1}{\beta} \right). \qquad (3)$$

When EG is confined, *chemical* and *physical* traps cause a slowing-down of the reorientational diffusive dynamics. The substitution of the H-atom of the Si-OH groups with the inert CH_3 groups allowed us to put into evidence the different role of active and non-active surface sites on the reorientational dynamics. The analysis of the system in unmodified gelsil reveals, in the far wing of the spectrum, a transition from a CILS[2] fluidlike exponential decay to a Lorentzian collective relaxation.

EGmE exhibits a behaviour analogous to EG: a frustration in molecular mobility is observed in connection with geometrical restrictions and surface interactions, reflected in the increased value of $\langle \tau \rangle$.

EGdE suffers only the retardation process related to *physical* traps: its diffusional dynamics in modified and unmodified Gelsil is, in fact, quite similar. It follows a single Debye decay in the time domain, represented by a Lorentzian line.

TABLE 1. Rayleigh wing bast-fit parameters for all investigated samples.

Samples	β, $\langle \tau \rangle$ (ps)	τ_{exp} (ps), τ_{Ltz} (ps)
EG bulk	0.42, 5.18	0.09, -
EG in unmodified Gelsil	0.81, 5.92	-, 0.1
EG in modified Gelsil	0.47, 5.38	0.1, -
EGmE bulk	0.5, 3.84	0.1, -
EGmE in unmodified Gelsil	0.73, 5.03	-, 0.1
EGmE in modified Gelsil	0.56, 5.85	0.1, -
EGdE bulk	1, 1.8	0.1, -
EGdE in unmodified Gelsil	1, 4.0	0.1, -
EGdE in modified Gelsil	1, 4.0	0.1, -

REFERENCES

1. Schuller, J., Mel'nichenko, Y. B., Richter, R., and Fischer, E. W., *Phys. Rev. Lett.* **73**, 2224 (1994).

2. Carini, G., Crupi, V., D'Angelo, G., Majolino, D., Mel'nichenko, Y. B., and Migliardo, P., *J. Chem. Phys.* **107**, 2292-2299 (1997).

3. Graener, H., Je, T. Q., and Laubereau, A., *J. Chem. Phys.* **90**, 3413 (1989).

4. Alvarez, F., Alegria, A., and Colmenreo, J., *Phys. Rev. B* **44**, 7306 (1991).

Fragility and Dynamical Properties of Glass-Forming Liquids above their Tg

M.Cutroni[a], A.Mandanici[a], R.Pelster[b], A.Spanoudaki[b]

[a]Dipartimento di Fisica – Università degli Studi di Messina and INFM – Unità di Ricerca di Messina,
ctr.Papardo salita Sperone, 31 – 98166 Messina, Italy
[b]II. Physikalisches Institut der Universität zu Köln, Zülpicher Straße 77, 50937 Köln, Germany

Abstract. A new broadband dielectric spectroscopy technique, operating from some Hz up to the microwave frequency region, has been used to study the collective dynamical aspects of some simple molecular liquids approaching their glass transition temperature. Accessing to the relaxational dynamics of metatoluidine ($CH_3C_6H_4NH_2$) on a mesoscopic timescale, a very high value of fragility has been obtained for this liquid in the classification scheme recently refined by Angell.

The scientific problem of glass transition is probably one ot the most debated in this part of the century and the fragility concept offers a fundamental key for the description of glass-forming liquids approaching their Tg. Usually several experimental methods can be used to study the dynamics of liquids as a function of temperature: by measuring the viscosity, η; the electrical conductivity, σ_{dc}; the frequency dependent permittivity, $\varepsilon^* = \varepsilon' - i \cdot \varepsilon''$; the attenuation, α, and the velocity, v, of longitudinal acoustic waves. The imaginary part of dielectric or mechanical response function, analyzed as a funtion of frequency, exhibits a loss peak corresponding to a relaxation process in the material . The relaxation time, dielectric or mechanical, at a certain temperature T, is defined by the equation

$$2\pi f_P \cdot \tau = 1 \tag{1}$$

where f_P is the frequency at which the maximum in the dielectric or mechanical loss spectra occurs. In liquids well above their melting temperature, the relaxation time, as well as viscosity and the inverse conductivity, obeys a simple Arrhenius law

$$\tau = \tau_0 \cdot \exp\left[E_A / (k_B T)\right] \tag{2}$$

with an activation energy E_A and a characteristic time τ_0. As the temperature is lowered, a deviation from the simple activated behaviour is observed: the apparent activation energy continuously increases and the relaxation time tends to diverge with respect to usual experimental timescales. Below the glass transition temperature, T_g, the main relaxation peak disappears. Liquids that notably deviate from the Arrhenius

CP513, *Nuclear and Condensed Matter Physics*, edited by A. Messina
© 2000 American Institute of Physics 1-56396-929-7/00/$17.00

behaviour can reach the supercooled phase even without a rapid cooling: they are called "fragile". A fragile behaviour reflects the ability of the system to perform thermally activated cooperative rearrangements of its microscopic units and corresponds to a considerable excess of specific heat in the metastable phase with respect to the stable crystal at the same temperature around T_g.

The deviations from the Arrhenius behaviour are quantitatively described by the fragility concept, according to a recent definition due to Angell [1],

$$F_{1/2} = 2(T_g / T_{1/2}) - 1 \qquad (3)$$

where the temperature $T_{1/2}$ is individuated by the relation $\tau(T_{1/2}) = 10^{-6} s$. Experimental values of τ on this mesoscopic timescale are directly accessible to dielectric measurements, so fragility $F_{1/2}$ can be evaluated independently of any data fitting function or procedure.

To achieve a better understanding of relaxational dynamics of glass forming liquids in their supercooled phase, the dielectric response of simple molecular liquids has been studied as a function of frequency between some Hz and 20GHz in a wide temperature range from +50°C down to –100°C [2]. We report here some preliminary results on the dielectric properties of *meta*-toluidine, a simple disubstituted benzene. At room temperature its main dielectric relaxation peak occurs in the microwave frequency region.

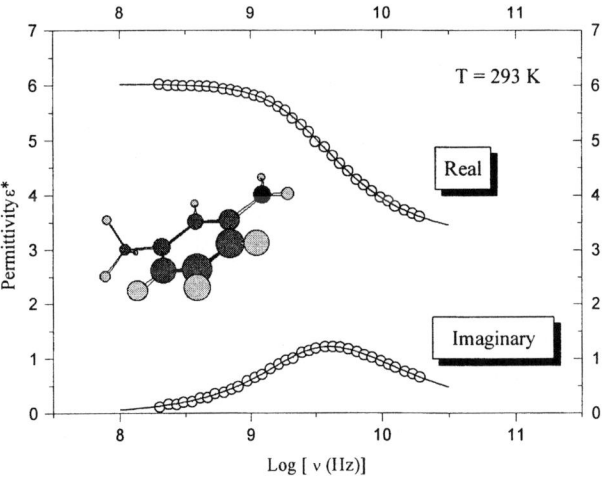

FIGURE 1. F Real and imaginary part oc ele tri al permittivity at room temperature, as a cun tion oc crequen y between 200MHz and 20GHz. The relaxational response is shicted towards lower crequen ies at lower temperatures.

The dielectric response can be well described by the Cole-Davidson empirical equation

$$\frac{\varepsilon^* - \varepsilon_\infty}{\varepsilon_s - \varepsilon_\infty} = \frac{1}{\left[1 + i \cdot \omega \tau_{CD}\right]^{\beta_{CD}}} \tag{4}$$

with an exponent β_{CD} very close to one, so the liquid has a nearly-Debye behaviour at this temperature. At lower temperatures the peak height increases and the frequency of the maximum dielectric loss is shifted towards lower values, as expected for a thermally activated process. The shape of the relaxation is progressively modified and the parameter β_{CD} decreases, in correspondence with a broader and asymmetric peak. The dielectric complete spectra between 5Hz and 2GHz, acquired with a broadband spectroscopic technique [3] at temperatures between 203K and 213K, can be used to evaluate the fragility of m-toluidine (Fig.2). A value $F_{1/2} \approx 0.80$ has been found, which denotes the highly fragile character of this liquid. For a comparison the characteristic parameters of selected glass-forming liquids are reported in Table 1.

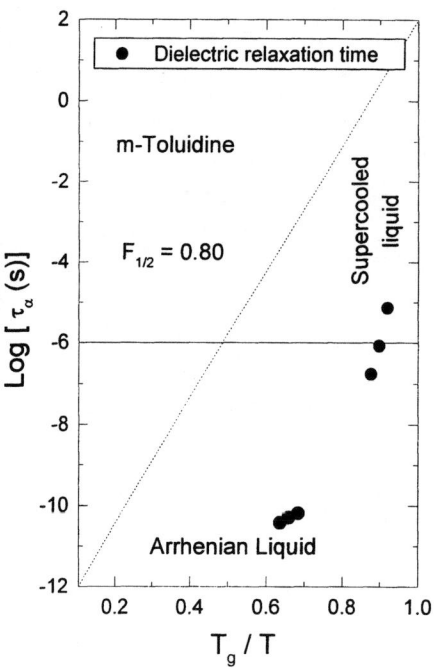

FIGURE 2. Angell's plot of the dielectric relaxation time for the m-toluidine at selected temperatures. The glass transition temperature of this liquid, determined by DSC at 10K/min, is 186.8K [2,4].

TABLE 1. Fragility of selected liquids [1] and m-toluidine [this work] in terms of $F_{1/2}$.

Substance	T_g (K)	$F_{1/2}$
SiO_2	1473	0.13
glycerol	190	0.62
OTP	246	0.74
m-toluidine	186.8	0.80

Looking at the structure of the simple molecular units, a mutual exclusion effect between benzene rings is expected [5] while the breaking and formation of hydrogen bonds could be the elementary steps of the observed macroscopic relaxation, as confirmed by an estimate of the activation energy in the Arrhenian region of the liquid around room temperature [2].

ACKNOWLEDGMENTS

We should like to thank V. Grunow for his help during experiments. DAAD and CRUI are also gratefully aknowledged for the support to the mobility of researchers involved in the Vigoni Program 1998-99.

REFERENCES

1. Richert, R., Angell, C.A., *J. Chem. Phys.* **108**, 9016 (1998).

2. Cutroni, M., Mandanici, A., Pelster, R., Spanoudaki, A., Nimtz, G., *in preparation*

3. Pelster, R., *IEEE Trans. Microwave Theory and Techn.* **43**, 1494 (1995).

4. Alba-Simionesco, C., Fan, J., Angell, C.A., *J. Chem. Phys.* **110**, 5262 (1999).

5. Morineau, D., and Alba-Simionesco, *J. Chem. Phys.* **109**, 8494 (1998).

Applications Of Thin Film Semiconductors

Dario della Sala

ENEA – Centro Ricerche Casaccia
Via Anguillarese 301, 00060 S. Maria di Galeria, ITALY

Abstract. The increasing use of thin film semiconductors in device applications has forced the change to new deposition and processing technologies, compared to the familiar wafer semiconductor technologies. In spite of the lower performance, the process flexibility allowed with low temperature processing has opened the way to quite new products, based on thin film semiconductors.

INTRODUCTION

Wafer semiconductor technologies generally end up with ~100 μm-thick polished wafers, obtained with high temperature growth process, close to the equilibrium temperature with the melt. Also the other process steps for device manufacturing require high temperature processing, except for lithographic patterning.

In recent years other semiconductor technologies based on thin films, deposited and processed at low temperature on supporting substrates, have gained popularity and market niches. The active semiconductor layer is deposited at much lower temperature compared to the melting temperature of the material.

The thin film semiconductors (TFS) are "as thin as needed" by effective sunlight absorption, field effect and radiation sensing. The advantages of thin film techniques are: low thermal budget, compatibility with cheap and flexible substrates (glass, metal and polymer foils, plastic sheets), very low atom inter-diffusion, abrupt electrical junctions, arbitrary stoichiometry and tuneable physical properties.

STRUCTURE AND ELECTRICAL PROPERTIES

At low deposition temperature, epitaxial growth is difficult even on mono-crystalline substrates: the bulk structure of the thin film ranges from amorphous to micro-crystalline, with the average grain size increasing with the film thickness. As a consequence, the roughness of the film free surface is systematic.

The bulk microstructure of the deposited material seriously affects the electrical properties and the device applications. In the case of silicon, the crystal state has well defined energy bands and forbidden energy gap; holes and electrons in the band

CP513, *Nuclear and Condensed Matter Physics,* edited by A. Messina
© 2000 American Institute of Physics 1-56396-929-7/00/$17.00

energy states are described by Bloch wavefunctions ("extended states"); the defect state concentration inside the energy gap is very low. In amorphous silicon (a-Si), on the contrary, the structure is aperiodic at the atomic scale, the Si-Si bonds arrange a network of distorted tetrahedral cells, and the internal strain generates dangling bonds. The strained bonds cause a large concentration of localized states close to the band edges ("tail states") and affect the carrier mobility because of trapping/detrapping events. The dangling bonds provide a large amount of localized states at midgap (ranging from 10^{16} to 10^{19} $cm^{-3}eV^{-1}$), and influence the minority carrier lifetime.

In a mixed-phase polycrystalline material the dangling bonds and the strained bonds are concentrated close to the grain boundaries, therefore the carrier lifetime and mobility are directly proportional to the average grain size. The electric mobility value in TFS ranges from 10^{-1} cm^2/Vs for modern polymeric materials to 1 cm^2/Vs for amorphous silicon and to 100-400 cm^2/Vs for the best polycrystalline silicon (obtained with the pulsed excimer laser irradiation of amorphous silicon) and is compatible to several product areas. Also other low temperature thin film materials crucial for device build-up are high quality. Semiconductor oxides with 10^6 V/cm breakdown voltage and 10^{11} cm^{-2} interface states, metal films with electrical resistivity comparable to bulk materials (less than 10^{-5} Ωcm), transparent conductive oxide films with resistivity of the order of 10^{-4} Ωcm, all play an essential role in solar cells and sensor devices.

SOLAR CELLS

In crystalline silicon (c-Si), the absorbance of the red and infrared solar radiation is not very high. Special 3-D textures should be realized on the wafer surface for improving the effective light path in the wafer by geometrical (refractive) effects. Thin films of amorphous silicon, CdTe, $CuInSe_2$ less or equal to 1 μm, absorb the whole solar spectrum better that ~100 μm-thick Si wafers.

However, the minority carrier diffusion length is very low in TFS (~0.1 μm in a-Si, compared to ~100 μm in c-Si). Therefore, the photogenerated carriers must be driven by the internal fields, and the use of undoped active layers is mandatory, because thin film doped layers are very rich of gap states. The thin film doped layers are rather resistive and require and additional transparent conductive oxide layer for external current collection. As a result, the typical thin film cell structure is: TCO/p-i-n/back metal, instead of the usual structure: metal grid/p-n/back metal, for wafer cells.

The low level of species inter-diffusion during processing allows the manufacture of multi-junction solar cell [1], where two or three complete cells are stacked on top of each other (for example: a-Si (thin)/a-Si (thick), a-Si/a-SiGe/a-SiGe, a-SiC/a-Si/a-SiGe). The rule of thumb is placing the more transparent (or wide band-gap) cells on the top, favouring the progressive absorption of sunlight in the stack.

Being able to deposit layers at a low temperature on flexible substrates without process-induced inter-diffusion of species, TFS technologies allows the "roll-to-roll" mass production, where a continuous flexible roll (stainless steel or polymer) winds up across a sequence of deposition chambers and laser scribing chambers. The reciprocal dopant contamination between adjacent chambers is maintained low, by intermediate chambers with a larger pressure of inert gases, that behave like "gas barriers". Such technologies provide flexible solar panels (for boats and homes) and photovoltaic tiles to be nailed on the roof-tops [1].

MICROCIRCUITS AND SENSORS

In the core of liquid crystal active matrix displays (AMLCD), integrated switches are useful to control the voltage state and the transparency the pixels, that are actually "voltage-driven filters". Such switches can be made as a thin film transistor (TFT) of CdSe, a-Si and poly-Si. A similar switch is required also in many modern sensor arrays, for driving the charge accumulated by microsensors in the sensing operation.

The TFT is therefore the natural "building block" for thin film microcircuits. In the AMLCD, TFTs also multiplex the pixel rows once-at-a-time, and in the time frame a row is active, the columns are made active in parallel, feeding the multiplexed video-composite signal to the pixels. With a symmetrical concept, the driving circuits of a sensor arrays multiplex and drive out the charge generated by microsensors, for serializing and amplifying the signal corresponding to the sensed images. If p-channel and n-channel TFTs can be realized (as in poly-Si technology), monolithic low power CMOS control circuitry are also available, for mobile applications.

The time available to the switching TFT for the capacitor charging in AMLCDs can be calculated of the order of a few tens microseconds. In this time window, an a-Si TFT provides an ON current not larger than $I_{MAX}=1$ μA, but the charge transferred (about 10^{-11} Coulomb) is large enough to provide the required aligning voltage to the liquid crystal (V=Q/C=few volts, being C~10^{-12} Farad). Problems arise in the multiplexing circuitry, realized on the same substrate instead of bonding thousands of high-resolution line pads to external circuits. The transit time in the a-Si TFT channel is $T=L^2/\mu V_{drain-source}$~0.25 μsec (L=5 μm, V_{ds} =1 Volt). The individual a-Si TFTs therefore cannot work at a clock rate 1/T longer than few megahertz. This cut-off frequency is compatible with row multiplexing, but not with column multiplexing.

A similar barrier limits the application of a-Si TFTs to driving high resolution light sensors [2]. Whenever high-speed integrated control circuits are required for thin film large-area 2-D devices, as in the 30 cm x 40 cm prototype X-ray imagers for radiography [3], the high-mobility of laser-recrystallized poly-Si is required.

Other voltage driven filters can be made with thin film technologies on large-area substrates. In sunlight blinders for bioclimatic applications, a semi-transparent p-i-n

SiC solar cell drives Li^+ ions from an ion storage material (V_2O_5) through an ion conductor (MgF_2) to the electrochromic layer WO_3. The transition to the opaque $WO_3:Li^+$ is reversible and proportional to the dose of Li^+ ions, and is fast enough (minutes) [4].

HYBRID ARCHITECTURES OF TODAY AND TOMORROW

The device functionality can be extended by applying TFS to more sophisticated "substrates". This is the case of "sensor-on-ASIC" [5], where smart pixels capable of performing on-chip image pre-processing can be arranged, based on crystalline silicon ASICs, that provide the signal conditioning, close to the pixel, at high speed. A further advantage is that three-color integrated image sensors can be arranged with a stack of thin film layers, the so-called "ni^{3}p" photodiodes, where a reverse-biased high speed nip photodiode, with different zones in the "i" layer allows to separate the contribution of the three fundamental colors [5].

In the past few years, the increasing performance of Organic Light Emitting Polymers, that can be applied on low cost-substrates by spin-on of a liquid, are making feasible display prototypes based on poly-Si circuits and polymeric light emitters, being all-polymer circuits still poor in performance.

The need for adding data storage to thin film electronic systems is being satisfied by floating-gate TFTs, where the memory effect is provided by silicon nanoparticles, that behave like injected charge absorbers. The nanoparticles are obtained during the same laser irradiation processing used for poly-Si formation [6]. These progresses pave the way to the sophisticated "systems-on-glass" or "systems-on-plastic" for mobile applications.

REFERENCES

1. http://www.ultraflexgroup.it/ute/index.html

2. Tomiyama, S., Ozawa, T., Ito, H., and Nakamura, T., *Journal Non-Cryst. Solids* **198-200**, 1087-1092 (1996).

3. Street, R. A., Apte, R. B., Granberg, T., Mei, P., Ready, S. E., Shah, K. S., and Weisfield, R. L., *Journal Non-Cryst. Solids* **227-230**, 1306-1310 (1998).

4. Bullock, J. N., Bechinger, C., Benson, D. K., and Branz, H. M., *Journal Non-Cryst. Solids* **198-200**, 1163-1167 (1996).

5. http://www.uni-siegen.de/dept/fb12/ihe/forschung/

6. Nomoto, K., Gosain, D. P., Noguchi, T., Usui, S. and Mori, Y., *MRS Spring Meeting 1999*, San Francisco, April 5-9, 1999, Materials Research Society, Symposium Proceedings (to be published).

Electromagnetic field Quantization in Quantum Confined Systems

Omar Di Stefano, Salvatore Savasta and Raffaello Girlanda

INFM and Dipartimento di Fisica della Materia e Tecnologie Fisiche Avanzate,
Università di Messina
Salita Sperone 31, I-98166 Messina, Italy

Abstract. We extend recently developed schemes for field quantization in absorbing dielectric media with local susceptibilities to dielectric systems described by a non local susceptibility .The method is applied to the cases of a semiconductor quantum well embedded in infinite barriers and embedded in planar semiconductor microcavities. As an application of the formalism, we analyze the effects of the propagation through a semiconductor microcavity (MC) on a continuous-mode squeezed coherent state.

The rapid growth in experiments on quantum-optical processes which take place inside material systems has stimulated the development of techniques for the quantization of the electromagnetic field in dielectrics. An usual approach to the problem of quantization in lossy dielectrics uses Langevin forces to represent the noise, and has been applied to the calculation of quantum-optical processes in dielectric slabs with local susceptibility [1–3]. Three-dimensional quantization schemes have been presented by Ho Trung Dung *et al.* [4] and by us [5]. These quantization schemes for dispersive and absorbing linear dielectrics, provide expressions which can be applied directly to such problems as the effects of propagation through absorbing dielectrics on light that initially displays nonclassical features [6], or to calculate the field vacuum fluctuations. The aim of the present paper is to extend the field quantization to those material systems whose interaction with light is properly described by a nonlocal susceptibility [7]. Quantum wells (QW,s) and superlattices are an important example of dielectrics driven by the field via a nonlocal susceptibility [8,9]. In order to quantize the electromagnetic field in a medium with a nonlocal susceptibility, we specialize to a realistic microscopic model, by considering a semiconductor medium with translational symmetry broken in the z-direction and derive the corresponding photon-electron interaction Hamiltonian, by using the multipolar form of the interaction in dipole approximation [7,8]. By using the Heisenberg-Langevin method, we obtain a Fredholm integral equation of second kind with a noise current operator as source term:

CP513, *Nuclear and Condensed Matter Physics*, edited by A. Messina

$$\left(\frac{\partial^2}{\partial z^2} + k_z^2 \right) \hat{E}_{\mathbf{p}}^+(z, \omega) = -\frac{\omega^2}{\varepsilon_0 c^2} \left(\int_{-\infty}^{\infty} dz' \chi_{\mathbf{p}}(z, z', \omega) \hat{E}_{\mathbf{p}}^+(z', \omega) + \alpha \hat{f}_{\mathbf{p}}(z, \omega) \right), \quad (1)$$

where $\mathbf{p}$ is the in-plane wave vector and it is a good quantum number for the systems that we take in consideration and $\alpha = \sqrt{\hbar\pi/\mathcal{A}}$ is a constant depending on the quantization surface $\mathcal{A}$. This Fredholm equation, together with the commutation relation of the noise operators, $\left[\hat{f}_{\mathbf{p}}(z, \omega), \hat{f}_{\mathbf{p}'}^{\dagger}(z', \omega') \right] = \mathrm{Im}[\chi_{\mathbf{p}}(z, z', \omega)]\delta_{\mathbf{p},\mathbf{p}'}\delta(\omega-\omega')$, is the starting point for field quantization in systems described by a nonlocal susceptibility, and can be considered a direct generalization of the corresponding results for field quantization in dielectrics with a local susceptibility [3]. The solution of Eq. (1) provides the expression for the quantized light field in a nonlocal absorbing dielectric medium. The solution of the Fredholm equation provides the expression for the electric-field operator . Owing to space limitations, we give here only the quantized field expressions in vacuum outside a symmetric arbitrary dielectric ($|z| > L$) with a QW inside and at normal incidence. These expressions provides all the information needed for describing input and output quantum fields. The positive components of the electric field operator at normal incidence outside the dielectric structure are given by

$$\hat{E}^+(z, \omega) = \left(\frac{\hbar\omega}{4\pi\epsilon_0 c\mathcal{A}} \right)^{\frac{1}{2}} [\hat{a}_R(\omega)e^{ik_z z} + \hat{b}_L(\omega)e^{-ik_z z}] \qquad z \leq -L,$$

$$\hat{E}^+(z, \omega) = \left(\frac{\hbar\omega}{4\pi\epsilon_0 c\mathcal{A}} \right)^{\frac{1}{2}} [\hat{a}_L(\omega)e^{-ik_z z} + \hat{b}_R(\omega)e^{ik_z z}] \qquad z \geq L. \quad (2)$$

In Eq. (2) $\hat{a}_{R(L)}(\omega)$ are the input photon operators related to the field traveling (at normal incidence) towards the dielectric structure and obeying the usual Bosonic commutation rules, while $\hat{b}_{R(L)}(\omega)$ are the output operators related to the field escaping from the dielectric structure. They are given by

$$\hat{b}_{L(R)}(\omega) = A(\omega)\hat{F}(\omega) + R(\omega)\hat{a}_{R(L)}(\omega) + T(\omega)\hat{a}_{L(R)}(\omega), \quad (3)$$

where the noise Langevin operators, $\hat{F}(\omega)$ describe the degrees of freedom of the reservoir, they have zero expectation values and satisfy the usual commutation relations. The coefficients $R(\omega)$, $T(\omega)$, and $A(\omega)$ are respectively the reflection, the trasmission, and the absorption coefficients [7,10]. An application of the quantized field operator is provided by the derivation of the spectrum of electric field fluctuations in the vacuum state of electromagnetic field. As it is well known, the spontaneous emission rate of an atom embedded in a dielectric medium, depends directly from the power spectrum of vacuum field fluctuations. Vacuum fluctuations also determine nonlinear spontaneous processes in semiconductors [11,12]. The value of the correlation function at a common spatial position determines the power spectrum $\mathcal{S}_{\mathbf{p}}(z, \omega)$ of the field fluctuations [3]. The spatial variations of the power spectrum of the field fluctuations for three different values of ω are shown in

107

Fig. 1. The oscillations are a consequence of the partial standing-wave character of the excitation, as the QW behaves as a mirror with a frequency-dependent reflectivity. The oscillation amplitudes reach their maximum at resonance ($\omega = \omega_0$), where, in the absence of absorption, the excitation is completely reflected by the QW. Fig. 1 has been obtained considering normal incidence and a QW 10 nm thick. We have used $\Gamma = 2 \cdot 10^{-3}$ meV for the radiative broadening, $\omega_0 = 1.5$ eV for the homogeneous non-radiative broadening and $\gamma = 0.5 \cdot \Gamma$ for the exciton energy level. We can now examine how the propagation through the MC influences the squeezing

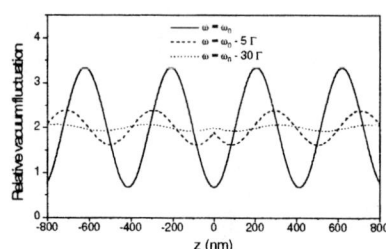

FIGURE 1. Spatial variations of the power spectrum of the field fluctuations for three different values of ω.

of an incident signal beam. We take the signal beam traveling rightward to the MC as a continuous-mode squeezed coherent state, such that produced by a degenerate parametric amplifier. We will analyze the case in which also a nonclassical input leftward signal is sent to the sample in addition to the rightward signal. We can chose for the state describing the input leftward a continuous-mode squeezed vacuum state. Quadrature squeezing occurs when the quantum fluctuations in one of the quadrature components of electromagnetic fields drop below the vacuum level, this is characterized by states of the field with no classical analogues. The effect of squeezing can be measured by a balanced homodyne detection scheme [6]. For sufficiently long detection time , it is possible to obtain a simple expression of the variance $\langle [\Delta \hat{a}_R(\phi_{LO}, \omega_{LO})]^2 \rangle$ of the homodyne measurements made on a specific field state e.g. $|R >$ as a function of the local oscillator (LO) frequency ω_{LO} and phase ϕ_{LO}, [6]. By using Eq. (3) and assuming a zero temperature reservoir, we can obtained the variance of the field transmitted through to the right of the MC as a function of the variances of the input fields,

$$\langle [\Delta \hat{b}_R(\phi_{LO}, \omega_{LO})]^2 \rangle - 1 = |R|^2 (\langle [\Delta \hat{a}_L(\phi_{LO} - \arg R), \omega_{LO}]^2 \rangle - 1)$$
$$+ |T|^2 (\langle [\Delta \hat{a}_R(\phi_{LO} - \arg T, \omega_{LO})]^2 \rangle - 1), \qquad (4)$$

with the coefficients evaluated at ω_{LO}. We now use these results to analyze the propagation of quadrature noise through a symmetric MC made of a λ layer of refractive index $n_c = 3.4$ with one QW embedded inside with a radiative decay rate of the bare exciton amplitude $\Gamma_0 = 0.064$ meV and with the bare exciton

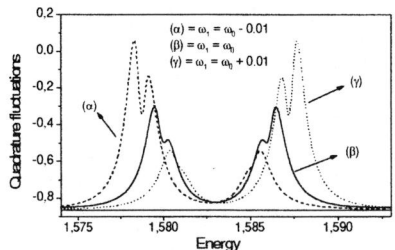

FIGURE 2. Transmitted quadrature variance as the local oscillator frequency is swept through the Rabi peaks of a semiconductor microcavity for different values of the microcavity detuning. The input signal beams are a rightward and a leftward propagating continuous-mode squeezed state. Parameters are given in the text and in the figure.

energy $\omega_0 = 1.583$ eV. The mirrors of the MC are two DBRs with 20 double layers of refractive index $n_1 = 2.95$, $n_2 = 3.32$. The squeezing strength σ and phases ϕ_σ of the input rightward signal are taken to be constant over the resonance region. In particular we chose $\sigma = 1$ and $\phi = 2\phi_{LO}$ producing a variance of the homodyne measurements on the input field $\langle [\Delta \hat{a}_R(\phi_{LO}, \omega_{LO})]^2 \rangle - 1 = -0.865$. The loss of squeezing by reflection can be compensated sending also an input leftward squeezed signal. We chose $\sigma' = \sigma$ and $\phi' = \phi$. The obtained quadrature fluctuations for different cavity detunings and for $\gamma = 0.8$ meV are displayed in Fig. 2. As a reference we have also displayed the constant input noise level $\langle [\Delta \hat{a}_R]^2 \rangle - 1 = \langle [\Delta \hat{a}_L]^2 \rangle - 1 = -0.865$. As expected squeezing is degraded at frequencies close to the Rabi peaks mostly in correspondence of the exciton-like Rabi-peak. Furthermore one can observe rapid output quadrature variations of the noise for energy near the Rabi peaks due to variations of the phase of the MC reflection and transmission coefficients.

REFERENCES

1. L. Knöll and U. Leonhardt, J. Mod. Opt. **39**, 1253 (1992).
2. U. Leonhardt, J. Mod. Opt. **40**, 1123 (1993).
3. R. Matloob, R. Loudon, S. M. Barnett and J. Jeffers, Phys. Rev. A **52**, 4823 (1995).
4. Ho Trung Dung L. Knöll and D. G. Welsch, Phys. Rev. A **57**, 3931 (1998).
5. O. Di Stefano, S. Savasta and R. Girlanda, to appear on Phys. Rev. A .
6. M. Artoni and R. Loudon, Phys. Rev. A **59**, 2279 (1999).
7. O. Di Stefano, S. Savasta and R. Girlanda, Phys. Rev. A **60**, 1614 (1999).
8. K. Cho, J. Phys. Soc. Jpn., **55**, 4113 (1986).
9. L. C. Andreani, Phys. Lett. A **192**,99, (1994).
10. O. Di Stefano, S. Savasta and R. Girlanda, to appear on Phys. Status Solidi .
11. S.Savasta and R.Girlanda, Phys. Rev. Lett. **77**, 4736 (1999).
12. S.Savasta and R.Girlanda, Phys. Rev. B , 15409 (1999).

Interaction Water-Polyethyleneglycols: Determination of Hydration Parameters from Hydrodynamic Data

Ines D. Donato*, Pasquale Agozzino° and Patrizia Perzia*

*Dipartimento di Chimica Fisica — Università di Palermo
°Dipartimento di Chimica e Tecnologie Farmaceutiche — Università di Palermo

Abstract. Densities, viscosities and refractive indexes of diluted aqueous solutions of polyethylene glycols (Mw = 200 - 6000) are determined at 25 °C. The non-linear increase of viscosity B and C coefficients with the number of ethylenoxide (EO) units are discussed. The solvation (hydration) becomes dominant when the number of ethylenoxide group increases. The hydration parameter, ω, the specific increment of density, dd_s/dc, and of refractive index, dn_{IV}/dc, for the various PEGs are the consequence of dimension and of conformation that assumes the polymer.

INTRODUCTION

Polyethylene glycols (PEGs) are frequently used in various fields: pharmaceutical and cosmetic preparations, biomedical applications and strengthening of waterlogged woods.

Despite the numerous studies on aqueous solution of those polymers, complete investigations on polymer-solvent and polymer-polymer interactions are still lacking.

In order to investigate these interactions we examined the coefficients of viscosity and other parameters from measurements of density and refractive index for aqueous solution of various PEGs.

EXPERIMENTAL

Polythyleneglycols were from Aldrich, with the following average molecular weights: 200, 400, 600, 1000, 1500, 2000, 3000, 3400, 4600 and 6000.

Viscosities were measured at 25 ± 0.02 °C by an Ubbelohde viscosimeter connected with AVS 440 (Schott-Geräte); the accuracy of absolute viscosity was ± 0.001 cp. Densities were measured at 25 ± 0.005 °C by a Sodev 03D densimeter, the accuracy of density was $\pm 3.0 \ 10^{-6}$ g cm^{-3}. Refractive indexes were measured at 25 ± 0.02 °C by a GPR 11-37 refractometer (Index Instrument), the accuracy was ± 0.0002. Polyethylene glycol solutions were prepared for weighing. The concentration of polymer was not exceeding 10%.

CP513, *Nuclear and Condensed Matter Physics,* edited by A. Messina
© 2000 American Institute of Physics 1-56396-929-7/00/$17.00

RESULTS AND DISCUSSION

For moderate concentrations solutions of non-electrolytes, the relations between the relative viscosity η_r and the B and C coefficients is generally defined by a polynomial.

$$\eta_r = \frac{\eta}{\eta_0} = 1 + Bc + Cc^2 + \ldots \ldots \tag{1}$$

The B and C coefficients are determined by a polynomial fit of η_r against concentration; table 1 reports the fitting parameters specific to every polymer.

B is a measurement of *long range* interactions; the higher term C involves structural solute-solute interactions.

TABLE 1. Coefficients B, C and Bsize, B_{solv}, values for PEGs in water at 25 °C.

M_W PEG	B dm^3 mol^{-1}	C dm^3 mol^{-1}	B_{size} dm^3 mol^{-1}	B_{solv} dm^3 mol^{-1}
200	0.73	0.13	0.43	0.30
400	1.48	2.16	0.84	0.64
600	2.96	4.09	1.26	1.70
1000	5.48	28.4	2.10	3.38
1500	10.8	77.7	3.15	7.70
2000	12.9	248.1	4.20	8.68
3000	33.0	568.5	6.30	26.72
3400	38.4	1034	7.12	31.28
4600	72.6	1350	9.64	63.01
6000	93.9	4415	12.55	81.35

B and C increase non–linearly with the number of ethylenoxide units; the values of C are lesser than those of B at low molecular weight, but are higher at high molecular weight; the variations become evident beginning from 34 EO units (Mw=1500) as shown in Fig.1.

The solute-solute interactions become stronger with chains lengthening. Increasing M_w, the chains approach each other, reducing the intermolecular distances between macromolecules.

The B variations can be related to two effects. The first effect, B $_{size}$, depends on the solute size giving a structural ordering of water molecules, through hydrophobic interaction and increase with EO. The second, B_{solv}, is the effect of solvation (hydration) due to hydrogen bonding between the ether oxygen and water molecules, solute-solvent interactions (1).

$$B = B_{size} + B_{solv} \tag{2}$$

$B_{size} = K_s \overline{V}_2$ is deduced from the Einstein theory; K_s is the shape factor; the value of K_s can be taken as 2.5, because previous studies showed that the polyethylene glycols considered in this work behave as impermeable spheres in aqueous solution (2). To determine $\overline{V}_2$, we measured the densities of PEGs solution at various concentrations, and we determined the partial molar volume. The values of B_{size} were independent of the solute concentration. The values of B_{size} and B_{solv} are reported in table 1.

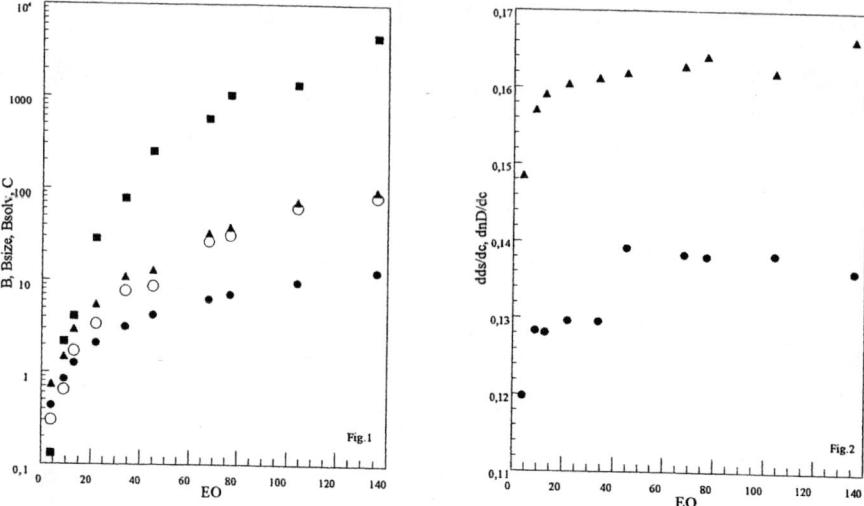

Figure 1. B (▲), B$_{size}$ (●), B$_{solv}$ (○), C (■) values as a function of EO units.

Figure 2. dd$_s$/dc (▲) and dn$_D$/dc (●) values as a function of EO units.

Fig 1 showed and confronted the two contributions to B; as EO units increase, B$_{solv}$ becomes the major effect.

The solvation depends on the solute concentration; the hydration parameter, ω, is deduced as a function of solute concentration by viscosity and density data. We evaluated η$_c$ and ω$_c$ by the relation:

$$[\eta_c] = K_s\left(\upsilon + \frac{\omega_c}{d_0}\right)$$

(3)

Where: $\upsilon = \dfrac{\overline{V_2}}{M_2}$; $[\eta_c] = \eta_{sp}\dfrac{\upsilon}{\Phi}$; $\Phi = c\overline{V}_2 \cdot 10^{-3}$

Increasing the concentration of the solute, the ω$_c$ values rise non-linearly and ω, the weight of water hydrating 1g of PEG, is determined by a polynomial fit of ω$_c$ against polymer concentration. Table 2 shows the ω values for every polyethylenoxide.

The density and refractive index increase linearly with the concentration.

In table 2 reportes the specific increment of density, dd_s/dc, and of refractive index, dn_D/dc, for the various PEGs. These values represent the slope of the linear trend for density and for refractive index respectively versus the concentration.

Fig 2 shows the plots of dd_s/dc and dn_D/dc versus the EO units number.

The trends of values dds/dc, dn_D/dc and ω versus the EO units number are non-linear at high abscissa values. The trends are connected with the interaction between the water and the polymer. These interactions are a direct consequence of the PEG conformation taking on water. A different hydration of PEGs, at high values of molecular weight, can be induced from variation of chain conformation: helicoidal, folding and random coils (3).

TABLE 2. Hydrodynamic data and specific increments of density and refractive index.

EO	ω g_{H2O}/g_{PEG}	n_{H2O} / EO	dd_s / dc	dn_{Ds} / dc
4	0.33	0.92	0.14846	0.1199
9	0.81	2.00	0.15700	0.1283
13	0.90	2.31	0.15905	0.1281
22	1.37	3.46	0.16040	0.1296
34	2.10	5.15	0.16117	0.1295
45	2.91	7.18	0.16188	0.1391
68	3.33	8.16	0.16285	0.1383
77	4.22	10.35	0.16409	0.1380
104	4.45	10.94	0.16207	0.1383
136	5.10	12.50	0.16637	0.1362

ACKNOWLEDGMENTS

Financial supports from Italian Ministry for University and Technological Research (MURST) and from C.R.R.N.S.M. are gratefully acknowledged.

REFERENCES

1. Bahri H., Güveli, Colloid e Polymer Sci, **266**, pp. 141-144 (1988).
2. Rigby D, Stepto RFT Macromolecules, **14**, 1808-1812 (1981).
3. Donato ID, Magazù S. Molecular Physics, **87**, 1463-1469 (1996).

Modelling the phase diagram of transition metal alloys by the embedded atom method

M. G. Donato, P. Ballone, P. V. Giaquinta

Istituto Nazionale per la Fisica della Materia
Università degli Studi di Messina
Dipartimento di Fisica, Contrada Papardo, C.P. 50, 98166 Messina, Italy

Abstract. The embedded atom method (EAM) is a simple yet powerful model for the potential energy of ordered and disordered transition metal systems. We demonstrate its versatility and reliability by a simulation study of the phase diagram for Cu_xPd_{1-x} as a function of concentration x and temperature. We discuss the limitations of the present method, and we outline a few directions for improvements of the basic model.

Introduction. The interpretation and prediction of thermodynamic and mechanical properties of transition metal alloys increasingly relies on computer simulation at the atomistic level, which, in turn, requires a simple and accurate model for the potential energy of an assembly of atoms as a function of their coordinates.

Metallic bonding is due to the delocalised cloud of valence electrons, and, therefore, it is intrinsically many-body. The embedded atom model mimics the many-body interactions in metals by introducing an explicit (although oversimplified) description of the valence charge distribution, and by relying on basic density functional concepts to account for the dependence of the interatomic interaction on the local environment.

The Cu_xPd_{1-x} phase diagram. A simple and representative example of a binary alloy of transition metals is given by the Cu_xPd_{1-x} system. On cooling from the liquid state, Cu_xPd_{1-x} crystallizes in a continuous solid solution, based on the fcc lattice. At low temperature, the phase diagram displays several intermetallic phases, of which the ordered L1$_2$ (at $x \sim 0.75$) and B2 (at $x \sim 0.5$) structures are the most prominent ones [2]. These two structures are both based on cubic lattices (the fcc for L1$_2$, and the bcc for B2), with a regular distribution of the Cu and Pd species on simple cubic sublattices.

The embedded atom model. We adopt the original approach, introduced in Ref. [1], together with the explicit parametrization of Ref. [4] for the Cu-Pd potential. The potential energy E as a function of the atomic coordinates $\{\mathbf{R_I}; (I =$

CP513, *Nuclear and Condensed Matter Physics*, edited by A. Messina
© 2000 American Institute of Physics 1-56396-929-7/00/$17.00

$1, N)\}$ is written as:

$$E[\mathbf{R_I}] = \frac{1}{2} \sum_{I \neq J} \phi_{IJ}(|\mathbf{R_I} - \mathbf{R_J}|) + \sum_I F_I[\rho(\mathbf{R_I})] \tag{1}$$

where ϕ_{IJ} is a repulsive pair potential, and $F_I[\rho(\mathbf{R_I})]$ is the energy gain in embedding the atom I into the valence charge density $\rho(\mathbf{R_I})$. In turn, $\rho(\mathbf{R})$ is given by the superposition of all the valence electron distributions associated to each atom.

The repulsive potential ϕ and the embedding function F are devised in order to reproduce the ground state structure, and to fit the elastic properties and vacancy formation energies of the pure elements.

The simulation method. We sample the phase space of the system by the Monte Carlo method. All simulations are performed at zero pressure for samples of 1024 atoms, adopting a cubic cell for the B2 phase, and a tetragonal cell for the phases based on the fcc lattice. Since the order-disorder transition is a crucial component of the phase transformation, we sample the atomic exchange processes by attempting to swap the position of a pair of atoms chosen at random. A detailed description of the simulation is contained in Ref. [5].

Results. The $T = 0\text{K}$ phase diagram of $Cu_xPd_{(1-x)}$ is determined as a function of composition $0.30 \leq x \leq 0.70$ by comparing the potential energy of the ordered B2 and random fcc alloy.

The results for the excess energy of mixing at $T = 0$, defined by:

$$E_{mix}(x) = E(x) - xE(Cu) - (1-x)E(Pd)$$

(where $E(x)$, $E(Cu)$ and $E(Pd)$ are the cohesive energies per atom of the alloy and of the pure metals, respectively) are reported in Fig.1.

It is apparent that the EAM predicts the stability of the B2 phase at $T = 0\text{K}$ for concentrations $0.31 \leq x \leq 0.63$. This results is in qualitative agreement, but apparent quantitative disagreement with the experimental phase diagram: according to the data reported in Ref. [2], the B2 phase is stable for x between 0.42 and 0.72

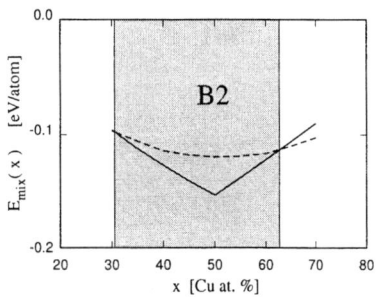

FIGURE 1. Excess energy of mixing at $T = 0$ K for the ordered B2 (full line) and disordered fcc (dash line) phases as a function of concentration.

As a second stage, we perform a series of simulations for the system at $0 \leq T \leq$ 1500 K, with the B2 and disordered fcc structures. In the case of the B2 structure, $U(T)$ has a clear deviation from linearity at ~ 700K, corresponding to the onset of the order-disorder transition. The fcc phase does not have such a transition, and the corresponding $U(T)$ is almost linear.

The relative stability of the B2 and fcc phases is determined by their free energy difference $\Delta G(T)$, which is reported in Fig. 2. With increasing temperature, the conspicuous potential energy advantage of the B2 phase at $T = 0$ is progressively compensated by the mixing entropy of the fcc alloy. The compositional disordering of the bcc-based alloy at $T \sim 700$K, slows down the free energy gain of the random fcc phase, but it does not prevent the crossing of the B2-random fcc free energy that occurs at $T_B = 750$K. Considering the simplicity of the EAM model, this result is in surprisingly good agreement with the experimental transition temperature of 770K at $x = 0.5$

We investigate the dependence of the transition temperature on composition x. On the one hand, it is apparent already from Fig. 1 that the potential energy advantage of the B2 phase decreases rapidly in moving away from the $x = 0.5$ composition. On the other hand, also the mixing entropy that stabilizes the random fcc structure has a maximum at the stoichiometric ($x = 0.5$) composition. Our simulations show that the energy dependence on x is, by far, the dominant factor, resulting in a transition temperature that follows closely the behavior of the potential energy difference. For instance, at $x = 0.58$, which corresponds to the highest transition temperature measured by experiments, the computed T_B is already reduced to 300K, in apparent disagreement with the experimental data.

As a final point, we investigated the transition from the L1$_2$ phase to the disordered fcc alloy for $x = 0.75$. In this case, the potential energy U displays, as a function of T, a clear linear behavior both at low and at high temperatures, with a cross-over starting at $T \sim 400$K and culminating at $T \sim 500$K. The specific heat, obtained by differentiating a Padé fit for $U(T)$, has a peak centered at $T = 470$K,

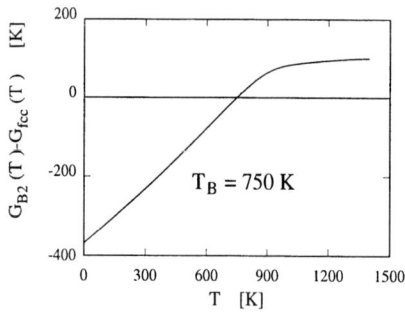

FIGURE 2. Difference in the Gibbs free energy (per atom) of the B2 and disordered fcc phases as a function of temperature at composition $x = 0.5$.

116

that we identify with the order-disorder transition temperature. This result has to be compared with the experimental transition temperature of 730K at $x = 0.75$, and a stability region for the L1$_2$ phase centered at $x \sim 0.82$, with a maximum transition temperature of 770K.

Discussion. The analysis of the computational results shows, at first, that the EAM is remarkably successful in reproducing the qualitative features observed in the experimental phase diagram: the B2 phase is stable at low temperature around $x = 0.5$, and transforms with increasing temperature to the disordered fcc structure. Moreover, the stability of the L1$_2$ phase at $x = 0.75$ is also reproduced by EAM, with a critical temperature for the order/disorder transition in qualitative agreement with the experimental result. In both cases, the driving force stabilizing the ordered structures is the positive heat of mixing of Cu and Pd, favoring the optimal alternation of these two elements in the alloy.

However, if we look more in detail into the comparison of the computational with the experimental data, we see that EAM is unable to reproduce the maximum in the transformation temperature at $x \sim 0.58$, and, moreover, the temperature range of stability for the B2 phase is somewhat underestimated.

The most crucial problem is related to the asymmetry of the stability region around $x = 0.5$, that is not reproduced by the EAM formulation. Previous *ab-initio* computations [3] reveal that the electronic structure of Cu$_x$Pd$_{1-x}$ alloys is fairly complicated, and the electronic energy is affected by a variety of factors, including Fermi surface effects, which are strongly dependent on the band filling (i.e., on x), relativistic effects (mainly spin-orbit interactions), etc. These features are not, and cannot be, fully included in a simple model like EAM, that implicitly assumes a spherical Fermi surface.

However, the inclusion of these effects into slightly more sophisticated models ([6], [7], [8]) is possible, at least to some degree of approximation. Then, the bainitic transformation in Cu$_x$Pd$_{1-x}$, described fairly well by the zero order model, but displaying also sizable differences with the experimental data, could provide an ideal testing ground for the extension of these methods to alloys.

REFERENCES

1. M. S. Daw, and M. I. Baskes, Phys. Rev. Lett. **50**.
2. M. Hansen, *Constitution of Binary Alloys*, (McGraw-Hill, New York, 1958).
3. See: E. Bruno and B. Ginatempo, Europhys. Lett. **42**, 649 (1998) for $x = 0.5$, and: Z. W. Lu, S.-H. Wei and A. Zunger, Phys. Rev. B **45**, 10314 (1992) for $x = 0.75$.
4. M. Foiles, M. S. Daw, and M. I. Baskes, Phys. Rev. B **33**, 7983 (1986).
5. M. G. Donato, P. Ballone, and P. V. Giaquinta, Phys. Rev. B *in press*.
6. M. I. Baskes, J. S. Nelson, and A. F. Wright, Phys. Rev. B **40**, 6085 (1989).
7. T. J. Raeker, and A. E. De Pristo, Phys. Rev. B **39**, 9967 (1989).
8. S. M. Foiles, Phys. Rev. B **48**, 4287 (1993).

Slow Dynamics Features in Aqueous Solutions of High Molecular Weight Poly(Ethylene Oxide)

A.Faraone, C.Branca, S.Magazù, G.Maisano, P.Migliardo and V.Villari

Dipartimento di Fisica and INFM, Università di Messina, C.da Papardo S.ta Sperone 31, P.O.55, 98166 Messina, Italy

Abstract. In this contribution results of Photon Correlation Spectroscopy measurements on aqueous solutions of high molecular weight Poly(Ethylene Oxide) in the semidilute regime are reported. They clearly reveal the existence of two distinct k^2-dependent processes on far different time scales: a collective diffusion process at short times and a slow dynamics at long times. The experimental findings suggest that the slow dynamics is likely due to the formation of "clusters" which crucially depends, besides on the fact that the overlap concentration is reached, on the polymer main chain length.

INTRODUCTION

Recently, semidilute and concentrated polymeric solutions have been drawing a lot of attention because they represent a challenging many-body problem with various interesting applications. In fact, for the existence of chain overlaps and entanglements, it is necessary to take into account not only the direct intermolecular interactions, but also the effect of long-ranged hydrodynamic components on the collective motions. For these reasons one of the most debated subjects nowadays concerns with the presence of a slow dynamics in the intermediate scattering functions of many entangled polymeric solutions.[1] The autocorrelation functions of many semidilute solutions, in fact, show an additional slow decay mode together with the fast diffusive one. The decay of these slow modes covers a wide time range (up to 3 orders of magnitude) and shows components with characteristic times of about ten or hundred seconds. The analysis is complex, both from the experimental and the theoretical point of view, and, up to now, an underlying microscopic understanding of the physical origin of these modes and in particular of their temperature dependence have not been provided. To address this problem we investigated semidilute aqueous solutions of high molecular weight (M_w) Poly(Ethylene Oxide) (PEO), a water soluble polymer of relevant theoretical and applicative importance in polymer physics.[2]

EXPERIMENTAL SECTION

The solutions were freshly prepared from standard PEO samples (Sigma Aldrich co.) and ultrapure H_2O. Great care was taken in order to obtain stable, clear, and dust

CP513, *Nuclear and Condensed Matter Physics*, edited by A. Messina
© 2000 American Institute of Physics 1-56396-929-7/00/$17.00

free samples. PEO solutions were filtered in ricirculation through 0.22 μ Millipore PTFE filter. Photon Correlation Spectroscopic (PCS) measurements were performed using a BROOKHAVEN BI-2030 correlator to collect the intensity autocorrelation data. As exciting source, the 4880 Å vertically polarized line of a unimode Ar^+ laser INNOVA 70 was used. Under the hypothesis of a large number of independent scatterers, the Siegert's relation can be applied:

$$g^{(2)}(k,t) = <I(k,0)I(k,t)>/<I>^2 = 1 + b \left| g^{(1)}(k,t) \right|^2 \qquad (1)$$

where b is an optical parameter, $k = 4\pi n/\lambda \; sin(\theta/2)$ the scattered wave vector and $g^{(1)}(k,t) = <E_S(0)E_S(t)>/< I>$ the normalized dynamic structure factor.

Considering that two decays contribute to $g^{(1)}(k,t)$, the experimental intensity autocorrelation functions have been fitted by the following relation:

$$g^{(2)}(k,t) = \left[A_1 \exp\left(-(t/\tau_1)\right) + A_2 \exp\left(-(t/\tau_2)^\beta\right) \right]^2 \qquad (2)$$

where β is the shape parameter of the Kohlraush-Williams-Watts function.

RESULTS AND DISCUSSION

For polymers with high M_w, like for example PEO 900000, the semidilute regime is reached at low concentration (c~1% by weight). This regime is defined by the overlap concentration, above which the chains begin to touch and overlap: $c^* = 3M(N_A 4\pi R_H^3)^{-1}$.

In figure 1 the normalized intensity auto-correlation functions of aqueous solutions of PEO 900000 at w=0.5% by weight are shown at different scattering angles, corresponding to a k range of $1.3 \cdot 10^5 \div 3.3 \cdot 10^5$ cm^{-1}. In all the measurements it is possible to distinguish two components on different time scales. The long time decay follows a stretched exponential law indicating that it originates from a distribution of relaxation processes characterized by different decay rates. In the insert, the k dependence of the two modes is reported. As it can be seen both the modes, in the limit of the experimental error, can be considered k^2 dependent; this finding clearly reveals that both the processes are diffusive in character. However the relative amplitude of the two decays is strongly dependent on the exchanged wave vector; in fact the analysis of the A_1/A_2 ratio indicates that the long time contribution is more and more dominant as the scattering vector increases. Moreover also the parameter β, which takes the value 1 in the case of a pure exponential decay, decreases strongly by increasing the scattering angle, from 0.80 at $\theta=45°$ to 0.48 at $\theta=150°$. These values of β state that a wide distribution of processes contributes to the slow relaxation process.

In figure 2 the correlation functions are reported at three different temperatures for the same scattering angle ($\theta=90°$). As it can be seen both the fast and the slow process decay rates decrease increasing temperature. This evenience can be rationalized in terms of the effect of viscosity which decreases increasing temperature.

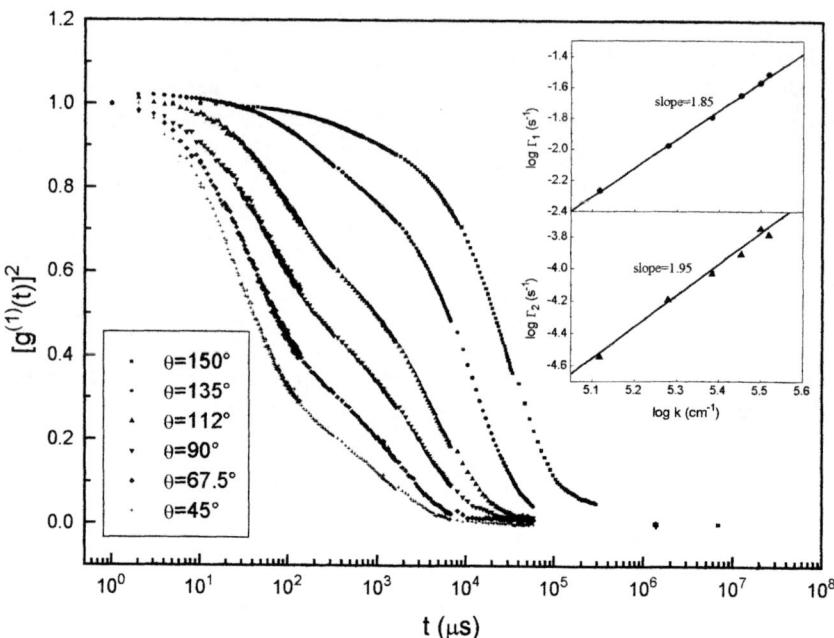

FIGURE 1. Intensity autocorrelation functions at c=0.5% for different scattering angles. The inserts indicates that the decay rates display a k^2-dependence.

To show this effect we calculated the effective dynamic correlation length corresponding to the fast and the slow modes: $\xi_i = k_B T/6\pi\eta D_i$ (i=1,2), η being the solvent viscosity. This quantity is intrinsically scaled for the ratio T/η so that it is not influenced by the temperature dependence of viscosity. The results are reported in the insert as a function of temperature. ξ_1 is smaller than R_H, as obtained from the scaling relation for PEO in aqueous solutions $R_H=0.145\,M_w^{0.57}$,[3] indicating that at w=0.5% the intermolecular interactions play a relevant role. As far as the slow mode is concerned, ξ_2 is much longer than R_H suggesting that the decay of this slow fluctuation is correlated on a wide space scale. From these findings we can attribute the fast relaxation process to the collective diffusion of the polymer blobs. On the other hand, the occurrence of a single relaxation mode, in the dilute regime at w=0.1%, suggests the hypothesis that the slow dynamics is likely due to the formation of aggregates of many polymer blobs which relax together and whose dimensions are of the order of the dynamic correlation length. Increasing temperature the dimension of these "clusters" decreases. Analogous measurements performed on PEO 35000 aqueous solutions[4] far above the overlap concentration, do not show the presence of either a slow dynamics, or the existence of aggregation phenomena[5] or viscoelastic effects on the concentration fluctuations decay. Therefore, we can argue that the formation of these "clusters" crucially depends, besides on the fact that the overlap concentration is reached, on the polymer main chain length.

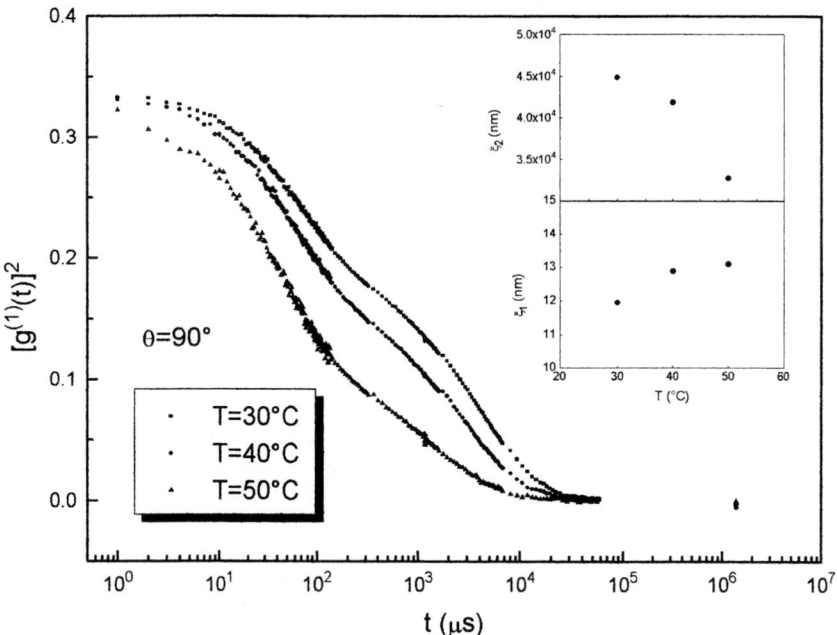

FIGURE 2. Temperature dependence of the diffusive relaxation processes. In the inserts the correlation lengths corresponding to both the decays are reported.

ACKNOWLEDGMENTS

V. Villari acknowledges a grant from Consiglio Nazionale delle Ricerche (CNR)

REFERENCES

1. Jian, T., Vlassoupoulos, G., Pakula, T., and Brown, W., *Colloid. Polym. Sci.*, **274**, 1033-1043 (1996).

2. Bailey, F.E., and Koleske, J.V., *Poly(Ethylene Oxide)*, New York: Academic, 1976.

3. Devanand, K., and Selser, J.C., *Macromolecules*, **24**, 5943-5947 (1991).

4. Faraone, A., Magazù, S., Maisano, G., Poneterio, R., and Villari, V., *Macromolecules*, **32**, 1128-1133 (1999).

5. Faraone, A., Magazù, S., Maisano, G., Migliardo, P., Tettamanti, E., and Villari, V., *J. Chem. Phys.*, **109**, 1801-1806 (1999).

Coupled Josephson Junction as Quantum Computing Devices

R.Fazio[1] G.M.Palma[2] and J.Siewert[1]

[1]Dipartimento di Metodologie Fisiche e Chimiche Universita' di Catania, & Unita' INFM
viale A.Doria 6, I-95125 Catania, Italy
[2]Dipartimento di Scienze Fisiche ed Astronomiche, Universita' di Palermo & Unita' INFM
Via Archirafi 36,I - 90123 Palermo, Italy

Abstract. The requirements that the candidate technologies have to fulfill in order to build a quantum computer are discussed with particular reference to a newly proposed implementation based on high capacitance coupled Josephson junctions.

REQUIREMENTS FOR A QUANTUM COMPUTER

Quantum computation is a new form of information processing which makes uses of purely quantum phenomena like quantum superposition and entanglement, to obtain an exponential gain in the resources (time, space, energy,..) needed in solving hard problems [1]. At an abstract level quantum computers are described as networks of gates operating on qubits [2], i.e. of a controlled coherent evolution of set of two level systems with time dependent couplings. It is evident that this theoretical frame imposes extremely stringent requirements on any experimental implementation of quantum information processing devices. Such requirements can be summarized as follows [3]:

1. It must be possible to control the computational Hilbert space. Such space should have the structure of a tensor product of subsystems, with a precisely enumerable set of states. As we will see below the dynamics of the system must be confined within such space.
2. it must be possible to prepare the system in a fixed starting state
3. it must be possible to change in time in a controlled way the energy of the individual subsystems as well as their mutual coupling to implement the required unitary gate operations
4. it must be possible to readout the final state of the system
5. it must be possible to isolate the system from the environment as such coupling would spoil the coherent computational unitary evolution introducing decoherence [4].

CP513, *Nuclear and Condensed Matter Physics*, edited by A. Messina
© 2000 American Institute of Physics 1-56396-929-7/00/$17.00

While the potentialities of this new form of computation are being explored at a theoretical level it is not yet clear which technology will be the best candidate for the physical implementation of a quantum computer. Several proposals have so far been put forward, ranging from cold trapped ions [5] to atoms in cavities [6], to nuclear magnetic resonance [7]. For large scale integration attention has been turned towards quantum dots [3], nuclear spin of phosphorous dopants in silicon [8], SQUIDS [9] and Josephson junctions [10]. In view of the recent experimental advances in the fields [11] it is on this latter that we will specialize our discussion.

JOSEPHSON QUBITS

Figure (1) represents schematically our system of two coupled Josephson junctions [10]. The electrostatic energy of the excess charge on each superconducting electron box is controlled by external bias voltages V though a capacitance C. Furthermore the currents flowing across the junctions are coupled through a common inductance L. The Hamiltonian describing this system is

$$H = \sum_{(i=1,2)} [E_c(n_i - n_{xi})^2 - E_J \cos\phi_i] + E_L(\cos\phi_1 + \cos\phi_2)$$

Where n_i is the number of Coper pairs on the ith island, ϕ_i is the corresponding conjugate variable $[n_i, \phi_j] = \delta_{ij}$, and the offset charge n_{xi} can be controlled by the external voltage V. The qubit is encoded in two adjacent charge states, say |0> and |1>. Truncated within this computational space the Hamiltonian becomes

$$H = \sum_{(i=1,2)} [\Delta E_c \sigma_{zi} - E_J/2\sigma_{xi}] - E_L/2\sigma_{y1}\, \sigma_{y2}$$

Where the σ are Pauli matrices and $\Delta E_c = E_c(n_{xi} - 1/2)$. During idle periods $\Delta E_c >> E_L, E_J$. One qubit gate operations are obtained by suddenly switching the offset charge to the degeneracy point $n_x = 1/2$ for a finite time. States |0> and |1> are then coupled by the Josephson tunneling. In a similar fashion two qubit gates are implemented by switching suddenly states |01> and |10> to degeneracy for a fixed time t , where they are mixed by the inductive coupling.

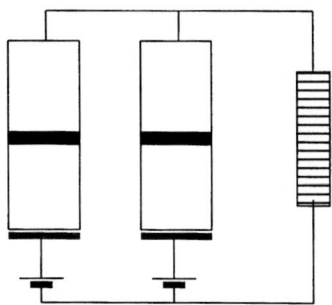

Fig.1: Two coupled junctions

123

FIDELITY AND LEAKAGE

The ideal dynamics described so far is restricted within the subspace spanned by states |0>, |1> of each junction. In reality this is a subspace of the much larger Hilbert space spanned by all the charge states. It is therefore to be expected that the presence of such states will cause a deviation from the ideal dynamics described in the previous section. A detailed estimate of such deviation is necessary in order to asset the possibility to implement error correcting protocols and fault tolerant computation with Josephson qubits and furthermore to optimize gate parameters [12].

The deviation from the ideal dynamics can be characterized by two quantities: the *leakage*, which measures the probability that the system will escape out of the computational state, and the *fidelity*, a positive real quantity smaller or equal to 1 (maximal fidelity) which measures how the real gate operations, including higher charge states, differs from the real one. For a quantitative definition of these two quantities and a related analysis see [12]. Here we will simply illustrate qualitatively our results, depicted in the figures below.

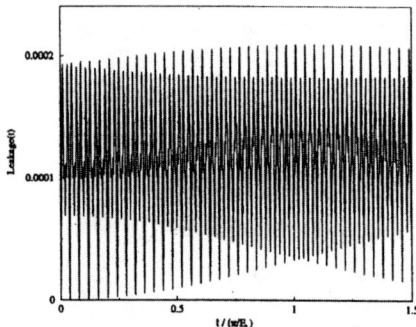

Fig.2 Leakage of the one -qubit gate as a function of time.

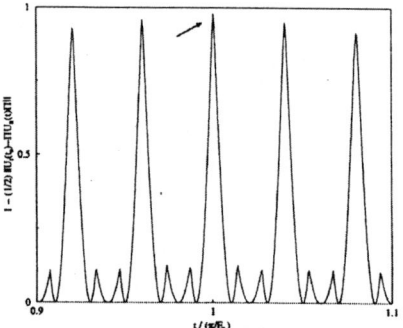

Fig.3 The gate fidelity is obtained by minimising the distance between the ideal and the real time evolution. For the single qubit case this corresponds to the point signed by the arrow.

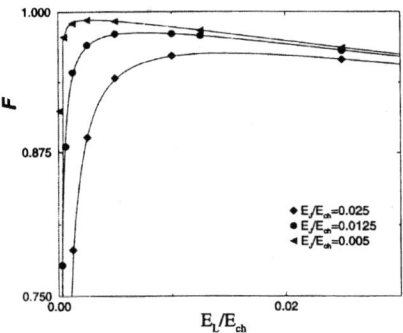

Fig.4 fidelity of the two-qubit gate as a function of gate patameters.

ACKNOWLEDGMENTS

This work was supported in part by the European TMR Research Network under contracts ERB 4061PL95-1412 and FMRX-CT-97-0143 and by the INFM under PRA *"Solid state implementations of quantum information processing devices"*.

REFERENCES

1. D.Deutsch, *Proc. R. Soc. London* **400**, 97 (1985);A. Ekert and R. Jozsa, *Rev. Mod. Phys.***68**, 733 (1996)

2. A.Barenco, C.H.Bennett, R.Cleve, D.DiVincenzo, N.Margolus, P.W.\Shor, T.Sleator, J.Smolin, and H.Weinfurther, *Phys.Rev.*A **52**, 3457 (1995).

3. D.Loss and D.DiVincenzo, *Phys. Rev.*A **57**, 120 (1998).

4. G.M.Palma, K.-A.Suominen and A.K.Ekert, *Proc. R. Soc. London* 567 (1996); W. Zurek, *Physics Today* **44**, 36 (1991).

5. J.I.Cirac and P. Zoller, *Phys. Rev. Lett.* **74** 4091 (1995).

6. Q.A.Turchette, C.J. Hood, W. Lange, H. Mabuchi and H.J. Kimble, *Phys. Rev. Lett.* **75**, 4710 (1997).

7. N.A.Gershenfeld and I.L. Chuang, *Science*, **275**, 350 (1995).

8. B.Kane, *Nature* **393**, 133 (1998).

9. J.E.Mooij, T.P.Orlando, L. Tian, C. van der Wal, L.Levitov, S.Lloyd, and J.J.Mazo, unpublished

10. A.Shnirman, G. Schoen and Z. Hermon, *Phys. Rev. Lett.* **79** 2371 (1997),Y. Makhlin, G.Schoen and A.Shnirman, *Nature*, **398**, 305-307 (1999).

11. Y.Nakamura, Yu.A. Pashkin, J.S. Tsai, *Nature* **398**, 786 (1999).

12. R.Fazio, G.M.Palma e J.Siewert, *Phys.Rev.Lett.*(submitted), e.print archive cond-mat/9906292, R.Fazio, G.M.Palma E.Sciacca e J.Siewert, *Physica B* (in press), R.Fazio, G.M.Palma E.Sciacca e J.Siewert, *J.L.Temp.Phys* (in press)

Implicit time integration procedure in Smoothed Particle Hydro codes

Gerardi G., Molteni D.

Dipartimento di Scienze Fisiche ed Astronomiche
Universita' di Palermo, Italia

Abstract. We propose an implicit way to solve the fluid dynamics equations discretized following the Smoothed Particles Hydrodynamics approach. We applied it only to ideal gas dynamics problems. The proposed algorithm is very stable and a gain of a factor 10 can easily achieved for any kind of problems. We foresee that the computational gain will be larger for cases heavily constrained by Courant condition, like the ones occuring in magnetohydrodynamics

INTRODUCTION

Smoothed particles Hydrodynamics (SPH) is numerical method used to integrate the fluid dynamic equations. It is essentially based on interpolation criteria. It is well known that a function can be expanded as sum of a set of basis functions. In SPH the basis function is always the same , for example a gaussian or a spline function. The coefficients are fixed and have the meaning of masses, related to the physical density of the fluid. The position of the interpolatin function changes with time following to the fluid motion. Therefore the interpolation point is frequently named 'particle' and the interpolation function is named 'kernel'. The method is therefore a lagrangean one. It is particularly useful in the case of large empty regions in the integration domain; since particles move and go were the density is not zero, the computation is limited to the 'full' regions. It is also very useful when mixing of different fluids occurs and for all problems which need information on the time history of each fluid parcel. For an extended review of SPH cfr. Monaghan 1985.

Standard SPH

The equations we integrate are the classical ones of compressible ideal fluid (Batchelor, 1969) Let us resume the equations for a two dimensional motion in cartesian space. With the Lagrangian derivative

CP513, *Nuclear and Condensed Matter Physics,* edited by A. Messina
© 2000 American Institute of Physics 1-56396-929-7/00/$17.00

$$D = \frac{\partial}{\partial t} + \mathbf{v} \cdot \nabla$$

we have for the mass conservation

$$\frac{D\rho}{Dt} = -\rho \nabla \mathbf{v} = -\rho \left(\frac{\partial v_x}{\partial x} + \frac{\partial v_y}{\partial y} \right)$$

For the momentum we have

$$\frac{D\mathbf{v}}{Dt} = -\frac{1}{\rho} \nabla P + \mathbf{g}$$

or

$$\frac{Dv_x}{Dt} = -\frac{1}{\rho} \frac{\partial P}{\partial x}, \frac{Dv_y}{Dt} = -\frac{1}{\rho} \frac{\partial P}{\partial y} \qquad (1)$$

respectively.

The energy equation describes the behaviour of the internal energy per unit mass ϵ

$$\frac{D\epsilon}{Dt} = -\frac{P}{\rho} \nabla \mathbf{v}$$

$$\frac{D\epsilon}{Dt} = -\frac{P}{\rho} \left[\frac{\partial v_x}{\partial x} + \frac{\partial v_y}{\partial y} \right]$$

So for the density we have the simple expression that identically satisfy the continuity equation in the form:

$$\rho(\mathbf{r}_i) \simeq \sum_{j=1}^{N} m_j W_{ij} \qquad (2)$$

$$m_k = \rho_k \Delta x \Delta y$$

Rewriting the fundamental equations in the formulation more suitable for the SPH evaluation (Monaghan, 1985), and applying the previous criteria we have the following expression;

The momentum satisfies:

$$\left(\frac{D\mathbf{v}}{Dt} \right)_i = -\sum_{j=1}^{N} m_j \left(\frac{P_i}{\rho_i^2} + \frac{P_j}{\rho_j^2} + \Pi_{ij} \right) \frac{\partial W_{ij}}{\partial \mathbf{r}_i} \qquad (3)$$

For the thermal energy per unit mass we obtain:

$$\left(\frac{D\epsilon}{Dt} \right)_i = -\frac{1}{2} \sum_{j=1}^{N} m_j \left(\frac{P_i}{\rho_i^2} + \frac{P_j}{\rho_j^2} + \Pi_{ij} \right) (\mathbf{v}_i - \mathbf{v}_j) \cdot \nabla_i W_{ij}$$

where Π_{ij} is the artificial term contribution.

Implicit method approach

No studies have been currently done to set up an implicit approach to SPH. Let us start with the study of the problem with only body forces. The lagrangean approach requires the following expression for the standard motion of one particle under the action of an acceleration $\mathbf{a}(x,y)$ due to a body force: The basic form to integrate is of the following type of equations

$$\frac{d\mathbf{r}}{dt} = \mathbf{v} \quad ; \quad \frac{d\mathbf{v}}{dt} = \mathbf{a}$$

The x component gives

$$\frac{dx}{dt} = v_x \rightarrow x^{n+1} = x^n + dt\; v_x^{n+1} \quad ; \quad \frac{dv_x}{dt} = a_x \rightarrow v_x^{n+1} = v_x^n + dt\; a_x^{n+1}$$

expanding a_x^{n+1}

$$a_x^{n+1} = a_x\left(x^{n+1}, y^{n+1}\right) = a_x\left(x^n, y^n\right) + \left(\frac{da_x}{dt}\right)^n dt + ...$$

we have

$$x^{n+1} = x^n + dt\left\{v_x^n + dt\left[a_x^n + \frac{\partial a_x}{\partial x}\left(x^{n+1} - x^n\right) + \frac{\partial a_x}{\partial y}\left(y^{n+1} - y^n\right)\right]\right\}$$

$$y^{n+1} = y^n + dt\left\{v_y^n + dt\left[a_y^n + \frac{\partial a_y}{\partial x}\left(x^{n+1} - x^n\right) + \frac{\partial a_y}{\partial y}\left(y^{n+1} - y^n\right)\right]\right\}$$

that is we have to solve the system

$$\left[\left(1 - \frac{\partial a_x}{\partial x}dt^2\right)x^{n+1} - \frac{\partial a_x}{\partial y}y^{n+1}dt^2\right] = N_1$$

$$\left[-\frac{\partial a_y}{\partial x}x^{n+1} + \left(1 - \frac{\partial a_y}{\partial y}dt^2\right)y^{n+1}dt^2\right] = N_2$$

with N1 and N2 known terms evaluated at the n-th time level.

Let us now take into account the pressure forces. The basic form to integrate is identical to the body force terms, but now

$$a_x = -\frac{1}{\rho}\frac{\partial P}{\partial x}$$

here is the **basic point** : since we are using a lagrangean approach, and the accelaration is a function of ρ, P, and therefore implicitly of the positions, we express the a_x^{n+1} as

$$a_x^{n+1} = a_x\left(x^{n+1}, y^{n+1}\right) = a_x\left(x^n, y^n\right) + \left(\frac{da_x}{dt}\right)^n dt + ...$$

$$a_x^{n+1} = a_x \left[\rho \left(r^{n+1} \right), P \left(r^{n+1} \right) \right]$$

we may proceed developing analytically the above expression

$$a_x^{n+1} = a_x^n + \left(\frac{\partial a_x}{\partial \rho} \frac{\partial \rho}{\partial x} \frac{dx}{dt} + \frac{\partial a_x}{\partial P} \frac{\partial P}{\partial x} \frac{dx}{dt} + \frac{\partial a_x}{\partial \rho} \frac{\partial \rho}{\partial y} \frac{dy}{dt} + \frac{\partial a_x}{\partial P} \frac{\partial P}{\partial y} \frac{dy}{dt} \right)^n dt$$

$$\left(\frac{da_x}{dt} \right)^n = \left(\frac{\partial a_x}{\partial x} \frac{dx}{dt} + \frac{\partial a_x}{\partial y} \frac{dy}{dt} \right)^n$$

Finally we have to equation for the x 'component':

$$x^{n+1} = x^n + dt \ \{v_x^n + dt \ [a_x (x^n, y^n) + T_x^n]\}$$

with

$$T_x^n = \left(\frac{\partial a_x}{\partial x} \frac{dx}{dt} + \frac{\partial a_x}{\partial y} \frac{dy}{dt} \right)^n dt$$

Analogous formula is valid for the 'y' component. So also in the case of pressure forces we have to solve the same type of system of two equations we had for plain body forces.

Finally we may discretize the energy equation as follows

$$\frac{\epsilon^{n+1} - \epsilon^n}{dt} = H^{n+1} \qquad where \qquad H = -\frac{P}{\rho} \nabla \mathbf{v}$$

$$H^{n+1} = H^n + \left(\frac{\partial H}{\partial x} \frac{dx}{dt} + \frac{\partial H}{\partial y} \frac{dy}{dt} \right) dt + ... =$$

$$H^n + \left(\frac{\partial H}{\partial x} \right)^n \left(x^{n+1} - x^n \right) + \left(\frac{\partial H}{\partial y} \right)^n \left(y^{n+1} - y^n \right)$$

So we have the same algebraic structure of the momentum equation. We may solve initially the momentum equantion and then use the new positions to evaluate the new thermal energy value.

Conclusions

We have therefore the following important result:
we have to invert only a small 2*2 (in the 2D cases, 3*3 for 3D) matrix for each particle at each implicit time step.

REFERENCES

1. Batchelor J.K., *An introduction to Fluid dynamics*, Cambridge University Press (1967).
2. Monaghan J., *Comp. Phys. Repts.* **3**, 71 (1985).

129

Distribution of A Substates in Saccharide Coated Carbonmonoxy-Myoglobin

Fabio Librizzi, Eugenio Vitrano, and Lorenzo Cordone

Dipartimento di Scienze Fisiche ed Astronomiche dell'Università di Palermo and INFM

Abstract. We report on preliminary experiments wherein we measured the distribution of A substates in Carbonmonoxy-Myoglobin (MbCO) partially dried either in the presence of a saccharide (glucose, sucrose, or trehalose) or in the absence of sugar. The results indicate that in our experimental conditions, glucose introduces the smallest spectral perturbation; this indicates a sizeable difference in protein-monosaccharide and protein-disaccharide interaction.

INTRODUCTION

Some organisms can survive extreme dehydration and high temperatures without suffering damages, in a state of suspended animation (anhydrobiosis). In the dry state these organisms contain large amounts of sugars, particularly trehalose, which has been found to be the most active saccharide for the preservation of biostructures (1-5). The properties that make trehalose unique with respect to other saccharides for biopreservation are still not clearly understood.

The CO stretching in Carbonmonoxy-Myoglobin (MbCO) exhibits, in the region $1900 \div 2000$ cm^{-1}, at least three distinguishable Voigtian bands, named in order of decreasing wavenumber A_0, A_1, and A_3 (6); these bands have been ascribed to different protein substates (7) and their spectral properties and relative intensity depend on external parameters such as pH, temperature, and pressure (6,8). We report here on preliminary experiments aimed at studying the different effects of disaccharides like trehalose or sucrose and of a monosaccharide (glucose) on the structure of MbCO, as measured from the distribution of A substates (6), in "*saccharide coated*" protein.

MATERIALS AND METHODS

Horse myoglobin and glucose were from Sigma (St. Louis, MO, USA); trehalose was from Hayashibara Shoij (Okayama, Japan), and sucrose from Fluka Chemie AG (Buchs, Switzerland). The above products were used without further purification. Measurements were performed on a FTIR Bio-Rad Digilab FTS-40A, with 2 cm^{-1} resolution. Trehalose (sucrose) sample was prepared by dissolving lyophilized ferric protein ($\sim 5 \cdot 10^{-3}$ M) in a solution containing $2 \cdot 10^{-2}$ M phosphate buffer pH 7 and $2 \cdot 10^{-1}$ M trehalose (sucrose).

CP513, *Nuclear and Condensed Matter Physics,* edited by A. Messina

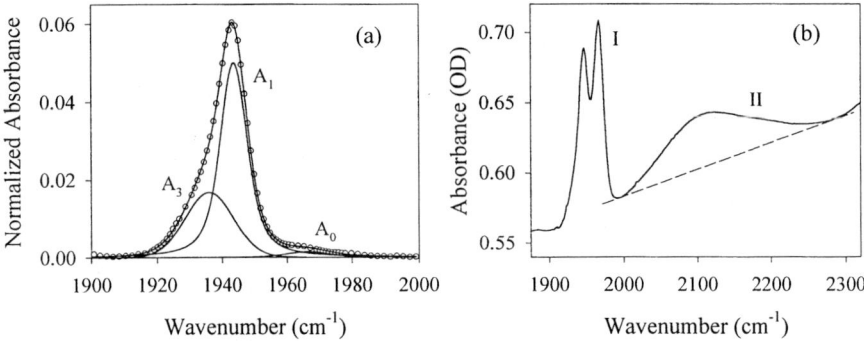

FIGURE 1. (a) A substates of MbCO; data refers to MbCO ~5mM in water solution pH 7 at room temperature. (b) CO stretching bands in MbCO (I) and Association band of water (II); we obtained information on the sample water content by drawing the tangent between the two minima. Data refer to partially dried MbCO sample without sugar.

The solution for glucose sample contained $4 \cdot 10^{-1}$ M glucose, in order to have all saccharide samples containing an equal concentration in monomers. Solutions were centrifuged, equilibrated with CO and reduced by anaerobic addition of sodium ditionite $(10^{-1}$ M); 0.5 ml of the above solutions were layered on CaF_2 windows of ~4.5 cm^2 surface and initially dried in a CO atmosphere in a silica gel desiccator. When samples were hard enough for staying in a vertical position, the drying process was allowed to continue inside the FTIR spectrometer, where a constant flux of dry nitrogen was maintained, until the desired water content was reached (see below). The sample without sugar was prepared by the same procedure and dried under CO flux until it was hard enough for staying in a vertical position. Fig. 1a shows the A substates of MbCO in water solution pH 7. Moreover, as shown in Fig. 1b, adjacent to the CO stretching bands, the water association band is evident (9). This band has been assigned to a combination of the bending mode of the H_2O molecule with intermolecular vibrational modes. We obtained qualitative information on samples' water content by measuring (in OD×cm^{-1}) the area delimited by the tangent to the curve at the two minima, as shown in Fig. 1b.

RESULTS AND DISCUSSION

Measurements were performed on samples wherein the water content is almost identical, thus enabling to get insights on the saccharide effects at analogous water content, in the glassy matrix. Fig. 2 shows the IR spectra of MbCO of partially dehydrated samples; as previously shown (10), the spectral properties of A substates in MbCO samples dried in the presence of trehalose (sucrose), or in the absence of sugar, strongly depend on the residual water in the sample. A comparison of data in Fig. 2 with the spectrum in Fig. 1a shows *i*) that, as already reported (10), water withdrawal affects the distribution of A substates, and that in particular it makes to increase the population of substate A_0, wherein the distal histidine is out of the heme pocket (11); *ii*) that the largest spectral distorsion with respect to the native distribution is observed in the sample dried in the absence of sugar. The increase of A_0 population, during the dehydration process, can be rationalized by considering that the free energy needed for exposing the hydrophobic residue histidine to the external medium decreases when water is progressively removed. Accordingly, the fact that the largest A_0 increase is found in the absence of sugar can be ascribed to the presence of direct protein-protein contacts in these samples. The lower A_0 increase in the presence of sugar suggests that saccharides screen proteins, thus preventing the formation of direct contacts.

Further interesting information can be obtained by comparing the data relative to the three saccharides. In the presence of glucose the perturbation on the distribution of A substates results to be the smallest and almost not dependent on the residual water; this suggests that the only relevant effect brought about by this sugar is the decrease of hydrophobic interactions between the "*exposed*" histidine and the external medium. Moreover, the sizeable difference among mono and disaccharides suggests that a relevant role is also played by propagation of saccharide structures that can take place in disaccharides and not in glucose. As already suggested (10), we ascribe the differences between sucrose and trehalose effects, evident in Fig. 2, to the symmetry differences between these two sugar molecules. Details will be reported in a forthcoming paper.

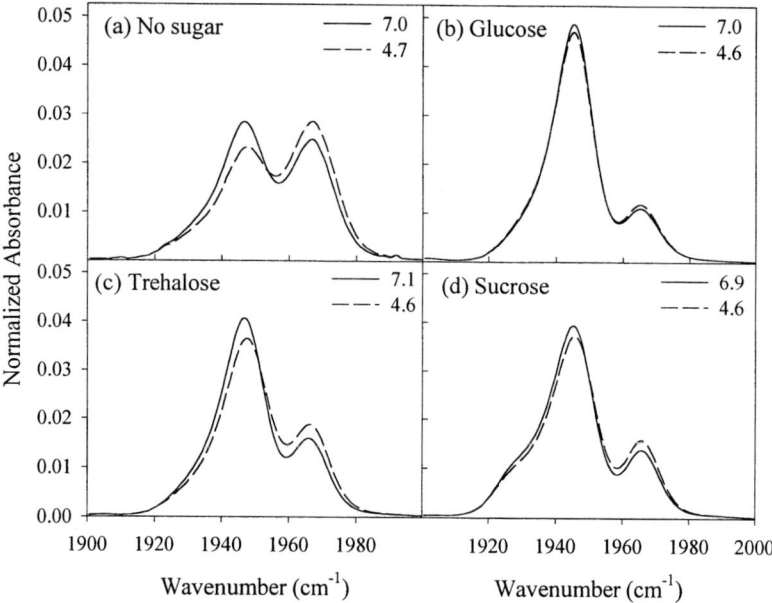

FIGURE 2. A substates of MbCO in our samples. The water content of the samples, measured in $OD \times cm^{-1}$, as described in the text, is reported within each panel, in the right upper corner. For the sake of comparison, spectra were normalized to unit area.

ACKNOWLEDGMENTS

This work is part of a project cofinancied by the European Community (European Funds for Regional Development). The technical assistance of Mr. G. Tricomi and S. Lo Cascio is gratefully acknowledged.

REFERENCES

1. Crowe, L.M., Reid, D.S., and Crowe, J.H., *Biophys. J.*, **71**, 2087-2093 (1996).

2. Panek, A.D., *Brazilian Journal of Medical and Biological Research*, **28**, 169-181 (1995).

3. Leslie, S.B., Israeli, E., Lighthart, B., Crowe, J.H., and Crowe, L.M., *Appl. Environ. Microbiol.*, **91**, 3592-3597 (1995).

4. Uritani, M., Takai, M., and Yoshinaga, K., *J. Biochem.*, **117**, 774-779 (1995).

5. Prestrelski, S.J., Tedeschi, N., Arakawa, T., and Carpenter, J.F., *Biophys. J.*, **65**, 661-671 (1993).

6. Ansari, A., Berendzen, J., Braunstein, D., Cowen, B.R., Frauenfelder, H., Hong, M.K., Iben, I.E.T., Johnson, J.B., Ormos, P., Sauke, T.B., Scholl, R., Schulte, A., Steinbach, P.J., Vittitow, J., and Young, R.D., *Biophys. Chem.*, **26** 337-355 (1987).

7. Frauenfelder, H., Sligar, S.G., and Wolynes, P.G., *Science*, **254**, 1598-1603 (1991).

8. Frauenfelder, H., Alberding, N.A., Ansari, A., Braunstein, D., Cowen, B.R., Hong, M.K., Iben, I.E.T., Johnson, J.B., Luck, S., Marden, M.C., Mourant, J.R., Ormos, P., Reinish, L., Scholl, R., Schulte, A., Shyamsunder, E., Sorensen, L.B., Steinbach, P.J., Xie, A.H., Young, R.D., and Yue, K.T., *J. Phys. Chem.*, **94**, 1024-1037 (1990).

9. Eisenberg, D., and Kauzmann, W., The structure and properties of water, London: Oxford University Press, 1969, ch. 4, pp. 228-231.

10. Librizzi, F., Vitrano, E., and Cordone, L., *Biophys. J.*, **76**, 2727-2734 (1999).

11. Morikis, D., Champion, P.M., Springer, B.A., and Sligar, S.G., *Biochemistry*, **28**, 4791-4800 (1989).

A Study of a Class of Power-Law Tail Quantum Wave Packets

Fabrizio Lillo and Rosario N. Mantegna

Istituto Nazionale per la Fisica della Materia, Unità di Palermo
and
Dipartimento di Energetica ed Applicazioni di Fisica, Università di Palermo, Viale delle
Scienze, I-90128, Palermo, Italia

Abstract. We study some properties of a class of quantum wave packets with power-law spatial tails analitically. Almost all packets of this class have the property that all the moments of the position operator are infinite. We prove that the free evolution of these packets presents an asymptotic decay of the wave packet maximum which is anomalous with respect to the behavior observed in the free evolution of customary quantum wave packets. We also study the temporal evolution of the tails and we show that the dominant term of the asymptotic expansion is conserved.

Power-law probability density functions [1] are receiving a lot of attention in the investigation of self organized [2] and complex systems [3]. One of the key aspects of the power-law probability distributions is that only a finite number of moments of the random variable are finite. In some cases the second moment is infinite and a typical scale for the variable is lacking. Recently we introduced the Power-Law Tail Wave Packets (PLTWPs) [4]. A PLTWP is a wave function $\psi(x)$ describing a non-relativistic spinless particle in one dimension, which decreases with x as

$$\mid \psi(x) \mid \sim \mid x \mid^{-\alpha} . \tag{1}$$

This class of wave packets is square-integrable only if $\alpha > 1/2$. In Ref. [4] we demonstrate some general physical properties of these wave packets and we study the behavior of these wave functions in the simplest dynamical evolution, namely the free evolution.

In this paper we study these properties for a specific class of PLTWPs defined as

$$\psi(x) = N \left[\frac{1}{(\gamma + ix)^\alpha} + \frac{1}{(\gamma - ix)^\alpha} \right], \tag{2}$$

where N is a suitable normalization constant and γ is a scale parameter. We restrict our attention to the values of $1/2 < \alpha \leq 1$. For these values of α the

CP513, *Nuclear and Condensed Matter Physics,* edited by A. Messina
© 2000 American Institute of Physics 1-56396-929-7/00/$17.00

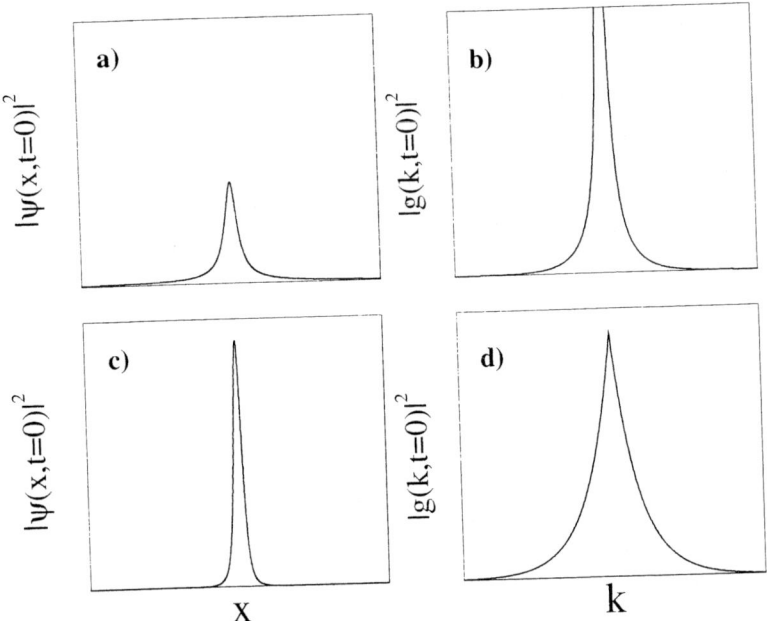

FIGURE 1. Probability density $|\psi(x,t=0)|^2$ for the wave packets of the class of Eq. (2) with $\alpha = 0.6$ (a) and $\alpha = 1$ (c). In (b) and (d) we show the corresponding momentum distribution $|g(k,t=0)|^2$ calculated from Eq. (3). In (b) the $|g(k,t=0)|^2$ diverges in $k=0$. Scales of (a) and (c) and of (b) and (d) are the same.

wave function of Eq. (2) is even, real and positive. The tails of $\psi(x)$ decrease asymptotically as $|\psi(x)| \sim |x|^{-\alpha}$ for $1/2 < \alpha < 1$ and as $|\psi(x)| \sim x^{-2}$ for $\alpha = 1$. In the general case of Eq. (1) the moments of position operator $\langle \hat{x}^m \rangle$ with $m \geq 2\alpha - 1$ are infinite. In Eq. (2) with $\alpha = 1$ only the first two moments are finite. On the other hand all the moments, including the mean, are infinite when $1/2 < \alpha < 1$. As a consequence the uncertainty in position Δx is infinite in these cases. This property reflects the fact that these packets are scale free in space. These packets can be considered as an intermediate case between usual wave packets (for example, the Gaussian wave packet), which have a typical scale and are square-integrable, and plane waves, which are scale free in space but are not square-integrable. The lack of finite moments of x has a counterpart in the properties of Fourier transform $g(k) = FT[\psi(x)]$ in $k=0$. We calculate the Fourier transform of Eq. (2), which gives the amplitude probability distribution of momentum, and we find

$$g(k) = \frac{N\sqrt{2\pi}}{\Gamma(\alpha)}|k|^{\alpha-1}e^{-\gamma|k|}. \tag{3}$$

When $\alpha < 1$ the function $g(k)$ is infinite in $k = 0$, whereas when $\alpha = 1$ $g(k)$ presents a discontinuity in the first derivative in $k = 0$. As an illustrative example

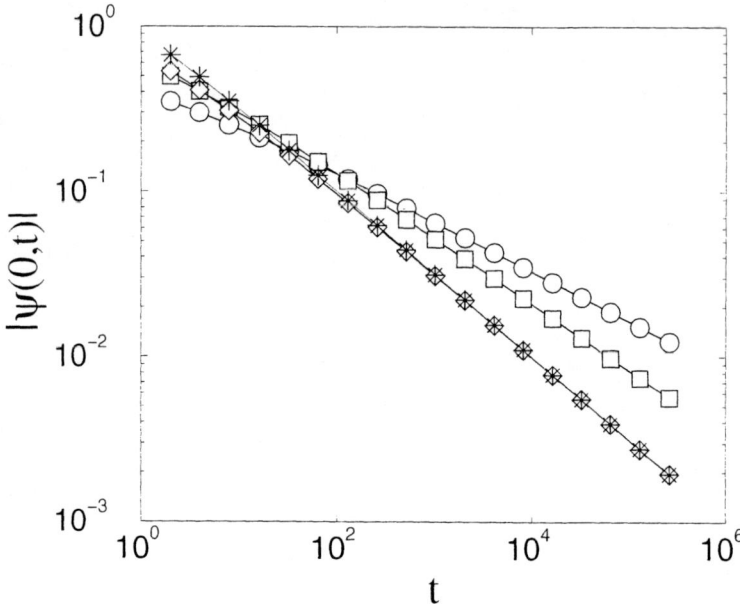

FIGURE 2. Log-log plot of the time evolution of the wave packet amplitude at $x = 0$ for four different wave packets. The usual asymptotic behavior $t^{-1/2}$ is observed for the Gaussian (star) and for the PLTWP of Eq. (2) with $\alpha = 1$ (diamond). The $t^{-\alpha/2}$ anomalous decay is observed for the PLTWPs of Eq. (2) with $\alpha = 0.6$ (circle) and $\alpha = 0.8$ (square). In our calculation we set $\gamma = 1$ and $\hbar/2M = 1$.

in Fig. 1 we show the probability distribution of position $|\psi(x)|^2$ and of momentum $|g(k)|^2$ for the wave function of Eq. (2) with $\alpha = 0.6$ and $\alpha = 1$.

Even if $g(k)$ diverges in $k = 0$, the probability of finding the particle with momentum between $\hbar k$ and $\hbar(k + \Delta k)$ is always finite. This is due to the fact that $g(k)$ is square-integrable. The exponential asymptotic form of $g(k)$ guarantees that all the moments of the momentum operator of the particle $\langle \hat{p}^m \rangle$ $(m \geq 1)$ are finite. Therefore these packets have finite mean kinetic energy and finite non zero uncertainty in momentum Δp. Since Δx is infinite, these packets are chararacterized by an infinite uncertainty product $\Delta x \Delta p$.

Hereafter we consider the properties of the PLTWPs of Eq. (2) in the simplest case of dynamical evolution, namely the free wave evolution. We assume that at $t = 0$ the wave packet is described by Eq. (2) and we focus our attention on the spreading of the wave packet. During the dynamical evolution in free space the wave function at time t is given by

$$\psi(x,t) = \frac{1}{\sqrt{2\pi}} \int_{-\infty}^{\infty} g(k)e^{i(kx - \hbar t k^2/2M)}dk =$$

136

$$= \frac{N}{(-2i\beta)^{\alpha/2}} \left(U(\frac{\alpha}{2}, \frac{1}{2}, \frac{(\gamma + ix)^2}{-4i\beta}) + U(\frac{\alpha}{2}, \frac{1}{2}, \frac{(\gamma - ix)^2}{-4i\beta}) \right), \qquad (4)$$

where $U(a, b, z)$ is the second solution of the confluent hypergeometric differential equation [5] and $\beta = \hbar t/2M$ is a scaled time.

The variance of the packets is infinite and it is not possible to characterize the free spreading of the wave packets by studying the temporal evolution of their variance. We characterize the spreading by investigating the evolution of the probability amplitude at the maximum of the packet $\psi(0, t)$. We briefly recall that in the case of a Gaussian wave packet the amount of spreading of the wave packet can be quantified either by considering the time dependence of the position variance or by determining the time dependence of the maximum of the wave function. In the Gaussian case, the variance of $|\psi(x, t)|^2$ is asymptotically proportional to t^2 and the maximum of the wave function $|\psi(0, t)|^2$ decreases as t^{-1} asymptotically. In Fig. 2 we show the dynamics of $|\psi(0, t)|$ for PLTWPs of Eq. (2) with $\alpha < 1$, with $\alpha = 1$ and for a Gaussian wave packet. In the figure is clear that the Gaussian and the PLTWP with $\alpha = 1$ soon converge to the usual $1/\sqrt{t}$ asymptotic behavior whereas the PLTWPs with $\alpha = 0.6$ and $\alpha = 0.8$ slowly converges to the anomalous asymptotic behavior of $1/t^{\alpha/2}$. This result is in agreement with a more general study [4], in which we show that for a wave packet with $|\psi| \sim |x|^{-\alpha}$ with $1/2 < \alpha < 1$ the maximum decreases in time in an *anomalous way* as $|\psi(0, t)| \sim t^{-\alpha/2}$.

In order to obtain a more complete description of the free wave spreading, we consider the time evolution of the tails of the packet. By using the asymptotic expansion of the confluent hypergeometric function we prove that at any time t the dominant term of the spatial asymptotic expansion of $\psi(x, t)$ is the term $|x|^{-\alpha}$. Also this result is valid for all PLTWPs [4]. From a physical point of view this result implies that during the free evolution the number of finite moments of position operator is a conserved quantity.

We thank INFM and MURST for financial support. We wish to thank Giovanni Bonanno for help in numerical calculations.

REFERENCES

1. P. Lévy, *Calcul des Probabilités* (Gauthier-Villars, Paris, 1925).
2. P. Bak, C. Tang and K. Wiesenfeld, Phys. Rev. Lett. **57**, 381 (1987).
3. F. Mallamace and H. E. Stanley (Eds.), *The Physics of Complex Systems* (IOS Press, Amsterdam, 1997).
4. F. Lillo and R. N. Mantegna, *Anomalous Spreading of Power-Law Quantum Wave Packets*, (submitted).
5. M. Abramowitz and I. A. Stegun, *Handbook of Mathematical Functons*, Dover, New York (1972).

Fractal Approach in Petrology: Combining Ultra Small Angle (USANS), and Small Angle Neutron Scattering (SANS)

F.Lo Celso[a], F. Triolo[a,b], A. Triolo[c], J.S. Lin[d], G. Lucido[e] and R. Triolo[a]*

[a] Department of Physical Chemistry, University of Palermo, Palermo, Italy, [b] Mount Sinai School of Medicine, New York, NY, USA, [c] Heriot-Watt University, Riccarton, Edinburgh, UK, [d] Oak Ridge National Laboratory, Oak Ridge, TN, USA, [e] Department of Chemistry and Physics of the Earth, University of Palermo, Palermo (Italy).

Abstract. Ultra small angle neutron scattering instruments have recently covered the gap between the size resolution available with conventional intermediate angle neutron scattering and small angle neutron scattering instruments on one side and optical microscopy on the other side. Rocks showing fractal behavior in over two decades of momentum transfer and seven orders of magnitude of intensity are examined and fractal parameters are extracted from the combined USANS and SANS curves.

1. INTRODUCTION

Many natural processes and objects have been shown to be fractals. Examples include Brownian motion, thermal convection, earthquakes, snowflakes, coastlines, rivers, faulting, folding and volcanic eruptions. Petrology is a branch of science that deals with the origin and evolution of rocks. So, its purpose is to study changes that occur naturally in rocky complexes, namely, sediments which undergo physical and chemical alterations, magmatic fluids which solidify, rocks which undergo partial or total melting. Fractal geometry is the natural mathematical language to describe much of what petrologists observe. Fractal geometry does not describe the mechanism that produces the fractal scaling, but it nonetheless helps to sort out possible mechanisms or explanations. Although much has been accomplished in quantifying rock and mineral chemistry, comparatively little effort has been made to quantify texture. Recent progress in ultra-small angle neutron (USANS) and X-ray scattering (USAXS) instrumentations[1] enables one to access the microstructure of rocks well beyond the limit of SANS intruments. In this paper we present new USANS, and SANS experiments performed on various samples.

2. THEORETICAL BACKGROUND

A particularly fortunate situation arises when rocks are investigated by means of neutron scattering. In this case the system, even if strictly multi-phase, can be treated as a two-phase system as most of the scattering originates from the contrast between the inorganic components and the voids. When a beam of neutrons illuminates a volume V in which scattering centers are distributed according to a distribution law $\rho(\mathbf{r})$, scattering events take place and a diffused beam will be generated, whose

CP513, *Nuclear and Condensed Matter Physics*, edited by A. Messina
© 2000 American Institute of Physics 1-56396-929-7/00/$17.00

differential scattering amplitude dA(**Q**)/dV as a function of the scattering variable **Q** will be given by

$$dA(Q)/dV = \rho(r) \exp(-iQr) \qquad (1)$$

In integrated form, equation (1) relates the scattering amplitude A(Q) to the Fourier transform of the scattering length (Q=$4\pi\lambda^{-1}\sin\theta$, 2θ being the scattering angle and λ the wavelength of neutrons).

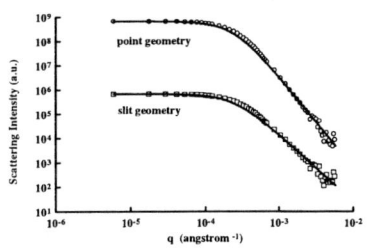

Figure 1. Usans data for a sample of rock. Lines are fits to the equation for surface fractals

In the late eighties there has been a strong interest to describe the scattering from natural systems (in primis rocks) in terms of fractal structures (mass and/or surface fractals). However, these attempts led to limits in the interpretation of data because most of the power law exponents found were in a range of uncertain attribution, and also because the Q range in which data could be obtained was too limited (often less than 1.5 decades). To extend the Q range we have modified the Oak Ridge National Laboratory USANS camera, which now incorporates the suggestions to suppress the wing effect by using two triple bounce channel cut ideal crystals in the Bonse-Hart set-up[1], and we have measured again some of the samples run earlier. In a few cases we have also extended the high Q portion of the scattering by performing our experiments at the TOF SANS instrument at the Rutherford Appleton Laboratory (UK), and by making use also of the high Q detector bank. By doing so we have been able to span almost seven decades of momentum transfer.

3. EXPERIMENTAL

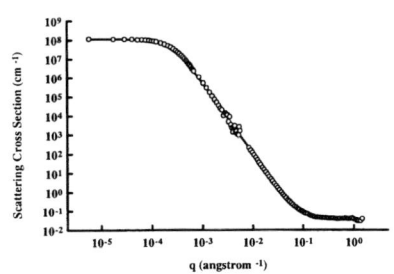

Figure 2. Combined USANS, SANS and IANS for the rock of figure 1.

The SANS and IANS results have been obtained at the ISIS pulsed neutron source of the EPSRC Rutherford Appleton Laboratory (UK) using the "LOQ" spectrometer[2] and at the Oak Ridge National Laboratory using the 30m SANS camera. Details of technical and experimental aspects together with data reduction procedures are given elsewhere[1,3].

The USANS results have been obtained at the HFIR USANS facility of the Oak Ridge National Laboratory USA). Details of technical and experimental aspects together with data reduction procedures are given elsewhere[1]. As the USANS camera becomes unreliable for $Q > 0.003$, there is a small region of Q not covered by both USANS and LOQ instruments.

4. DATA ANALYSIS

We shall now present the scattering equation for mass fractals and for surface

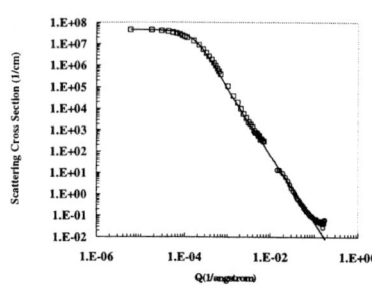

fractals. Bale and Schmidt[4] derived an expression for the correlation function of perfect surface fractals, i.e. fractals whose surface is self-affine and whose surface roughness scales according to a power law of the kind $S(r) = S_o (r/l_{max})^{2-D_s}$, where l_{max} is the upper limit for the scale invariance (i.e. the largest size for which the surface is still a fractal). In terms of the surface fractal dimension D_s the scattering function $I(Q)$ has been derived by Mildner and Hall [5]:

Figure 3. USANS and SANS data for a sample of clay mineral. See text for details.

$$I(Q)= Q^{-1} \Gamma(\beta)l_{max}^{\beta} [1+(Ql_{max})^2]^{-\beta/2}\sin[(D_s-1) \arctan(Ql_{max})], \text{ where } \beta =5-D_s. \quad (2)$$

When $l_{max}>>1/Q$ (i.e. in the range of linearity in a log $[I(Q)]$ vs. log Q plot) we have a power law dependence of $I(Q)$ from Q with exponent $\beta+1$. The existence of a lower limit for the scale invariance, l_{min}, is indicated in a double log plot by a flattening of the scattering curve at high Q values, following the linear portion. Fig. 1 shows the USANS result obtained for a sample of rock (benmoreitic inclusion in hyaloclastite, Linosa Island, Italy). Symbols indicate slit smeared and point desmeared experimental intensities, while the line indicates fit to the above scattering equation for surface fractals with a fractal dimension $D_s =2.3$ and a cut-off dimension l_{max} of $3.9x10^3$ Å. An entirely analogous equation can be derived[6] for mass fractals:

$$I(Q)= Q^{-1} \Gamma(\beta)l_{max}^{\beta} [1+(Ql_{max})^2]^{-\beta/2}\sin[(\beta) \arctan(Ql_{max})], \quad (4)$$

where D_m is the mass fractal dimension and $\beta =D_m-1$. Also in this case the scattering equation in the range of linearity in a log $[I(Q)]$ vs. log Q plot gives a power law with exponent $\beta+1$.

5. RESULTS AND CONCLUSION

Figure 2 shows USANS, SANS+IANS data for the same rock shown in figure 1. The rock shows fractal behavior in slight more than two order of magnitude in Q (and therefore in length scale) and seven orders of magnitude in intensity. Diffraction peaks are barely shown in the highest region of scattering variable Q, where the flattening of the scattering curve is due to l_{min}, while the flattening in the small Q region is due to the cut-off distance l_{max}.

Figure 3 shows combined USANS, SANS and IANS data for a sedimentary rock for which the best fits were obtained with the mass fractal equation. Fits yield a fractal dimension D_m =2.85 and a cut-off distance l_{max} of 6.0 x 10^3 Å. The rock shows fractal structure in slight less than three orders of magnitude in Q (and therefore in length scale), while intensity varies about eight orders of magnitude. Again, the existence of a lower limit l_{min} for the scale invariance can be seen in the high Q portion of the scattering pattern.

Figure 4 shows the transition from a surface fractal structure to a mass fractal structure. The sample is a dinosaur bone. Symbols are point geometry USANS and SANS data. Lines are fits to equation 2 and equation 3 of the text. Data are well reproduced except in the transition region from surface to mass fractals.

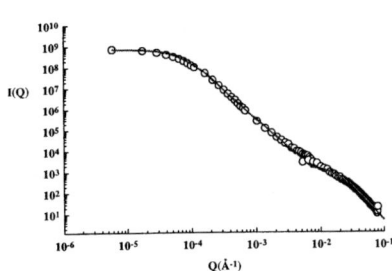

Figure 4. USANS and SANS data for a sample of dinosaur bone. See text for details.

In conclusion, examples of rocks showing mass and surface fractal geometry have been shown. These indicate the importance of combining USANS and SANS data to extend the range of momentum transfer explored.

ACKNOWLEDGMENTS

The authors thank ISIS and ORNL for provision of beamtime. MMA and JSL acknowledge support from the Divisions of Advanced Energy Projects and Materials Sciences, U. S. Department of Energy under contract No. DE-AC05-96OR22464 with Lockheed Martin Energy Research Corporation. AT, FT, GL and RT acknowledge financial support from the Italian National Research Council and The Italian Department of Education and Research (MURST).

REFERENCES

1) Agamalian, M.M., Wignall, G.D. & Triolo, R. (1997). J. Appl. Cryst., **30**, 345-349.
2) Heenan, R. K., Penfold, J. and King S. M. (1997). J. Appl. Cryst., **30**, 1140-1147.
3) Heenan, R. K., King S. M., Osborn, R. and Stanley,H. B. (1989) Rutherford Appleton Laboratory Report RAL-89-128.
4) Bale, H. & Schmidt, P.W. (1984). Phys. Rev. Lett., **53**, 596-599.
5) Mildner D.F.R. & Hall, P.L. (1986). J. Phys. D, **19**, 1535-1545.
6) Schaefer, D.W. & Keefer, K.D. (1984). Phys. Rev. Lett., **53**, 1383-1386.

Integrated Charge Preamplifier For Silicon Detectors With Very Low Capacitance

Ludovico Lo Nigro on behalf of the Microelectronic Group

Physics Department, University of Catania and INFN of Catania, Catania, Italy

Abstract. Low capacitance detectors, as microstrip, pixel e silicon drift (SDDs), require strongly front-end realised as integrated circuits. The main features of these preamplifier are: low noise, low power, high gain, good stability and low cost. In this paper we describe the design of charge preamplifiers in CMOS and BiCMOS technology both for GaAs pixel detector, used as a high-efficiency X-ray imaging detector, and for SDDs, used in ALICE experiment.

Charge preamplifier for GaAs pixel detector

A novel digital radiology device with inherent high spatial and energy resolution is based on a GaAs pixel detector and two custom-made VLSI electronic circuits which process each photon pulse and classify it according to its energy.

For low noise analog chip we have adopted a 0.8 μm BiCMOS design of a charge preamplifier and a CR-CR^4 shaper. The performance, with a detector capacitance of the order of 1 pF and a peaking time of the shaper of 1 μs, should guarantee a sufficient figure of ENC = 200 e⁻ in 10 μs readout time, with a power consumption less than 7 mW/ch.

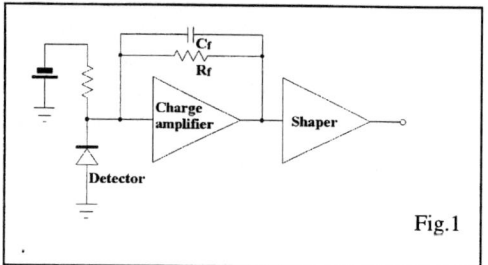

Fig.1

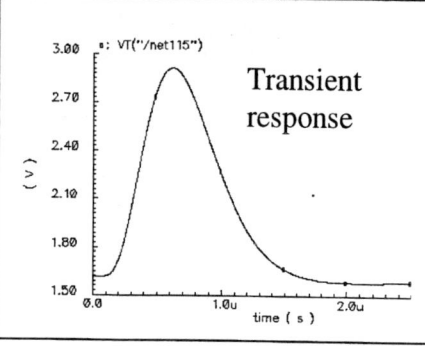

Transient response

FIGURE 1. and **2.** show the block diagram of circuit and the simulated output response vs time

CP513, *Nuclear and Condensed Matter Physics,* edited by A. Messina
© 2000 American Institute of Physics 1-56396-929-7/00/$17.00

FIGURE 3. The photo of the chip with 16 channels on board is shown.

Charge preamplifier for SDDs

The ALICE's Inner Tracking System consists of five planes of position sensitive detector. The 3rd and 4th planes are of Silicon Drift Detectors. The charge collected by the anodes implanted at the edge of the detector has a gaussian shape due to the diffusion: a rise time up to 70 ns is expected.

The first prototype was designed in 0.8 μm CMOS technology. We have adopted active feedback made with a Mos transistor instead of a resistor. This solution has led to a non linearity for high values of input charge.

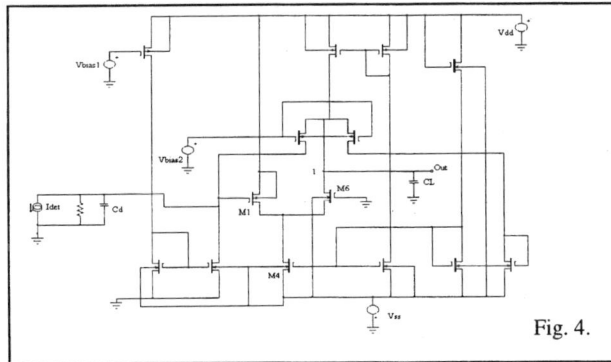

Fig. 4.

FIGURE 4.
Schematic diagram of the amplifier we have realised.

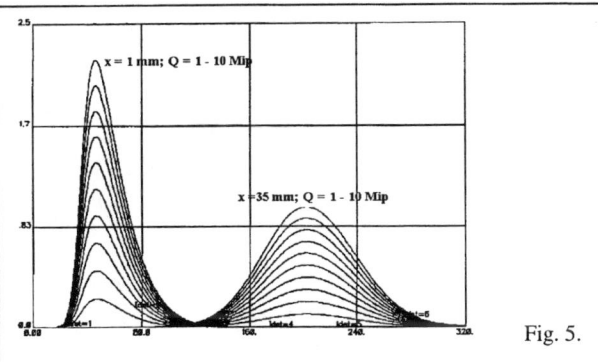

Fig. 5.

FIGURE 5. shows the simulated output response vs. time of the amplifier for several input charge with different shape due to the distance of the impact point of the particle in the detector.

143

With a current of 1.8 μA we have 150 kΩ transresistance gain in a wide input range over 25 MIP.

The precision to reconstruct the impact point depend on noise. Requirements for ALICE detector are ENC < 500 e⁻ rms to obtain spatial resolution less than 20 μm. Measured noise is 340 e⁻ rms.

An other key point is power dissipation, since the drift time and the resolution of the SDDs strongly depends on the temperature. The power consumption is less then 1 mW/ch.

A second prototype for SDDs was designed in according of new requirements of the read-out, with the following characteristics: gain of 900 kΩ, bandwidth > 25 MHz, noise < 2mV rms, driving capability of 7 pF, power dissipation < 2 mW/ch.

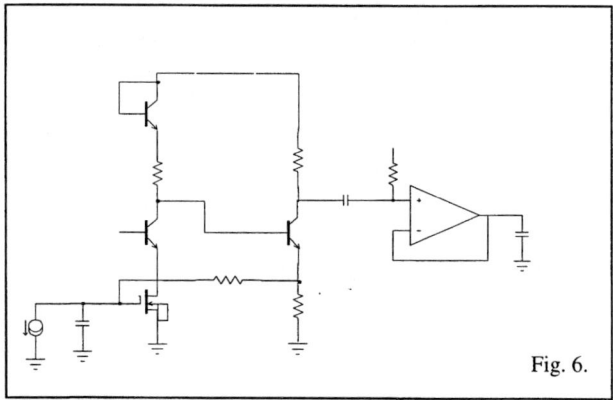

Fig. 6.

FIGURE 6. The schematic circuit of the amplifier.

Both high gain and good stability were achieved using a feedback resistance R_L = 300 kΩ. The technology used is BiCMOS 0.8 μm.

We have chosen a NMOS transistor as the input device, because it has input zero current. To eliminate the offset of the amplifier, the blocking capacitance C_B was set to 10 pF.

The third prototype of preamplifier-shaper for SDDs has been designed. It has 8 channels. Each channel consists of a charge preamplifier and CR-RC shaper to limit noise. Thermal and shot noise has been simulated as 300 e⁻ rms. Since the foreseen leakage current is not more than 10 nA, the preamplifier has a sensitivity of 155 mV/MIP, in order to achieve the assigned 7 MIP range. The shaper has a gain 1 V/V and a driving capability of 7 pF as required by its load, that is the 256-cell Analog Memory.

Analog front-end has been designed with a traditional scheme, as shown fig. 1. It consists of an integrator and CR-RC shaper plus an output buffer to drive the load.

144

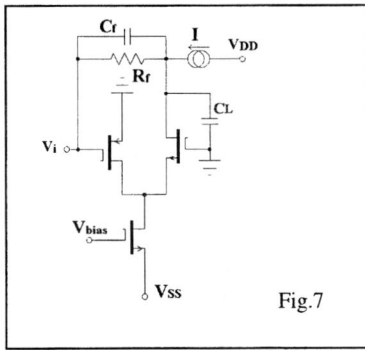

FIGURE 7. Shaper stage.

Both preamplifier and shaper stages are based on a single-ended folded cascode as shown fig. 7. This solution offers better performances in term of stability and gain compared with standard cascode. For preamplifier stage a choice of PMOS input transistor was addressed for better result in term of flicker noise.

The main characteristics of the amplifier are: noise < 500 e⁻ rms; sensitivity of 155mV/MIP; shaping time of 55ns; power dissipation of 2.2 mW; input range of 7MIP; output range of 1 V.

Accurate noise analysis was performed to sizing input transistor taking into account the detector capacitance and power dissipation. Coupling capacitor of 20 pf was implemented inside the chip. A 8-channels amplifier has been realised in 0.8 μm AMS CMOS technology.

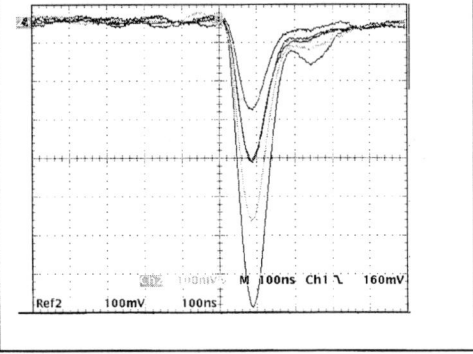

FIG. 8 and **9** . Microphotograph of the chip and measured output transient response of the amplifiers up to 7 MIPs.

We have some problem with noise performance that is worse compared with simulations, probably due to the AC coupling between the detector and the input of the preamplifier. This fact suggests a DC coupled amplifier with the compensation of the leakage current.

REFERENCES

1. L. Lo Nigro, Graduate Thesis, Catania University.

2. N. Randazzo, Gratuate Thesis, Catania University.

3. 4th Workshop on Elect. For LHC Experiments, N. Randazzo, L. Lo Nigro, G.V. Russo, Nucl. Instr. And Meth., vol 420, Roma, 21-27 Sept, 1998, pp 606-609.

On The Aggregation of Poly(Ethylene Oxide) in Water

S.Magazù, C.Branca, A.Faraone, G.Maisano, P.Migliardo, V.Villari

Dipartimento di Fisica and INFM, Università di Messina, C.da Papardo S.ta Sperone 31, 98166 Messina Italy

Abstract. The topic dealt concerns with a study of the structural and dynamical properties of aqueous solutions of Poly(Ethylene Oxide) (PEO) by means different spectroscopic techniques. All the findings show that aggregation processes do not constitute an inherent property of the PEO-water system. In particular, the polymer-water interaction strength increases with the polymerization degree, indicating that the hydrophylic character of PEO weakens at low molecular weight and that the aggregative processes could be, eventually, activated by inhomogeneities in the system.

INTRODUCTION

On recent years there has been a notable increase in interest, not only academic but applicative too, on Poly(Ethylene Oxide), [PEO: $OH-(CH_2CH_2O)_m-H$], aqueous solutions [1]. There are a number of factors behind this. One is undoubtedly due to the simpleness of its structure. That is the reason why these solutions provide an interesting model for studying the interaction mechanisms of water with hydrophilic surfaces, macromolecules and biological structures, so having implications on the field of Biophysics, Biochemistry and Biotechnology. From the applicative point of view, they finds an employment as lubricants, dispersants, plasticizers, recovery agents of tertiary oils, in the delicate area of restoration and conservation of structures, and as friction reducers in the flow of water through ducts and channels, when high Molecular Weight (M_w) polymer is present at concentrations even less than 10ppm [1]. This effect could be due to an interfacial layer of PEO between the water flow and the tube, or to the capacity of PEO to form long filaments in water flow, which reduce local turbulences. The first hypothesis would be supported by evidence of aggregation whereas a good solvent behaviour of water would support the second one. For this, also the aggregation behaviour of PEO in water, starting from very dilute concentrations and from oligomers up to high M_ws, is a controversial topic [1]. To address the difficult questions concerning whether aggregation is an inherent property of the PEO/water system, in the present work we will consider the findings obtained by the integrated employment of NMR, light and neutron scattering. All the presented experimental evidences show that at low concentration water is a good solvent for

CP513, *Nuclear and Condensed Matter Physics,* edited by A. Messina
© 2000 American Institute of Physics 1-56396-929-7/00/$17.00

PEO which behaves like a typical *swollen* random coil, with no evidence of aggregation.

EXPERIMENTAL SECTION

We examined aqueous (H_2O and D_2O) solutions of high purity PEO samples from Ethylene Glycol (EG) to PEO with average molecular weight, M_w, of 4000000 Da, purchased from Aldrich-Chemie. Raman Polarized (I^{VV}) and depolarised (I^{VH}) spectra were obtained by a SPEX Ramalog 5 triple monochromator in a 90° scattering geometry in the (293 K÷333 K) temperature range. Quasi Elastic Light Scattering (QELS) measurements were performed by means of Photon Correlation Spectroscopy (PCS) technique, using a standard scattering apparatus with a photon counting optical system and a BROOKHAVEN BI-2030 correlator to analyse the scattered light. Self-diffusion coefficient measurements were performed by the ^1H Pulse Gradient Spin-Echo Nuclear Magnetic Resonance (PGSE-NMR) technique, using the low resolution pulsed NMR spectrometer Bruker SXP 4-100 MHz in the temperature range 293÷333 K. QENS experiment was carried out using the IRIS backscattering high-resolution spectrometer at ISIS facility (DRAL, UK). The detectors used give Q-values ranging from 0.3 to 1.9 Å^{-1} with a mean energy resolution $\Gamma=15\mu eV$ of Full Width at Half Maximum (FWHM). The instrumental resolution function was determined by reference to a standard vanadium plate. Details on the experimental set-up are reported in Refs. [2-5]

RESULTS AND DISCUSSION

The most remarkable points which emerge from the employment of different complementary spectroscopic techniques [2-5], can be summarized in the following points:

i)   PEO chain presents in the crystalline state a helical conformation with a conformation assignment to the internal rotations which follows a trans-gauche-trans sequence. In the molten state, Raman and Infrared data indicate the existence of a highly disordered helix which approaches a random coil conformation. In water solution, however, the shifts in Raman frequencies from the crystalline state are not appreciable. To show the ordering effect of water on PEO let us focus the attention on the D-LAM contribution. The insert (a) of fig. 1 shows, for PEG 600, that by increasing the water content, the D-LAM contribution evidences a remarkable frequency increase (over 10 cm^{-1}) towards values corresponding to the crystal ones, and the significative sharpening of the spectral contribution. Such evidences confirm that PEO in water tends to assume, in respect to the melt case, a more ordered conformation, closer to the crystalline. The conformation ordering, which is further

147

promoted by a lowering of temperature, can be associated to the hydration with environmental water molecules.

ii)   PCS and Raman scattering evidence different conformational properties and a different temperature behaviour of the PEO/H$_2$O and PEO/D$_2$O systems. As an example, from a glance to the insert (a) of fig.1, one observes that, notwithstanding the similar behaviour, by adding heavy water molecules, the D-LAM centre frequency tends to an asymptotic value smaller than that relative to the PEO/H$_2$O system corresponding to the crystalline one. Such evidences confirm that PEO in water tends to assume, in respect to the D$_2$O solutions, a more ordered conformation, closer to the crystalline one. This clearly indicates that the PEO chain conformation in the presence of heavy-water is locally less ordered. Therefore the present work shows that the isotopic substitution technique, nowadays widely employed in a plenty of structural and dynamical studies, for PEO gives rise to different conformational arrangements.

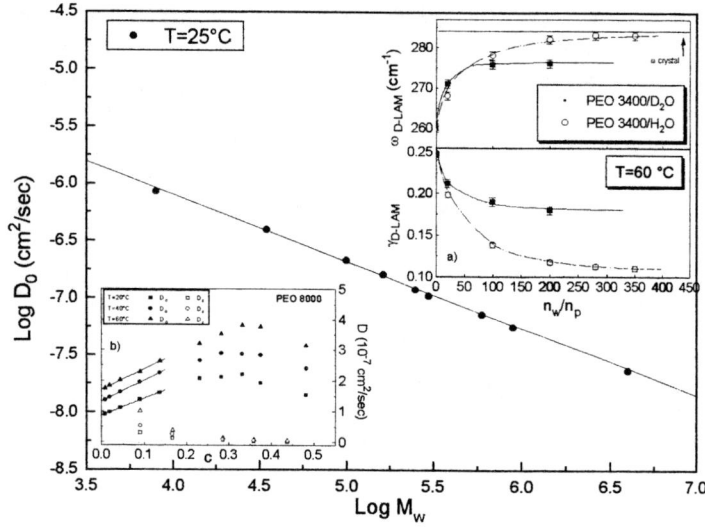

FIGURE 1. Scaling law relating D$_0$ and M$_w$; the scaling exponent is 0.57. In the insert a) the D-LAM center frequency and bandwidth are reported as a function of the ratio n_w/n_p, n_w and n_p being the water and polymer moles respectively. In the insert b) the effective diffusion coefficient (closed marks), as evaluated by PCS, and the self diffusion coefficient (open marks), as evaluated by NMR, are reported as a function of concentration.

iii) PCS, Raman scattering and NMR clearly indicates a swollen state of the polymer molecules and that in the temperature range investigated water behaves as a good solvent for PEO. In particular, see the insert (b) of fig. 1, in the very dilute regime the concentration dependence of the diffusion coefficient is normally expressed by $D=D_0(1+k_D c)$, where K_D, marks the initial slope of the effective diffusion coefficient, D_{eff}, as evaluated by PCS. . The initial increase of D_{eff} at low c, with always $K_D>0$, clearly indicates a swollen state of the polymer coil. The value of the diffusion coefficient, measured by PCS and NMR, at infinite dilution as a function of

M_w range 3400-4000000 Da satisfy, see fig. 1, a scaling law typical of polymer in a good solvent.Finally the increase of K_D in the 20°C-40°C temperature range and the successive saturation at higher temperatures supports the conclusion that the solvent power of water increases up to 45°C and then decreases at higher temperatures.

iv)   QENS data on PEO 80000 aqueous solutions suggests that the addition of water gives rise to the destroying of interactions between polymer chains because of the saturation of the "active" sites, generating as a consequence, less hindered unities. Furthermore PEO affects significantly the dynamics of water molecules in its neighborhood, consistently with the presence of a series of hydration layers or shells each of which is characterized by a relatively slow diffusion.

In conclusion, all the presented experimental evidences show that at low concentration water is a good solvent for PEO which behaves like a typical swollen random coil, with no evidence of aggregation. It clearly emerges that the unaggregate state provides the starting reference point for clarifying many of the friction properties of such systems. On the basis of these findings we can formulate an hydration model, see fig. 2, in which we assume that the water molecules at any instant in time can be partitioned in three different classes: the first one corresponds to the water molecules tightly bound to the backbone of the polymer (i.e. on a time scale longer than the characteristic time scale of the experiment); ii) the second one concerns with water molecules close to the polymer chain (i.e. second, third,..., hydration shells) whose hydrogen bonding network, coordination, energetics, local dynamics, both translation and rotation, are affected by the presence of the polymer; iii) finally the third one corresponds bulk water.

ACKNOWLEDGMENTS

V. Villari acknowledge a grant from Consiglio Nazionale delle Ricerche (CNR).

REFERENCES

1. Devanand, K., and Selser, J. C., *Nature*, **343**, 739 (1990) and references therein; Bieze, T. W. N., Barnes A. C., Huige, C. J. M., Enderby J. E., and Leyte, J. C., *J. Phys. Chem.*, **98**, 6568 (1994)

2. Magazù, S., *Physica B Condensed Matter*, **226**, 92 (1996)

3. Faraone, A., Magazù, S., Maisano, G., Migliardo, P., Tettamanti, E., and Villari, V., *J. Chem. Phys.*, **110**, 1801-1806 (1999).

4. Branca, C., Magazù, S., Maisano, G., Migliardo, P., and Villari; *J. Phys.: Condensed Matter*, **10**, 10141, (1998).

5. Branca, C., Faraone, A., Magazù, S., Maisano, G., Migliardo, P., Triolo, A., Triolo, R., and Villari; V., *J. Phys.: Condensed Matter*, **10**, 10141, (1998).

Microwave Dielectric Spectroscopy and Dynamical Processes in Superionic Glasses

A.Mandanici [α], M.Cutroni [α], C. Cramer [β], K.Funke [β], P.Mustarelli [γ]
and C. Tomasi [γ]

[α]*Dipartimento di Fisica – Università degli Studi di Messina and INFM – Unità di Ricerca di Messina, ctr.Papardo salita Sperone, 31 – 98166 Messina, Italy*
[β]*Institut für Physikalische Chemie der Westfälischen Wilhelms-Universität, Schlossplatz 4/7, D48149 Münster, Germany*
[γ]*Dipartimento di Chimica-Fisica, Università di Pavia, via Taramelli 16, 27100 Pavia, Italy*

Abstract. The dynamical response of ionic glasses can be accurately characterized, in terms of frequency dependent conductivity, by dielectric spectroscopy techniques. Performing measurements on superionic vitreous conductors in the microwave frequency region, it has been discovered that the dispersive behaviour of the conductivity obeys a double power law. Model considerations associate the two power law contributions at low and high frequencies respectively with translational and localized ionic hopping processes, characterized by different activation energies. In the present study on pure and doped silver phosphate glasses, experimental evidence is found that the power law exponents assume constant values, independent of the exact glass composition, and remarkably similar to those obtained for other superionic glasses.

Ionic glasses are widely studied because of their potential applications in new electrochemical devices as solid electrolytes. In fact the dc conductivity of vitreous ionic conductors, essentially due to migration of mobile ions, can be comparable to that of liquid electrolytic solutions. But what happens when an ionic glass is subjected to the action of an external oscillating electric field?

In the frequency domain, the linear response function relating the external disturbance (the electric field $\underline{E}(v)$) to the consequent effect (the current density $\underline{j}(v)$) is the frequency dependent conductivity $\sigma(v)$:

$$\underline{j}(v) = \sigma(v) \cdot \underline{E}(v) \tag{1}$$

As earlier pointed out by Jonscher [1], the dynamical response of ionic glasses, in terms of frequency dependent conductivity, obeys a simple power law,

$$\sigma(v) = \sigma_0 + A \cdot v^p, \qquad \text{with } 0 < p < 1 \tag{2}$$

CP513, *Nuclear and Condensed Matter Physics,* edited by A. Messina
© 2000 American Institute of Physics 1-56396-929-7/00/$17.00

i.e., a constant plateau value at low frequencies is observed, while, at higher frequencies, $\sigma(v)$ continuously increases with frequency. This behaviour, so-called "universal dielectric relaxation" (UDR), is well accounted for within the framework of the Jump Relaxation Model [2]. Considering the structural disorder and the progressive relevance of mutual interactions between mobile ions at increasing frequencies, the model reproduces the smooth transition between the dc plateau and the dispersive region and also provides for the existence of a high frequency plateau, avoiding the unphysical divergence of conductivity which would result in extrapolating the Jonscher power law to high frequencies.

The validity of the UDR behaviour has been largely tested in a frequency range extending from a few Hz to about 100 kHz, but recent experiments in the microwave frequency region [3, 4, 5], have revealed that for a good description of the conductivity behaviour in the whole frequency range investigated a "new" power law contribution with an exponent $q > 1$ is required:

$$\sigma(v) = \sigma_0 + A \cdot v^p + B \cdot v^q \qquad (3)$$

In this work the dielectric response of some pure and doped silver phosphate glasses in the frequency range 8.0-60 GHz has been studied using several waveguide equipments [6]. With this extension of previous measurements [4,7], broadband conductivity spectra spanning over about ten decades in frequency have been obtained.

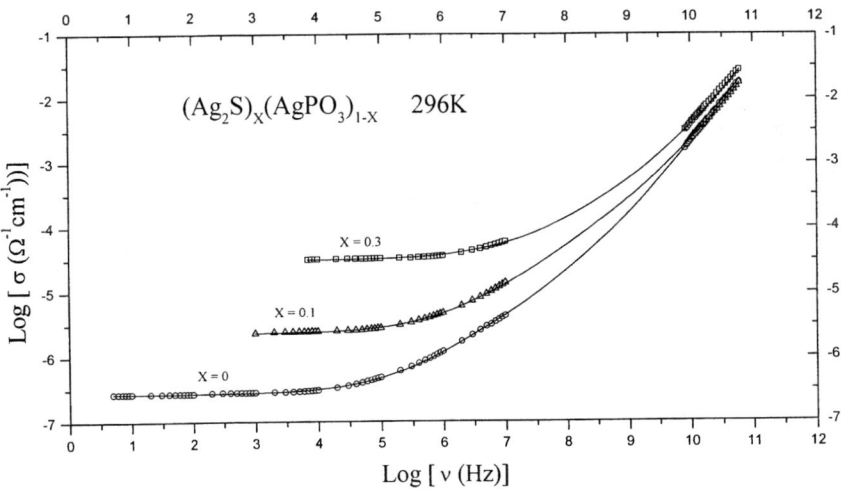

FIGURE 1. Log-log plot of frequency dependent electrical conductivity data, $\sigma(v)$, up to 60 GHz for the $(Ag_2S)_x(AgPO_3)_{1-x}$ glasses with different compositions, x=0, 0.1, 0.3, at room temperature. Conductivity data from 1Hz to 10MHz are taken from refs. [4,7].

These broadband datasets have allowed for a systematic and accurate evaluation of the power law exponents as a function of the glass composition. The power law exponents are found to have constant values $p \cong 2/3$ and $q \cong 4/3$, which are closely similar to those recently reported for other superionic glasses [8]. A question arises: is there a reason to have power law exponents with the same numerical value in glassy ionic conductors of different composition? Further work is in progress to check if the power law exponents are constant also as a function of temperature [6].

To account for the double power law behaviour, a suitable extension of the Jump Relaxation Model has been proposed: the Unified Site Relaxation Model [9]. The basic mechanism responsible for the response experimentally observed is the ion hopping motion. Just after the initial hop that an ion has performed from some site X to some neighbouring site Y, two different relaxation processes come into play: the ion may perform a correlated back hop to the site X, or the neighbourhood may relax with respect the new site Y. The relaxation towards a new equilibrium configuration can be accomplished by the shifting of the Coulomb cage of nearest neighbouring mobile ions or by local adjustments of the glassy network.

The power law exponent of the resulting hopping conductivity will be given by the ratio between the initial back-hop rate and the site relaxation rate. In this scheme the exponent $q > 1$ is due to a backward hopping "faster" than the site relaxation and it corresponds to a mean square displacement which approaches a finite value in the long-time limit. This is the case of a localized dynamical process that will not contribute to the dc conductivity. Instead if the ion, after the initial hop, tends to stay in the new site ("good site"), a successful hop is performed, which contributes to the long range translational transport, i.e. to the σ_0 term and to the Jonscher power term of conductivity.

The different kinds of dynamical processes could correspond to the existence of different local configurations, so one site could be favourable or less favourable for an ion hop depending on the size and charge distribution in its surroundings. On the other hand, the analysis of conductivity data as a function of temperature [7], supported by results of ultrasonic measurements [10], has revealed the existence of several activated processes with typical energies of about 0.5eV, 0.1eV and 0.01eV respectively. While the energy $\Delta E_1 \sim 0.5$eV is strictly related to the dc conduction process, the other values ΔE_2, ΔE_3, are associated to the dispersive regimes characterized by the exponents p and q, then essentially corresponding to translational and localized processes.

From a microscopical point of view it should also be considered that, in a glassy material characterized by topological disorder, several energetically accessible positions can exist for an ion to perform a localized hopping in a given neighbourhood without leaving it. Instead, a hop of greater length and larger activation energy is required to leave the neighbourhood, contributing to the translational component of the frequency dependent conductivity. Anyway, it has to be noted that the occurrence of the double power law, here discussed in terms of different dynamical processes, has

152

been observed only in ion-conducting glasses but not in crystalline ionic conductors ($RbAg_4I_5$ [11]) nor in polaron conducting glasses [12]. On this basis, the new power law contribution found at microwave frequencies seems to be directly related to the coexistence of two factors: the disorder of the glassy state and the ionic mobility. A localized hop in a neighbouring site induces a slight rearrangement of chemical bonding and local structure of the glassy network, making the potential experienced from the ion time dependent. Otherwise a localized hopping in a rigid potential would lead to a power law with an exponent exactly equal to 2, i.e. a Debye response [11].

Further experiments are in progress to achieve a larger and more detailed knowledge of the high frequency dynamical process in a variety of vitreous ionic conductors. At the same time new theoretical approaches are attempted to understand the strong similarities between the numerical values of the power law exponents in different superionic glasses.

REFERENCES

1. Jonscher, A. K., *Nature* (London) **267**, 673 (1977).

2. Funke, K., *Prog. Solid St. Chem.* **22**, 111-195 (1993).

3. Cramer, C., Funke, K., Saatkamp, T., *Phil. Mag. B.* **71**, 701-711 (1995).

4. Cutroni, M., Mandanici, A., Piccolo, A., Fanggao, C., Saunders, G.A., Mustarelli, P., *Phil. Mag. B.* **73**, 349 (1996).

5. Pradel, A., Ribes, M., *J. Non-Cryst. Solids.* **172-174**, 1222 (1994).

6. Mandanici, A., Cutroni, M., Cramer, C., Funke. K., Tomasi, C., Mustarelli, P., *in preparation.*

7. Cutroni, M., Mandanici, A., Piccolo, A., Fanggao, C., Saunders, G.A., Mustarelli, P., *Solid State Ionics* **90**, 167-172 (1996).

8. Cramer, C., Funke, K., Saatkamp, T., Wilmer, D., Ingram, M.D., *Z. Naturforsch.* **50a**, 613 (1995); Cramer, C., Funke, K., Roling, B., Saatkamp, T., Wilmer, D., Ingram, M.D, Pradel, A., Ribes, M., Taillades, G., *Solid State Ionics* **86-88**, 481 (1996)

9. Bunde, A., Funke, K., Ingram, M.D., *Solid State Ionics* **86-88**, 1311 (1996); Funke, K., Roling, B., Lange, M., *Solid State Ionics* **105**, 195 (1998).

10. Cutroni, M., Mandanici, A., *Solid State Ionics* **105**, 149 (1998).

11. Funke, K., *Phil. Mag. A* **68**, 711 (1993).

12. Funke, K., Cramer, C., Roling, B., Saatkamp, T., Wilmer, D., Ingram, M.D., *Solid State Ionics* **85**, 293 (1996); Roling, B., Funke, K., *J. Non-Cryst. Solids* **212**, 1 (1997).

Engineering quantum superpositions of an ion confined in a 2D-Paul trap

S. Maniscalco, A. Messina and A.Napoli

INFM and MURST, Dipartimento di Scienze Fisiche ed Astronomiche dell'Università di Palermo, via Archirafi 36, 90123 Palermo, Italy

Abstract. A method for engineering cat-like states of an ion confined in a bidimensional isotropic Paul-trap is presented. We briefly discuss their importance in connection with the problem of the individuation of a border between quantum and classical worlds.

Over the last few years there have been developed sophisticated techniques of laser cooling and trapping of atoms opening a new research field for testing fundamental features either of atomic physics and quantum optics [1]. It is possible, in fact, to demonstrate that an ion confined in an electromagnetic trap is describable as a particle in a harmonic potential in the sense that its center of mass (c.m.) can be quantized as harmonic oscillator [2]. Thus, we have practically at hand a single material harmonic oscillator which turns out to be very weakly coupled to the external environment [1]. This feature makes it possible the realization of a class of experiments, still considered as gedanken experiments only a few years ago, aimed at testing fundamental features of quantum mechanics.

A well known controversial question of quantum mechanics is its universality or, stated another way, the existence and the characterization of the border between quantum and classical worlds. In fact, a lot of troubles come out when one applies the superposition principle, which is at heart of quantum physics, to macroscopic systems. These difficulties were brought to the light for the first time by E. Schroedinger in 1935 [3]. In his paper Schroedinger proposed a gedanken experiment in which an unlucky cat were placed in a quantum superposition of the states |living *cat*⟩ and |dead *cat*⟩. This situation cannot be described neither saying that "the cat is living or dead" nor that "the cat is not living nor dead". There is no mental category for classifying this situation because it is something we have never experienced! Now, if quantum mechanics is a complete theory, why don't we see macroscopic superpositions like the cat's one?

According to one of the most qualified theories quite recently developed, the coupling of the system to the environment is responsible for decoherence phenomena converting quantum superpositions into statistical mixtures [4]. The decoherence

CP513, *Nuclear and Condensed Matter Physics,* edited by A. Messina
© 2000 American Institute of Physics 1-56396-929-7/00/$17.00

times were shown to be strongly dependent on the system's size, thus denying the possibility of detecting quantum features of macroscopic systems except than for extremely short times. It is then clear why a system like a trapped ion, which behaves in every respect as a quantum harmonic oscillator and is very weakly coupled to the external environment, can be very useful for studying the quantum-classical border.

In this paper we present a method for the generation of two different classes of cat-like vibrational ionic states. With the term cat-like states we mean superpositions of macroscopically distinguishable states, that is states possessing very different classical-like properties.

Consider the quantized motion of a two-level ion of mass m confined in a 2D isotropic harmonic potential characterized by the trap frequency ν. The annihilation operators of vibrational quanta in the X and Y directions respectively, are defined as:

$$\hat{a}_x = \frac{1}{\sqrt{2}}\left(\sqrt{\frac{m\nu}{\hbar}}\hat{X} + i\frac{1}{\sqrt{m\nu\hbar}}\hat{P}_x\right) \qquad \hat{a}_y = \frac{1}{\sqrt{2}}\left(\sqrt{\frac{m\nu}{\hbar}}\hat{Y} + i\frac{1}{\sqrt{m\nu\hbar}}\hat{P}_y\right) \quad (1)$$

Let's denote with $|n_x, n_y\rangle = |n_x\rangle|n_y\rangle$ the simultaneous eigenstates of the number operators $\hat{n}_x = \hat{a}_x^\dagger \hat{a}_x$ and $\hat{n}_y = \hat{a}_y^\dagger \hat{a}_y$ such that

$$\hat{n}_x|n_x n_y\rangle = n_x|n_x n_y\rangle; \qquad \hat{n}_y|n_x n_y\rangle = n_y|n_x n_y\rangle \qquad (2)$$

In what follows we will call the eigenvalues n_x and n_y *linear quanta* along the X and Y directions respectively. It is important to underline that, since the potential energy is invariant under rotation about z, we could just as well have chosen another system of orthogonal axes $\bar{X}$ and $\bar{Y}$ instead of X and Y. Therefore, in order to take better advantage on the symmetry of the problem under scrutiny, it is convenient to introduce the component $\hat{L}_z$ of the angular orbital momentum, defined as:

$$\hat{L}_z = \hat{X}\hat{P}_y - \hat{Y}\hat{P}_x = i\hbar\left(\hat{a}_x \hat{a}_y^\dagger - \hat{a}_y \hat{a}_x^\dagger\right) \qquad (3)$$

Introducing the two bosonic operators

$$\hat{a}_l = \frac{1}{\sqrt{2}}\left(\hat{a}_x + i\hat{a}_y\right) \qquad \hat{a}_r = \frac{1}{\sqrt{2}}\left(\hat{a}_x - i\hat{a}_y\right) \qquad (4)$$

the angular momentum $\hat{L}_z$ can be recast in the form:

$$L_z = \hbar\left(\hat{a}_r^\dagger \hat{a}_r - \hat{a}_l^\dagger \hat{a}_l\right) \qquad (5)$$

Following reference [5] it is reasonable to call the excitations associated to the number operators $\hat{n}_r = \hat{a}_r^\dagger \hat{a}_r$ and $\hat{n}_l = \hat{a}_l^\dagger \hat{a}_l$, *right and left circular quanta* respectively. For the scope of this paper it is useful, at this point, to observe that the simultaneous eigenstates $|\hat{n}_l, \hat{n}_r\rangle$ of the operators $\hat{n}_l$ and $\hat{n}_r$ are eigenstates of $\hat{L}_z$ belonging to the eigenvalue $\hbar(n_r - n_l)$.

155

A very interesting feature of the system under scrutiny is that it is possible to manipulate in a controlled way the oscillatory ionic motion coupling it to a pair of appropriately chosen electronic states by means of laser beams. The physical origin of this possibility stems from the momentum exchange, between the ion and the classical laser beams, associated to the absorption-emission events. By changing the external laser parameters, it is possible to realize, very easily and efficiently, a huge class of effective interactions involving vibrational and electronic ionic degrees of freedom. This circumstance makes it possible on one side simulating of many physical phenomena wherein bosonic and fermionic modes are coupled and on the other one engineering a great number of oscillatory and/or internal ionic states.

In particular, it has been demonstrated that, irradiating the trapped ion with two laser beams applied along the two orthogonal directions, $\bar{X}$ and $\bar{Y}$, with an angle of $\pi/4$ relative to the X and Y axes respectively and both tuned to the second red vibrational sideband, the physical system under scrutiny can be described by the following effective Hamiltonian model [6]

$$\hat{H} = \hbar\nu(\hat{a}_x^\dagger\hat{a}_x + \hat{a}_y^\dagger\hat{a}_y) + \frac{\hbar\Delta}{2}\hat{\sigma}_z + g\left[(\hat{a}_x\hat{a}_y)\hat{\sigma}_+ + (\hat{a}_x^\dagger\hat{a}_y^\dagger)\hat{\sigma}_-\right] \qquad (6)$$

The effective coupling constant is $g = 2\Omega\frac{\eta^2}{2}\hbar e^{-\frac{\eta^2}{2}}$ and $\eta = \sqrt{\frac{k^2\hbar}{2\nu m}}$ is the Lamb-Dicke parameter. The detuning $\omega_0 - \omega_L = 2\nu$ of the laser beams frequency from the electronic transition frequency has been denoted by Δ. Moreover $\hat{\sigma}_z = |+\rangle\langle+| - |-\rangle\langle-|$, $\hat{\sigma}_+ = |+\rangle\langle-|$, $\hat{\sigma}_- = |-\rangle\langle+|$ describe the internal degrees of freedom, $|+\rangle$ and $|-\rangle$ being the ionic excited and ground states respectively.

We have analytically shown [9] that, preparing the system in the initial state $|\Psi(0)\rangle = |n_{\bar{x}} = N, n_{\bar{y}} = 0\rangle|-\rangle$ [7] with $N \gg 1$, two simultaneous laser pulses realizing Hamiltonian (6) and having a duration of $t = \frac{\pi N}{4g} \equiv t_e$ if N is even or $t = \frac{\pi N}{4g} - \frac{\pi}{4gN} \equiv t_o \simeq t_e$ if N is odd, leave the ion in the following states:

$$|\Psi(t_o)\rangle = \frac{1}{\sqrt{2}}\left(|n_l = N, n_r = 0\rangle - i(-1)^{\frac{N+1}{2}}|n_l = 0, n_r = N\rangle\right)|-\rangle \qquad \text{odd N} \quad (7)$$

$$|\Psi(t_e)\rangle = \left(|n_{\bar{x}} = N, n_{\bar{y}} = 0\rangle + (-1)^{\frac{N}{2}}|n_{\bar{x}} = 0, n_{\bar{y}} = N\rangle\right)|-\rangle +$$
$$- i\sum_{k=1}^{N-1} P_k \sin(f_k t_e)|N-k-1, k-1\rangle|+\rangle \qquad \text{even N} \quad (8)$$

where $f_k = 2g\sqrt{k(N-k)}$ and $P_k = \frac{1}{2^{N/2}}\left(\begin{array}{c} N \\ k \end{array}\right)^{1/2}$. Let's focus, first of all, our attention on the state (7). This state is a superposition of the two eigenstates of the orbital angular momentum $\hat{L}_z$ corresponding to the maximum $(+\hbar N)$ and minimum $(-\hbar N)$ eigenvalue respectively, with $N \gg 1$. The state generated by means of our procedure is then a quantum superposition of two macroscopically distinguishable states describing ionic circular right or left motion respectively.

Consider now the state described by equation (8). According to the projection postulate, a measurement of the electronic state of the ion, currently performed by means of quantum jumps techniques [8], resulting in $|-\rangle$, leaves the ion in the following state

$$|\Psi(t_e)\rangle_M = \frac{1}{\sqrt{2}} \left(|n_{\bar{x}} = N, n_{\bar{y}} = 0\rangle + (-1)^{\frac{N}{2}} |n_{\bar{x}} = 0, n_{\bar{y}} = N\rangle \right) |-\rangle \qquad (9)$$

$|\Psi(t_e)\rangle_M$ is a superposition of two states describing ionic oscillations along the two orthogonal directions $\bar{X}$ and $\bar{Y}$ respectively, which is clearly a cat-like state, if $N \gg 1$.

Summing up we have proposed a method for the realization of two different classes of bidimensional cat-like vibrational states. We wish to emphasize that, in the context of the scheme outlined in this paper, the total number of excitations N present in the initial state of the c.m. motion, plays the role of an adjustable parameter allowing the generation of very different bimodal vibrational Schrödinger cat-like states.

ACKNOWLEDGEMENTS

The financial support from CRRNSM-Regione Sicilia is greatly acknowledged. This work was also supported by MURST 40% co-financial support in the framework of the research project "Amplificazione e Rivelazione di Radiazione Quantistica".

REFERENCES

1. W.M. Itano and D.J. Wineland Phys. Rev. A **25**, 35 (1982); D.J. Wineland et al., Phys. Rev. A **36**, 2220 (1987); F. Diedrich et al. Phys. Rev. Lett. **62**, 403 (1989); C. Monroe et al. Phys. Rev. Lett. **75**, 4011 (1995)
2. C.A. Blockley et al. Europhys. Lett. **17**, 509 (1992); J.I. Cirac et al., Phys. Rev. A **49**, 1202 (1994)
3. E. Schrödinger, Naturwissenschaften **23**, 807 (1935)
4. W.H. Zurek, Phys. Today **44** (10), 36 (1991)
5. Cohen-Tannoudji C.,Diu B. and Laloë F., 1977, Quantum Mechanics Vol.1 (New York: John Wiley and Sons), pp. 727-741
6. C.C. Gerry et al., Phys. Rev. A **55**, 630 (1997)
7. D.M. Meekhof et al., Phys. Rev. Lett. **76**, 1796 (1996)
8. J. F. Poyatos et al., Phys. Rev. A **54**, 1532 (1996)
9. S. Maniscalco , A. Messina and A. Napoli submitted to Phys. Rev. A

Blowing Up Reversibility
In Biomolecular Self-Assembly

Mauro Manno (1,2) and Antonio Emanuele (1)

(1) INFM, CRRNSM and Dept. of Physics, Univ. of Palermo, via Archirafi 36, I-90123 Palermo, Italy.
(2) CNR-IAIF, via U. La Malfa 153, I-90153 Palermo, Italy

Abstract. The self-assembly of hydrogels involves the interplay of many processes: thermodynamic demixing, molecular crosslinking and molecular conformational changes. Interactions among these processes lead to complex structures encompassing many length-scales. Here, we monitor experimentally changes of the structure of biomolecular aggregates during the melting of an agarose gel, that is during the inverse process of gelation, which exhibits a large hysteresis. Observations on different lengthscales clarify the role of the processes involved in sustaining irreversibility.

INTRODUCTION

Self-assembly of biomolecular structures in solution often involves an interplay among different processes, such as phase-separation, molecular crosslinking and molecular conformational changes [1]. This mechanism of process interaction is a common feature of several systems [1-3]. Its extension to important pathological coagulations is of wide interest specially in those cases involved in amyloidosis and Prion diseases [4].

In the present work we are concerned with the inverse process of self-assembly. That is, the dissolution of individual polymers from ordered supramolecular structures previously obtained by gelation. We study the melting of an hydrogel, whose formation was the object of previous papers [1]. By monitoring the structural changes at different length-scale we gain insights in the role of the different processes in the stability of supramolecular structure and in the thermal hysteresis of gels.

EXPERIMENTS

Present experiments concern the self-assembly and disassembly of hydrogels of Agarose, an uncharged polysaccharide [Seakem HGT(P), from BioProducts, Marine Colloids Division, Rockland, Maine]. Powder was dissolved in Millipore (Super Q) water for 20 min at 100 °C and filtered (0.22 μm filter) directly in measuring cells [1]. The chosen concentration is 2% (wt/wt).

For light scattering experiments we used a ILT argon laser tuned at 514.5 nm. Data were automatically collected at wide angle using a Brookhaven BI-200SM goniometer. When dealing with non ergodic samples, such as gels, a motor driven cell holder was used to scan different regions of the specimen, thus allowing ensemble averaging. For small angle ligth scattering measurements we used a CCD Panasonic Camera as in Ref.1.

We follow the procedure of Ref.1 to obtain a structured, spinodal-assisted gel. An Agarose solution 2% (wt/wt) was quenched from 90 °C to 43 °C. At this temperature and at the chosen concentration the solution is in its instability region and a spinodal demixing occurs. As a consequence, a pattern of lower and higher local concentration domains develops. Domains have a characteristic correlation length L_m of about 5 μm, as shown by peak in the structure function $S(q)$ at a scattering vector $q_m \sim 13000$ cm^{-1} (Fig.1). After 30 min the early stage of spinodal demixing is over and the macroscopic gelation occurs, as monitored by the ball-drop method. Gelation freezes the size distribution of demixed droplets. As pointed out in Ref. 1, a third process contributes to the gel formation, the process of conformational change at the molecular level on a length-scale characteristic of polymer sizes $L_p \sim 0.1$ μm.

In the interval $L_p < q^{-1} < L_m$ there is no other characteristic length: the system is self-similar and the mass M of molecules within a radius r can be expressed by a power law: $M \sim r^D$, where D is called mass fractal dimension. The radial correlation function is essentially proportional to the mass density. Therefore we obtain the following scaling expression for the correlation function and for the structure function, which is its Fourier transform: $S(q) \sim q^{-D}$. In the fractal regime, $L_p < q^{-1} < L_m$, the structure function carries information about the structure and texture of the solutions or gels within demixed domains or within clusters of molecules. The fractal dimension is a measure of the space filling properties of the branched structure of the gel. In addition to direct structural information, fractal dimension can be sometimes related to the kinetic mechanism of aggregation. In the present case in which the final structure is a result of the kinetic competition of different processes, the fractal dimension has been shown to grow in the course of self-assembly up to a stationary value (of about 1.4), at a time which corresponds to the time of macroscopic gelation [1]. Moreover, the value of 1.4 is a characteristic feature of this type of molecular crosslinking and it is rather independent of the kinetics. Thus, by measuring the fractal dimension one obtains information on both the gel structure and the progress of crosslinking [1].

In the first part of Fig.2, we observe the growth of the peak of the structure function at q_m and the growth of the fractal dimension in the self-similar length-scale. Within 30 min. we observe the onset of macroscopic gelation and the freezing in of the geometrical properties of the system (the correlation length L_m and the fractal dimension D), as in Ref.1.

Hydrogels samples obtained at 43 °C in 30 min. were heated quickly to 70 °C. At this temperature freshly homogeneous solution of agarose does not form gels, since the system is neither in the region of direct gelation nor in the region of thermodynamic instability. Nevertheless, previously formed gels do not melt. Long range (Fig.1) and fractal structure do not change at this temperature (Fig.2). Raising the temperature to 92 °C caused gel melting. The structure of demixed domains is lost quickly, but a longer range correlation fades off slowly (Fig.1). Also, the fractal structure disappears, as monitored by the dicrease of the fractal dimension.

DISCUSSION

In the present work, we monitored the formation and destruction of a hydrogel structure at different length-scale. In previous papers [1], the gelation was found to be the result of the interaction among processes. Spinodal demixing provides a canvas of domains of given correlation length, and molecular crosslinking, which involves a conformational change, freezes this correlation length by building a fractal structure on a smaller length scale. Now, we heat back the system at 70 °C in a region where the solution is stable as such. Unaffected value of fractal dimension evidences that the geometrical constraints imposed by the percolating net of crosslinks do not allow the destruction of the demixed domain structure. The loss of supramolecular order and return to the homogeneous solution state occurs at 92 °C [5].

These experiments stress the interest of studying kinetics in addition to equilibrium properties. The actual role of each process is different during the gelation or the melting. In fact, it is context- (and time-) dependent, that is it depends on the relative progress of each of the processes involved. In particular, self-trapping in the gel state, evidenced by large hysteresis, points out the highly non-linear nature of interactions, markedly of those relating fractal and topological properties.

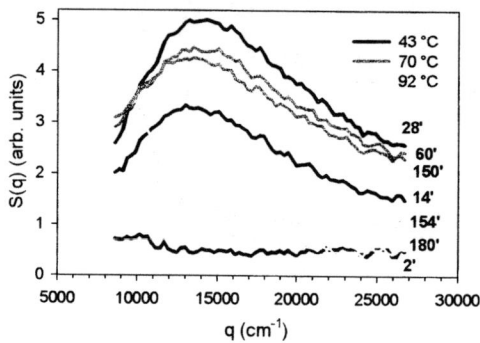

FIGURE 1. Structure functions S(q) at different times and temperatures after quenching. Solid black lines: 43 °C. Gray lines: 70 °C and 92 °C (gel melting).

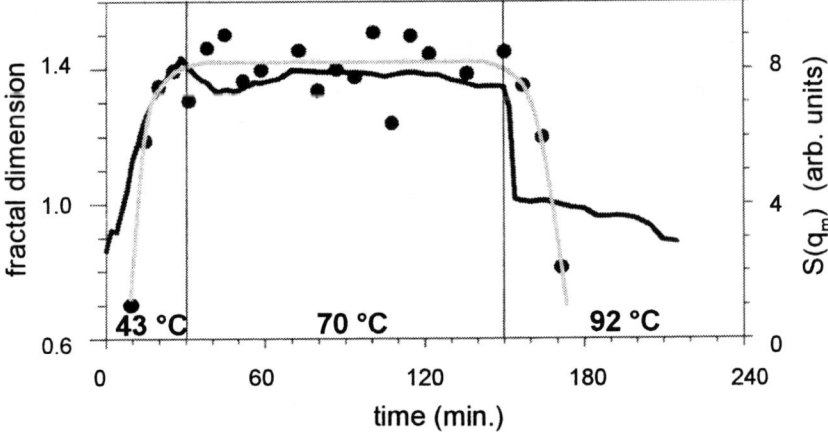

FIGURE 2. Kinetics relative to the three temperature steps: S(qm), solid black line; fractal dimension (black circles), the gray line is a guide for eyes for the fractal dimension..

ACKNOWLEDGMENTS

We thank M.U. Palma, M.B. Palma-Vittorelli, P.L. San Biagio, D. Bulone, and V. Martorana for discussions and long-term collaboration, and M. Lapis and R. Megna for technical assistance.

REFERENCES

1. Manno, M., and Palma, M. U., *Phys. Rev. Letters* **79**, 4286-4289 (1997); Manno, M., Emanuele, A., Martorana, V., Bulone, D., San Biagio, P. L., Palma-Vittorelli, M. B., and Palma, M. U., *Phys. Rev. E* **59**, 2222-2230 (1999).

2. San Biagio, P. L., Martorana, V., Emanuele, A., Vaiana, S. M., Manno, M., Bulone, D., Palma-Vittorelli, M. B., and Palma, M. U., *Proteins* **37**, 116-120 (1999).

3. San Biagio, P. L., and Palma, M. U., *Biophys. J.* **33**, 743-752 (1991); Sciortino, F., Prasad, K. U., Urry, D. W., and Palma, M. U., *Biopolymers* **33**, 743-752 (1993).

4. Kelly, J. W., *Curr. Opin. Struct. Biol.* **6**, 11-17 (1996); Safar, J., "The folding intermediate concept of prion protein formation and conformational links to infectivity" in *Prions, prions, prion*, edited by S. B. Prusiner, Berlin: Springer, 1998, pp. 69-76.

5. Vento, G. , Palma, M. U., and Indovina, P., *J. Chem. Phys.* **70**, 2848-2853 (1979).

Near Field Optical Spectroscopy of Interacting Electron Systems

Giovanna Martino, Salvatore Savasta and Raffaello Girlanda

INFM and Dipartimento di Fisica della Materia e Tecnologie Fisiche Avanzate,
Università di Messina
Salita Sperone 31, I-98166 Messina, Italy

Abstract. We present a numerical study of the local optical properties of electrons and holes interacting via the Coulomb potential. The local linear optical susceptibility is calculated with Lanczos iteration techniques for a quasi-one dimensional extended Hubbard semiconductor model. We apply this numerical method to analyze the combined effects of disorder-induced localization, Coulomb correlations and local probing in semiconductors. Calculations are carried out for different spatial resolutions ranging from diffraction limited spectroscopy to near field spectroscopy.

The achievement of high spatial resolutions in optical spectroscopy offers many opportunities for directly probing the world of mesoscopic dimensions. In particular the possibilities offered by the near-field optical spectroscopy in confining the optical excitation to a correspondingly small volume below the diffraction limit, appears to be crucial for probing the properties of nanostructures, as it can permit a direct observation of quantum confinement [1], of effects related to the Coulomb correlation [2] and of exciton localization due to random potential fluctuations at the interfaces [3].

We present a numerical study of local optical properties of electrons and holes in a quasi-one dimensional lattice interacting via the Coulomb potential. Here we employ a two-band, tight-binding semiconductor model with long-range Coulomb interaction and with broken translation simmetry. The Hamiltonian is given by $H = H_0 + V_d + V_C$. H_0 contains the usual single particle and translationally invariant terms,

$$H_0 = \sum_m E_g c_m^\dagger c_m + \left(t^e \sum_m c_{m+1}^\dagger c_m + t^h \sum_m d_{m+1}^\dagger d_m + \text{H.c.} \right) , \qquad (1)$$

where $c_m^\dagger$ and $d_m^\dagger$ create respectively an electron and a hole at site m. V_d is the single particle potential breaking the translational invariance. It can be a confining potential or it can be a random potential modeling the disorder potential originating from interface roughness felt by electrons and holes. Here we consider the following

CP513, *Nuclear and Condensed Matter Physics,* edited by A. Messina
© 2000 American Institute of Physics 1-56396-929-7/00/$17.00

diagonal potential, $V_d = \sum_m V_m^e c_m^\dagger c_m + \sum_m V_m^h d_m^\dagger d_m$. The Coulomb interaction is given by

$$V_C = \frac{1}{2} \sum_m U_{m,m'} \rho_m \rho_{m'} , \tag{2}$$

with the charge densities $\rho_m = c_m^\dagger c_m - d_m^\dagger d_m$. For the long-range Coulomb interaction we use $U_{m,m'} = \pi U / [N \sin(\pi |m - m'| /N)]$, for $m \neq m'$, being N the number of sites in the one dimensional semiconductor ring. The on-site Coulomb interaction is a different parameter, $U_{m,m} = \eta U$.

The Hamiltonian describing the interaction of the electronic system with a monochromatic classical light field of frequency ω, in the dipole and rotating wave approximations, is given by $H_I = -\mu \sum_m E(m,\omega) c_m d_m + \text{H.c.}$, where $E(m,\omega)$ is the electric field at site m, c_m and d_m are the annihilation operators for the electron and the hole at site m and μ is the interband dipole moment.

Considering a light field of given profile $E(m,\omega) = E(\omega)g(m - M)$, centered around the beam position M, we can define a generalized linear optical response function which is function of the beam position and shape,

$$\chi(M,\omega) = \langle 0| P_M \frac{1}{\omega - H + i\gamma} P_M^\dagger |0\rangle , \tag{3}$$

where $|0\rangle$ is the semiconductor ground state, $P_M = \sum_m g(m-M)c_m d_m$, and γ is the homogeneous broadening for the electron system. This local optical susceptibility takes into account properly the nonlocal character of the light-matter interaction in extended systems [2]. At this point we can define a local absorption coefficient $\alpha(M,\omega) = \text{Im}\chi(M,\omega)$, that is a function of the beam position and shape, and relates the total absorbed power to the power of a local excitation (illumination mode) [2]. Of course spatial resolution depends on the given beam profile. In the following we will describe the narrow light beam by a Gaussian profile $g(m) = \exp -(m^2/\sigma^2)$. We use the *Lanczos* algorithm to calculate the local optical susceptibilty and hence the local absorption $\alpha(M,\omega)$. The numerical evaluation proceeds in the usual way. The total number of basis states is N^2. The Lanczos algorithm tridiagonalizes the semiconductor Hamiltonian in the one electron-hole pair subspace starting with the (normalized) initial state $P_M^\dagger |0\rangle$. Each iteration step produces one new basis state. The iteration is extremely fast due to the sparsness of the Hamiltonian matrix. We truncate the iteration after the spectrum $\chi(m,\omega)$ stabilizes. The relevant spectrum comes from the inversion of the resolvent matrix $(\omega + i\gamma - H)^{-1}$ which is simple in the tridiagonal form of the Hamiltonian H. The main usefulness of this procedure is that no eigenvalues and eigenvectors have to be calculated to obtain the local absorption coefficient. In this way a system size of $N = 2000$ with a number of basis states $N^2 = 4 \cdot 10^6$ can be easily treated. We observe that the scheme here presented solves the problem of two interacting particles (the electron and the hole) in N sites illuminated by a local probe. The solution contains no approximations, e.g. the contributions from all the exciton states are automatically included, the usual approximation assuming that the e-h relative motion is not perturbed by the disorder potential is here relaxed. As a consequence this model appears suitable to analyze the combined effects of disorder-induced localization and Coulomb correlations for different spatial resolutions of the optical probe. Now we breafly present two prototype applications. First we analyze the case of a quasi-one dimensional ordered nanostructure under a local electromagnetic excitation polarized along the free z axis of the structure and including the Coulomb correlation Eq. (2). This prototype example gives indications on what happens when a homogeneous extended system is locally probed and gives indications on the interpretation of high resolution near field spectra. Fig. 1 displays the local absorption spectra for different spatial resolutions. The figure clearly shows a deformation and a broadening of the spectra as the spatial resolution (i.e. the variance σ of the gaussian beam) approaches the exciton Bohr radius a_0. Typical exciton Bohr radii in GaAs systems are of the order of 10 nm. The spectra in Fig. 1 have been obtained for a system size $N = 2000$ using the following parameters, $E_g = 1.585$ meV, $t^e = -0.23$ eV $t^h = 0.034$meV, $U = 0.03$ meV $\eta = 1.5$ and $\gamma = 0.7$ meV. The asymmetric lineshape of the high-resolution spectra compared with the global spectrum is a consequence of the localized excitation. An analogous asymmetric shape is usually observed as a consequence of the exciton localization due to interfacial disorder [4]. We now take into account the presence of a random potential V_m such that $V_m^e = (2/3)V_m$, $V_m^h = (1/3)V_m$, we consider the simple case of a quantum wire with coulomb correlation and with a white noise random potential satisfying the correlation relation $\langle V_m V_{m'} \rangle = \Gamma^2 \delta_{m,m'}$. Fig. 2 displays absorption spectra for different spatial resolutions ranging from the diffraction limited spectroscopy to near-field spectroscopy. The variance of the gaussian beam is expressed directly in number of sites. We have used $\Gamma = 2.5$ meV for the disorder strength. The global spectrum shows the typical disorder-induced inhomogeneous broadening resulting in an asymmetric gaussian lineshape. Increasing the spatial resolution, the inhomogeneous line start to reveal its fine structure until it splits into several very sharp and homogeneously broadened peaks. Of course this fine structure depends on the

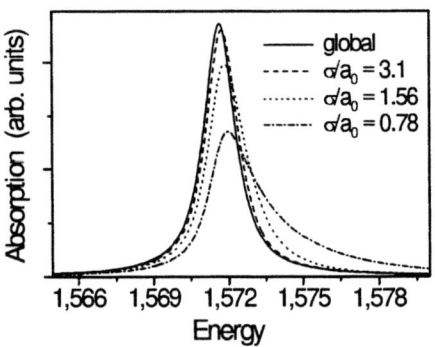

FIGURE 1. Global and local absorption spectra (including electron-hole correlation) for an ordered quantum wire. Parameters are given in the text and in the figure.

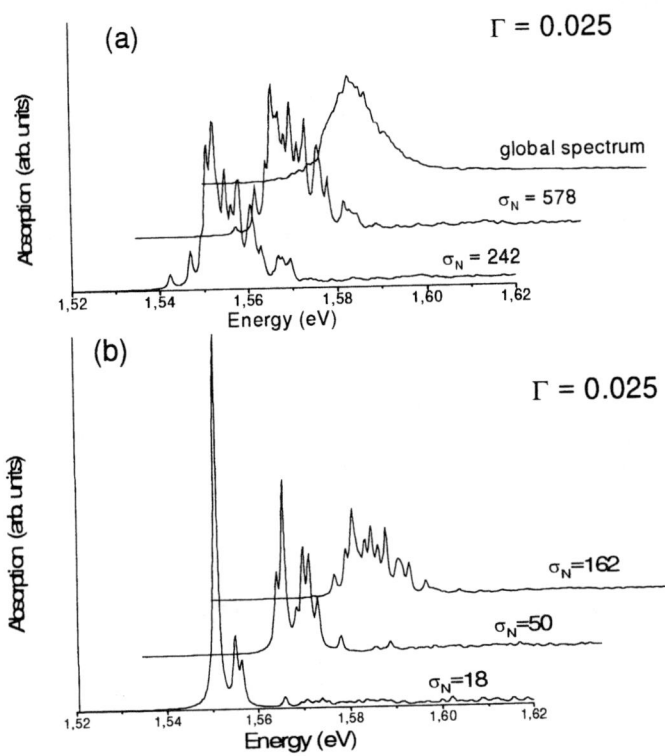

FIGURE 2. Global and local absorption spectra for a quantum wire including electron-hole correlation and a white noise random potential. Parameters are given in the text and in the figure.

position of the beam and on the particular disorder realization. We further observe that, increasing the spatial resolution, the higher peaks are mainly at low energy. This confirm that states in the low energy tail are more localized. This behaviours have been observed in several quantum wire samples [1,5] and gives indications on the possibility of this method of modeling the quantum confined structures under current investigations.

We acknowledge the support of INFM through the project *NANOSNOM*

REFERENCES

1. A. Gustafsson *et al.*, J. Appl. Phys. **84**, 1715 (1998).
2. O. Mauritz *et al.*, Phys. Rev. Lett. **82**, 847 (1999).
3. D. Gammon *et al.*, Phys. Rev. Lett. **76**, 3005 (1996).
4. S. Glutsch and F. Bechstedt, Phys. Rev. B. **50**, 7733 (1994).
5. J. Bellessa *et al.*, Phys. Stat. Sol. **164**, 273 (1997).

Superconducting dot in a magnetic field

A. Mastellone, G. Falci, Rosario Fazio, and G. Giaquinta

Dipartimento di Metodologie Fisiche e Chimiche per l'Ingegneria,
Università di Catania, viale A. Doria 6, 95125 Catania - Italy
Istituto Nazionale di Fisica della Materia (INFM), Unità di Catania , Italy

Abstract. We study superconducting correlations in small metallic dot with fixed particle number in an applied magnetic field. We determine the tunneling spectra finding a good qualitative agreement with experimental data. For utrasmall dots we report an example of "softening" of the crossover by a "superconductive" state to a normal one due to the magnetic field.

SUPERCONDUCTIVITY IN METALLIC DOTS

In a series of recent experiments Ralph, Black and Tinkham (RBT, [1]) found evidence of superconducting correlations and parity effects [2] in nanometer-scale Al dots. In these systems effects of a finite level spacing $\delta \sim 1/N(0)V$ ($N(0)$ is the density of states at the Fermi energy and V the volume of the dot) are important so the interplay between the physics of a quantum dot and the effect of superconductivity have to be studied. The standard BCS theory of superconductivity can no longer be applied as the dots become smaller [3] and the problem of how to characterize superconducting correlations, if any, is technically complicated because the large charging energy of the dots requires an analysis in the canonical ensemble. This problem has been recently studied using scaling techniques and Lanczos diagonalizations [4–6]. It has been shown that due to pairing correlations a new energy scale Δ appears, and various low energy features of superconducting dots are described in terms of universal functions of δ/Δ. This allow to distinguish between the regime of superconducting dots ($\delta/\Delta > 1$) and the regime of large dots ($\delta/\Delta < 1$) and to describe quantitatively the physics of pairing correlations as a Kondo-like crossover phenomenon, where Δ plays the role of the Kondo temperature and the low-energy cutoff is related to δ. These results have been fully confirmed by a Density Matrix Renormalization Group study [7] and from the analysis of the exact solution of Richardson and Sherman [8]. An important result is that pairing correlations are effective even in the superconducting dot regime. For instance they determine the reentrant behavior of the spin susceptibility in odd dots [9], which could be experimentally detected with present day technology [10].

CP513, *Nuclear and Condensed Matter Physics,* edited by A. Messina
© 2000 American Institute of Physics 1-56396-929-7/00/$17.00

In this contribution we focus on the spectral features of a superconducting dot in a magnetic field. Experiments [1] have been carried out by measuring tunneling spectra. They can be associated to tunneling events which change the electron number $N \to N \pm 1$. The initial state of the dot is the even (odd) ground state with N electrons whereas the final state is a generic state of an odd (even) dot with $N \pm 1$ electrons. Then the tunneling spectrum can be easily derived from the spectrum of isolated even and odd dots which is the central subject of our investigation.

In small metallic dots the charging energy is much larger than all other energy scales (bulk gap, typical values of transport voltage and temperature) and strongly suppresses fluctuations of the electron number, so we study the problem in the canonical ensemble. We study the dynamics of the electrons in the Debye shell ($|\epsilon - \epsilon_F| < \omega_D$, the Debye energy) starting from the hamiltonian [11,4,5]

$$H = \sum_{\substack{n=1 \\ \sigma=\pm}}^{\Omega} \epsilon_n \, c_{n,\sigma}^\dagger c_{n,\sigma} - \alpha \delta \sum_{m,n=1}^{\Omega} c_{m,+}^\dagger c_{m,-}^\dagger c_{n,-} c_{n,+} - 2 \sum_{\substack{n=1 \\ \sigma=\pm}}^{\Omega} \mu_B H \, \sigma \qquad (1)$$

where the indices m and n label the single electron energy levels with annihilation operator $c_{m,\sigma}$. Time reversed states are labelled by $\sigma = \pm$. We assume equally spaced noninteracting doubly degenerate levels, $\epsilon_n = n\delta$, so their number is $\Omega = 2\omega_D/\delta$. We consider a half filled Debye shell, $N = \Omega$. The pairing interaction has dimensionless coupling constant α. Other states in the conduction band do not appear in the Hamiltonian eq.(1) because they are not coupled by the pairing interaction and give a trivial contribution to the energy. The last term accounts for the Zeeman effect. This is the only relevant term in the experiments [1] where the tunneling spectra depend linearly on the magnetic field. Finally we introduce the energy scale related to pairing correlation which in this model is given by $\Delta = \omega_D/[2\sinh(\delta/\alpha)]$.

EXCITATION SPECTRA IN A MAGNETIC FIELD

Unpaired electrons in the Debye shell are not scattered by the interaction. The configuration of these "frozen" electrons is a good quantum number for this problem. Eigenvalues of the Hamiltonian eq.(1) have to be found by diagonalizing it in each subspace determined by the configuration of frozen levels.

In the odd case we find the ground and first three excited states from the lowest-lying eigenstates in the subspaces obtained moving the upper electron from the Fermi level (ground state) to first, second and third single-particle energy level above and under it (it's present a degeneracy due to simmetry of the half-filling scheme). In the even case, in a weak magnetic field we obtain the first excited state from the lowest-lying eigenstate in the subspace obtained breaking a couple, the second one from the first excited state in the ground subspace and the third

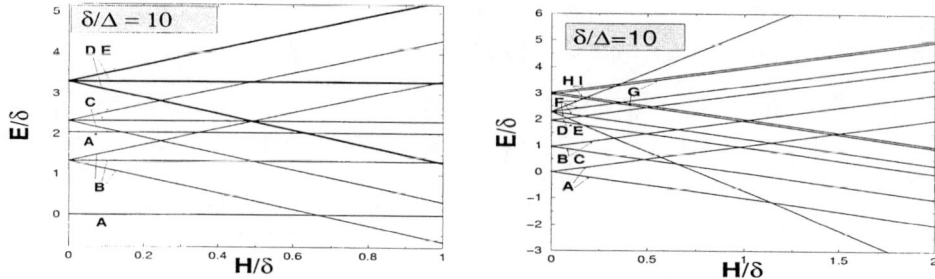

FIGURE 1. Low-lying energies in function of H for dots with an even (a) and an odd (b) number of electrons. Different caps letters denote subspaces with different configurations of blocked levels. Notabily A and A^* denote the ground and first excited states when no level is blocked.

one from the subspace obtained moving the upper electron in next available single-particle level. Instead, in a strong magnetic field regime, the subspaces of excited states are the same as in the odd case: the upper (lower) (unpaired, due to strong magnetic filed) electron moves one, two and three single-particle levels upward and downward. The same qualitative behaviour holds for different values of δ/Δ also. In the figure 1 we plot the energies versus the applied magnetic field.

The paramagnetic breakdown in small dots does not occur via a first order transition like in films but rather as a sort of avalanche process in which the Cooper pairs are progressively broken [13]. We find that (fig. 2) near BCS regime ($\delta/\Delta = 0.2$) several pairs break at once; instead in ultrasmall regime ($\delta/\Delta = 1.0$) pairs break one by one. This cleary suggests a softening of the transition (crossover) from a superconductive state to a normal one when δ/Δ grows.

We focus now on the tunneling spectra. The experimental data [1] show various features. Apart from the linear dependence of the energies on H, the spectrum shows a structure of kinks and discontinuities. These features can be explained by the interplay of the existence of selection rules and the smooth paramagnetic breakdown effect which implies a "cascading" of the ground state as H is increased. To obtain the spectra, we first fix the initial number of electrons in the dot N and spin sector of the lowest-lying eigenstate s_i. Then, since the dot's large charging energy ensures that only one electron can tunnel once a time, we choose the final states according to the "spin selection rule" $|s_f - s_i| = \frac{1}{2}$. In the figure 2 we report the tunneling for an odd-even transitions. We stress here that all spectra are *universal*: different N spectra collapse all on the same curves apart an additive value ΔE_{min}, the $H = 0$ lowest energy transition, provided that one plots $\Delta E/\Delta$ versus H/Δ.

Our result agree qualitatively with the results obtained in Ref. [15] where a variational after projection method was applied. We obtain also a "doublet" in the odd-even spectrum that a variational method cannot accunt for because one of lines of the doublet is the first excited eigenstate of the ground subspace. This doublet disappears at higher magnetic field due to minor pairing effects. We report

168

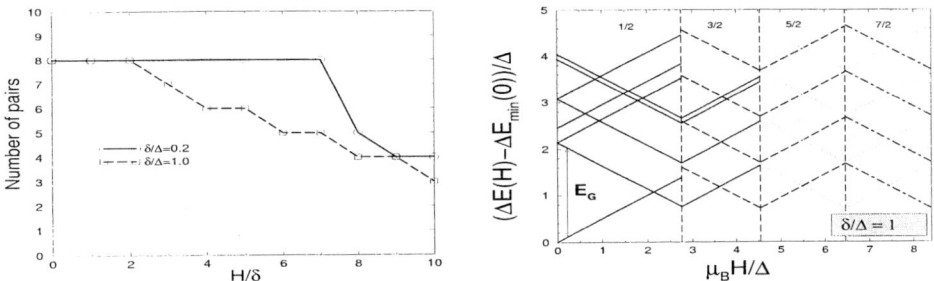

FIGURE 2. (a) Number of pairs in an applied magnetic field: as dots become smaller (larger δ/Δ) pairs break one by one; (b) Tunneling spectra for odd to even transitions. The numbers in upper part show the spin sector of the ground state.

also a good qualitative agreement with BRT experimental results [1]: the structure of spectra is well reproduced and our data display the parity effects and the kinks due to the ground state cascading with the magnetic field. We notice that more than qualitative agreement between theory and experiment cannot be expected due to neglecting nonequlibirium effects [14].

Acknowledgements. The authors thank L. Amico, A. Di Lorenzo and F.W.J. Hekking for stimulating discussions and comments.

REFERENCES

1. D.C. Ralph, C.T. Black, and M. Tinkham, Phys. Rev. Lett. **74**, 3241 (1995); C.T. Black, D.C. Ralph, and M. Tinkham, Phys. Rev. Lett. **76**, 688 (1996); D.C. Ralph, C.T. Black, and M. Tinkham, Phys. Rev. Lett. **78**, 4087 (1997).
2. D.V. Averin and Yu. V. Nazarov, Phys. Rev. Lett. **68**, 1993 (1992).
3. P.W. Anderson, J. Phys. Chem. Solids **11**, 28 (1959).
4. K.A. Matveev and A.I. Larkin, Phys. Rev. Lett. **78**, 3749 (1997).
5. A. Mastellone, G. Falci, and Rosario Fazio, Phys. Rev. Lett. **80**, 4542 (1998).
6. S.D. Berger and B.I. Halperin, *cond-mat* 9801286.
7. J. Dukelski and G. Sierra, Phys. Rev. Lett. **83**, 172 (1999)
8. R.W. Richardson and N. Sherman, Nuclear Physics **52**, 221 (1964).
9. A. Di Lorenzo, R. Fazio, F.W.J. Hekking, G. Falci, A. Mastellone and G. Giaquinta, subm. to Phys. Rev. Lett.
10. A review of the thermodynamics of a superconducting dot can be found in G. Falci, R. Fazio, F.W.J. Hekking, and A. Mastellone, to be publ. in Journ. Low temp. Phys., 2000.
11. J. von Delft, D.S. Golubev, W. Tichy, and A.D. Zaikin, Phys. Rev. Lett., **77** 3189 (1996).
12. B. Janko, A. Smith, and V. Ambegaokar, Phys. Rev. B **50**, 1152 (1994).
13. F. Braun, J. von Delft, D. Ralph, and M. Tinkham, Phys. Rev. Lett. **78**, 4542 (1997).
14. O. Agam, *cond-mat* 9812315.
15. F. Braun and J. von Delft, *cond-mat* 9801170.

Detection of Low-Intensity Far-Infrared Coherent Fields by Mesoscopic Josephson Junctions

R. Migliore, A. Messina, A. Napoli

Istituto Nazionale di Fisica della Materia and Dipartimento di Scienze Fisiche ed Astronomiche, University of Palermo, via Archirafi, 36 I-90123 Palermo, Italy.

Abstract. Here we consider the effects of exposing ultrasmall dc voltage-biased Josephson junctions to a single-mode quantized far-infrared field prepared in a coherent state. We prove that the I-V characteristic of the mesojunction exhibits quantum Shapiro steps (QSS) whose amplitude and quantum noise reflect the intensity and the phase of the non classical coherent state. In this way, such a nanodevice may be exploited to probe the parameters of an external quantized e.m. field, detecting the amplitude and the quantum fluctuations of the first QSS only.

The I-V characteristic of a mesoscopic Josephson junction irradiated by non classical electromagnetic fields exhibits current spikes, presents in a classical framework too [1,2], known as quantum Shapiro steps (QSS) [3-5]. In references [6-8] we have shown that their structure displays an interesting sensitivity to the quantum coherences of the e.m. field which the mesojunction is coupled to. This means that a JJ operating in the mesoscopic regime, defined in accordance with Likharev [9], seems to possesses appropriate potentialities to play an important technological role in the conception of a new class of quantum electromagnetic field detectors.

In this paper, we report on a simple scheme by which the intensity $|\alpha|$ and the phase θ_α of a low-intesity ($|\alpha| \sim 1$) far-infrared ($\omega_1 \sim 2\pi \cdot 10^{14}\, Hz$) coherent quantum state $|\alpha\rangle = \left||\alpha|e^{i\theta_\alpha}\right\rangle$ might be extracted from measurements of the amplitude and the quantum noise characterizing the first QSS only.

A mesoscopic JJ is practically realized requiring very low capacitances ($C \sim 10^{-15}\, F$) and temperatures ($T \sim (10 \div 100)\, mK$) [10]. In such conditions, we cannot neglect the quantum mechanical nature of conjugate variables phase φ and supercharge q, which is taken into account adopting the commutation rule $[\varphi, q] = 2ei$ [9,11]. In accordance with the Voltage Bias Model, we suppose that $V_F = |\vec{E}|d = i\sqrt{\hbar\omega_1/2C_F}\,(a - a^\dagger)$, is the non classical electromotive force externally applied to the junction, $C_F = \dfrac{\varepsilon_0 \varepsilon_r}{d^2} \left(\dfrac{\pi c}{\omega_1} \right)^{\frac{1}{2}}$ being a capacitive parameter dependent on the relative dielectric constant ε_r of the insulating barrier and on its tickness d.

If V_0 is the dc voltage applied across the junction barrier, the hamiltonian for the total system can thus be cast in the following form [7,8]:

$$H = \frac{[q + C(V_0 + V_F)]^2}{2C} + E_J(1 - \cos\varphi) + \hbar\omega_1 \left(a^\dagger a + \frac{1}{2} \right). \tag{1}$$

Here, the Josephson energy E_J is defined through the relation $2eE_J = \hbar I_{cr}$, I_{cr} being the critical current of the junction. It is possible to show that, since $\omega_1 \sim 2\pi \cdot 10^{14}\, Hz$, so that $C \sim 10^{-15}\, F \ll C_F \sim 10^{-12}\, F$, the field operators evolve approximately freely

$$\frac{\partial a}{\partial t} \approx -i\omega_1 a \qquad \frac{\partial a^\dagger}{\partial t} \approx i\omega_1 a^\dagger \tag{2}$$

and that the junction operators evolve as follows:

$$\varphi \approx \frac{2e}{\hbar}(V_0 + V_F) \tag{3}$$

$$I \equiv -\dot{q} = I_{cr}\sin\varphi \tag{4}$$

After the explicit integration of eq. (3), we obtain the following equation describing the time evolution of the supercurrent operator as

$$I(t) = I_{cr}\sin[\omega_0 t - \xi(ae^{i\omega_1 t} + a^\dagger e^{-i\omega_1 t}) + \xi(a + a^\dagger) + \varphi_0], \tag{5}$$

where $\xi\sqrt{\hbar\omega_1 C_F} = \sqrt{2}e$, $\hbar\omega_0 = 2eV_0$ and φ_0 is the Schödinger operator for the initial phase of the junction. Let's choose the initial condition for the total system in the factorized form

$$\rho(0) = \rho_J(0) \otimes \rho_F(0), \tag{6}$$

$\rho_J(0)$ and $\rho_F(0)$ being the initial density matrices of the junction and the field respectively. In such conditions, it is immediate to derive the following expression for the expectation value $\langle I(t) \rangle$ of the supercurrent operator

$$\frac{\langle I(t) \rangle}{I_{cr}} = \mathrm{Im}\left\{ e^{\left(i\omega_0 t + i\frac{\pi\omega_0}{2\omega_1} \right)} \langle \alpha | D(\xi e^{i\omega_1 t}) | \alpha \rangle \right\} Tr_J \left(\rho_J(0) e^{i\varphi_0} \right) \tag{7}$$

where

$$D(\xi e^{i\omega_1 t}) = \exp[\xi(a^\dagger e^{i\omega_1 t} - ae^{-i\omega_1 t})] \tag{8}$$

and Tr_J means trace on the junction states. After the explicit calculation of the Weyl function $\langle \alpha | D(\xi e^{i\omega_1 t}) | \alpha \rangle$ appearing in eq. (7) and exploiting the Fourier-Bessel expansion of $\exp[izsin\theta]$, i is easy to show that

$$\frac{\langle I \rangle_{dc}^{(n)}}{I_{cr}} = e^{-\xi^2}(-1)^n J_n\left(2\xi|\alpha|\right)sin\left[n\left(\theta_\alpha + \frac{\pi}{2}\right)\right]Tr_J\left(\rho_J(0)e^{i\varphi_0}\right). \tag{9}$$

In eq. (9), valid under the nth resonance condition $\omega_0 = \pm n\omega_1$, J_n is the Bessel function of order n and $\langle I \rangle_{dc}^{(n)}$ is the dc component of expectation value of the supercurrent operator (the nth QSS).

In principle, as in the classical case, we can extract the field parameters $|\alpha|$ and the θ_α from the knowledge of the two quantities $\frac{\langle I \rangle_{dc}^{(2)}}{\langle I \rangle_{dc}^{(1)}}$ and $\frac{\langle I \rangle_{dc}^{(3)}}{\langle I \rangle_{dc}^{(1)}}$ as given by eq.(9).

However, since the amplitude of steps of order n≥2, in the case of low-intensity fields, is very small (typically < nA), it is in practice very hard to realize such an experiment. Fortunately, the quantumness of the supercurrent crossing a mesojunction suggests the idea of investigating on the field parameter dependence of the quantum noise characterizing a QQS. The aim is that of constructing analytical links other than eq. (9) between the irradiating field and relatively easier-to measure physical quantities.

Our main result is that, in accordance to such an approach, it is possible to extract the intensity $|\alpha|$ and the phase θ_α of our field state taking as a starting point *measurements on the first QSS only*. In fact, neglecting instrumental errors and exploiting the fact that, at $T \sim (10 \div 100)mK$, quantum fluctuation are predominant with respect the thermal ones, we can relate also the dc supercurrent quantum noise

$$\left(\Delta I^{(n)}\right)_{dc}^2 = \langle I^2 \rangle_{dc}^{(n)} - \left[I_{dc}^{(n)}\right]^2 \tag{10}$$

to the field parameters $|\alpha|$ and θ_α.

This approach leads to the following analytical expression of the dc component of the expectation value of $I^2(t)$

$$\frac{\langle I^2 \rangle_{dc}^{(n)}}{I_{cr}^2} = 2\left\{1 - Tr_J\left[\rho_J(0)e^{i2\varphi_0}\right]e^{-2\xi^2}J_{2n}\left(4\xi|\alpha|\right)cos\left[2n\left(\theta_\alpha + \frac{\pi}{2}\right)\right]\right\} \tag{11}$$

which together with eq. (9) allows the evaluation of $|\alpha|$ and θ_α.

In view of eq. (9) and (11), it is however necessary to fix the initial condition of the Josephson junction. It is very difficult to prepare the junction in a prefixed pure state but, due to the very low values of the operating temperature, it is reasonable to assume that, in the harmonic regime,

$$\rho_J(0) = \sum_{k=0}^{N} p_k|k\rangle\langle k|, \tag{12}$$

where p_k is the probability that the junction is in its energy level $|k\rangle$ and where we can put $N = 2$ because the Boltzmann weight p_0, as it is easy to check, is much larger than the other ones.

Summing up, starting from the knowledge of the experimental value of the amplitude and the quantum noise characterizing the first QSS and substituting the Bessel functions with its asymptotic expansion for low arguments ($\xi|\alpha| \approx 10^{-3} \ll 1$), we obtain the following approximated equations

$$|\alpha|^2 \approx \frac{\left(1 - 2\frac{\left\langle I^2 \right\rangle_{dc}^{(1)}}{I_{cr}^2}\right)e^{2\xi^2}}{2\xi^2 Tr_J\left[\rho_J(0)e^{i2\varphi_0}\right]} + \frac{2e^{\xi^2}\left[\left\langle I \right\rangle_{dc}^{(1)}\right]^2}{Tr_J\left[\rho_J(0)e^{i\varphi_0}\right]\xi^2 I_{cr}^2} \qquad (13)$$

$$\cos\left[2\left(\theta_\alpha + \frac{\pi}{2}\right)\right] \approx \frac{\left(1 - 2\frac{\left\langle I^2 \right\rangle_{dc}^{(1)}}{I_{cr}^2}\right)e^{2\xi^2}}{2\xi^2 Tr_J\left[\rho_J(0)e^{i2\varphi_0}\right]}\frac{1}{|\alpha|^2} \qquad (14)$$

that can be directly related to the value of the field amplitude $|\alpha|$ and the field phase θ_α. Our results confirm the potentialities of the mesoscopic Josephson junctions as probe for quantized coherent fields and their versatility both in the context of fundamental researches and in the field of applied physics.

ACKNOWLEDGMENTS

The financial support of CRRNSM and Murst 60% is gratefully acknowledge. The authors wish to thank M. Cirillo, G. Compagno and A. Vourdas for stimulating discussions. R. M. expresses gratitude to M. Cirillo for his hospitality.

REFERENCES

1. S. Shapiro, J. Appl. Phys. 38, 1879 (1967).
2. S. Shapiro, A. R. Janus and S. Holly, Rev. Mod. Phys. 36, 223 (1964).
3. A. Vourdas, PR B 49, 10040 (1994).
4. A. Vourdas, Z. Phys. B 100, 455 (1996).
5. A. Vourdas and T. P. Spiller, Z. Phys. B 102, 43 (1997).
6. R. Migliore, A. Messina, A. Napoli, *Quantum interference effects in mesoscopic Josephson junction*, in *Mysteries, Puzzles and Paradoxes in Quantum Mechanics*, Edited by R. Bonifacio (Am. Inst. Of Phys., NY 1999).
7. R. Migliore, *Risposta di una giunzione Josephson mesoscopica irradiata con sovrapposizioni quantistiche di stati elettromagnetici coerenti*, Thesis, University of Palermo 1999.
8. R. Migliore, A. Messina, A. Napoli, *Detecting quantum signatures of optical fields by ultrasmall Josephson junctions*, to be published in EPJ B (1999).
9. G. Schön, A. D. Zaikin, Phys. Rep. 198, 237 (1990).
10. K. K. Likharev, A. B. Zorin, J. Low Temp. Phys. 58, 347 (1985).
11. R. Fazio, A. Tagliacozzo, *Quantum Fluctuation and superconductivity*, in *Advances in Quantum Phenomena*, Edited by E. G. Beltrametti and J. M. Levy-Leblond (Plenum Press NY 1995).

Active Site Conformation In The αH87G Mutant Hemoglobin: An Optical Absorption And FTIR Study

Valeria Militello°, Maurizio Leone^, Clara Fronticelli*
and Antonio Cupane^

° INFM and Istituto di Fisiologia Umana, University of Palermo, 90134 Palermo, ITALIA
^ INFM and Dipartimento Scienze Fisiche ed Astronomiche, University of Palermo,
90123 Palermo, ITALIA
*Department of Biological Chemistry, Medical School of UMAB, Baltimore, 21201 Maryland, USA

Abstract. We have studied the active site conformation in the carbonmonoxy derivative of the αH87G mutant hemoglobin by means of optical absorption and FTIR spectroscopies. A red shift (≈ 30 cm^{-1}) of the Soret band peak frequency, together with a concomitant red shift (≈ 2 cm^{-1}) of the bound CO stretching frequency has been observed for the mutant protein. This indicates an altered electrostatic environment of the heme group in the mutated subunits. In view of the FTIR data showing that the bound CO molecule experiences an increased positive electrostatic field, we attribute the observed effects to a closer interaction of the CO ligand with the partially positively charged imidazole side chain of the proximal histidine.

INTRODUCTION

Site directed mutagenesis has proven to be a powerful approach to study the effect of single aminoacid on the structure-function relation in proteins. In particular, the αH87G human mutant hemoglobin, in which the α chains proximal histidines are substituted by glycines has been studied in some detail (1,2). In fact, in the α chains of this mutant hemoglobin, the covalent linkage between heme iron and protein is missing, so that the possibility is offered to investigate the role of the iron-proximal histidine bond on the conformational/functional equilibria and on the dynamics of this protein.

In the mutated chains the sixfold iron coordination is preserved by adding imidazole as an exogenous ligand; imidazole is a close homologue of the histidine side chain and has been shown by X-ray crystallography to bind at the proximal side (1); however the presence of imidazole as the sixth ligand and the lack of the covalent iron-proximal histidine bond may introduce conformational alterations.

CP513, *Nuclear and Condensed Matter Physics*, edited by A. Messina
© 2000 American Institute of Physics 1-56396-929-7/00/$17.00

In a preceding work, a temperature independent peak frequency shift of the Soret band had been observed for the αH87G mutant and was attributed to a different geometry of the iron-imidazole with respect to the iron-proximal histidine linkage (2); however, possible effects concerning the distal side of the heme pocket were not investigated.

In this paper, the information obtained with optical absorption spectroscopy in the Soret region has been complemented with data obtained with FTIR spectroscopy, which is a fine probe of the electrostatic environment of the bound CO and therefore of the distal side of the heme pocket.

MATERIALS AND METHODS

Mutant Hemoglobin Preparation

Bacterial growth, expression and purification of the recombinant protein, reconstitution and assembly into tetrameric hemoglobin have already been described (3). Site directed mutagenesis for replacing the proximal histidine with glycine was carried out using the pAlter Mutagenesis System of Promega Corporation.

Optical Absorption and FTIR Spectra

Optical absorption spectra in the Soret region (500-350 nm) were recorded in digitized form at 0.5 nm intervals with a PC-IBM controlled Cary Varian 2300 spectrophotometer. The scan speed was 0.5 nm/sec, the integration time was 0.5 sec and the spectral bandwidth was less than 0.2 nm in the whole wavelength range, corresponding to a spectral resolution of about 15 cm^{-1} at 420 nm. Samples for optical absorption measurements were prepared by diluting concentrated protein stocks in a 65% v/v glycerol/water mixture saturated with CO; they contained 0.1 M phosphate buffer pH 7, 10mM imidazole and about 3×10^{-4} M sodium dithionite (to ensure full reduction of the protein); the final protein concentration was about 10^{-5} M in heme. The spectra at each temperature were analyzed using an approach described in detail elsewhere (4); this enabled an unambiguous determination of the 0-0 electronic transition frequency (peak frequency of the band).

IR spectra were measured at room temperature using a Bio-Rad FTS-40A FTIR spectrophotometer equipped with a PbS detector. Samples were placed in a Specac cell with CaF$_2$ windows and a 0.025 mm spacer. The single beam spectra in the wavenumber range 1800-2400 cm^{-1} were measured with 256 scans at 1 cm^{-1} resolution. The absorption spectra of native and mutant hemoglobin were obtained by the subtraction between sample spectrum and solvent spectrum, after suitable normalization. The protein concentration used was about 5mM.

175

RESULTS AND DISCUSSION

Figure 1 shows the temperature dependence of the peak frequency of the Soret band for the αH87G mutant, in comparison with the native protein. As can be seen, for both proteins a peak frequency blue shift occurs as the temperature is lowered; this indicates the presence of quadratic coupling between the electronic transition and the low frequency modes of the system (2). However, a temperature independent red shift of about 30 cm^{-1} is observed for the mutant with respect to the native protein. This has to arise from a static rather than from a dynamic effect suggesting that in the mutated α chains the chromophore experiences an altered local electric field.

Figure 2 shows the FTIR spectra at 296 K of the mutant and native hemoglobin. As can be seen, the stretching band(s) of the CO molecule are shifted by about 2 cm^{-1} with respect to native hemoglobin. A red shift of the CO stretching band is indicative of a decreased C-O bond order and, in view of the well established negative correlation of an increased Fe-C bond order (4). According to the interpretation of CO stretching frequencies, this effect has to be attributed to an increased positive electrostatic field which enhances π back-bonding from the dπ−orbitals to the strongly antibonding COπ* molecular orbitals and therefore decreases the order of the C-O bond and the C-O stretching frequency.

Data obtained with optical absorption and FTIR spectroscopy are therefore in agreement, in that they consistently indicate an altered electrostatic environment of the heme in the mutated subunits; moreover FTIR spectroscopy shows that this is due to an increased positive electrostatic field experienced by the bound CO molecule. In view of the new FTIR data and at difference with our previous interpretation (2), we suggest that increased interaction of the bound CO with the (partially) positively charged

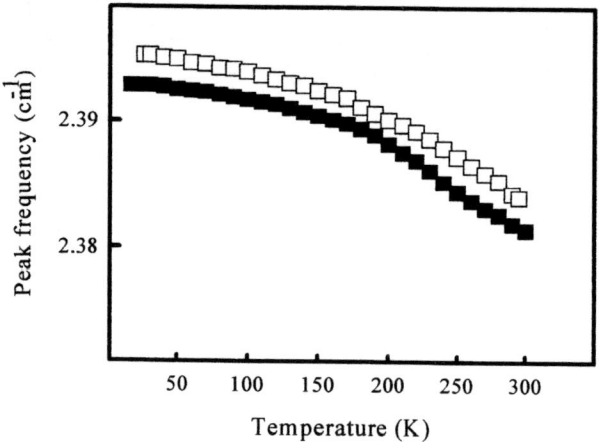

FIGURE 1. Temperature dependence of Soret band peak frequency for the CO derivatives of αH87G (■) and native hemoglobin (□).

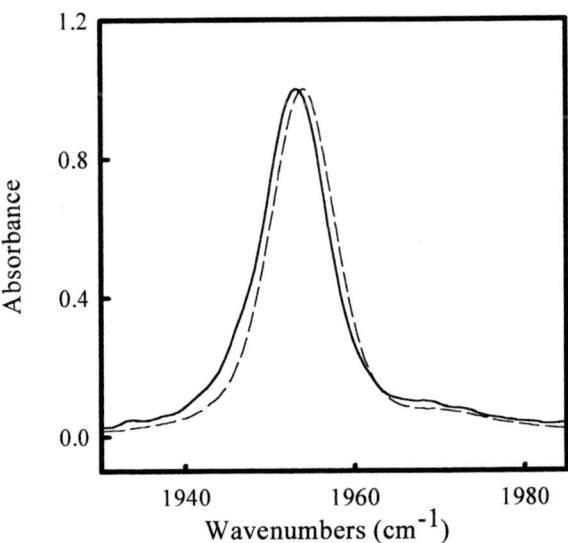

FIGURE 2. FTIR spectra of mutant (continuous line) and native hemoglobin (dashed line) at room temperature.

imidazole side chain of the distal histidine may be responsible for the observed effects. This confirms the relevance of the non covalent interactions between the heme and the aminoacid residues in the distal side of the heme pocket in determining the conformational and dynamic properties of hemoglobin.

ACKNOWLEDGMENTS

This work has been supported by Italian CNR grants n° 97.04350CT14.

REFERENCES

1. Barrick, D., Ho, N.T., Simplaceanu, V., Dahlquist, F.W. and Ho, C., *Nature Struct. Biol.* **4**, 78-83 (1997).
2. Cupane, A., Leone, M., Militello, V. and Fronticelli, C., "Effect of the Covalent Fe-Protein Linkage on the Iron-Porphyrin Dynamics" in *Biological Physics*, edited by H. Frauenfelder et al., AIP Conference Proceedings 487, New York: American Institute of Physics, 1999, pp. 139-146.
3. Sanna, M. T., Razynska, A., Karavitis, M., Koley, A. P., Friedman, F. K., Russu, I. M., Brinigar, W. S., and Fronticelli, C., *J. Biol. Chem.* **272**, 3478-3486 (1997).
4. Karavitis, M., Fronticelli, C., Brinigar, W. S., Vasquez, G. B., Militello, V., Leone, M., and Cupane A, *J.Biol. Chem.* **273**, 23740-23749 (1998).

Miniaturised Optical Devices Produced By Electron Beam Lithography In Lithium Fluoride Films

R. M. Montereali

ENEA, Centro Ricerche Frascati, P.O.Box 65, 00044 Frascati (RM), Italy

Abstract. The use of versatile, well-assessed and low-cost fabrication techniques consisting in physical vapor deposition of LiF films combined with an electron-beam direct writing lithographic process allows the realization of optically confined active structures, like broad-band emitters, channel waveguides and optical microcavities operating in the visible.

INTRODUCTION

In recent years an increasing interest has been growing for novel materials exhibiting active optical properties [1]. The miniaturisation of lasers is a key objective of optoelectronics and new configurations allowing a higher integration with the integrated optics technologies are under investigation.

Alkali halide crystals containing colour centres (CCs) are well known active media in optically pumped tunable solid state lasers [2] with efficiency close to unity. Among them lithium fluoride (LiF) stands apart because is practically not hygroscope and it can host different types of point defects laser active in the visible and in the near-infrared [3] even at room temperature (RT). They can be produced only by bombardment with ionising radiations.

Solid state lasers based on these crystals, however, even if compact, do not lend themselves to an easy integration with optical fibers and waveguides. In the frame of development of innovative miniaturised coherent light sources we investigated the optical properties of LiF films thermally evaporated on amorphous and crystalline substrates optically activated by low energy electron beam bombardment [4].

ELECTRON BEAM LITHOGRAPHY ON LITHIUM FLUORIDE

Low energy (2 – 20 keV) electron irradiation induces the efficient formation of stable CCs located at the surface of LiF films and crystals. Among them, the primary F centre (an electron trapped in an anion vacancy) and the F_3^+ and F_2 (two electrons bound to three and two anion vacancies, respectively) aggregate ones. These laser active defects show intense green and red emissions when excited in their almost overlapping absorption bands located around 450 nm.

CP513, *Nuclear and Condensed Matter Physics,* edited by A. Messina
© 2000 American Institute of Physics 1-56396-929-7/00/$17.00

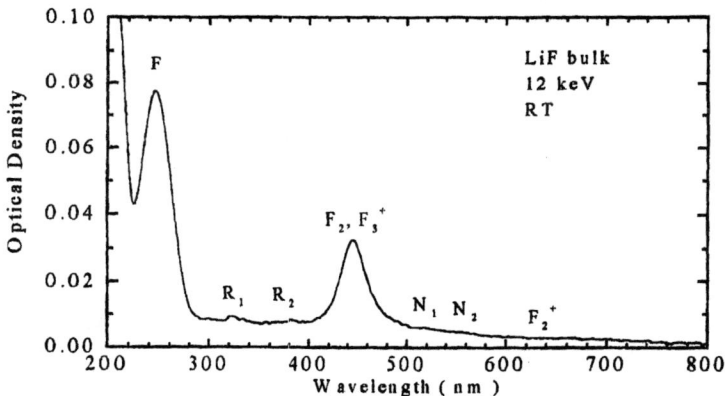

FIGURE 1. RT optical absorption spectrum of a LiF crystal irradiated by a 12 keV electron beam.

A typical RT absorption spectrum of a LiF crystal irradiated at RT by a 12 keV electron beam is reported in Fig.1. Two main absorption bands peaking at 245 nm (attributed to F centres) and at 450 nm (due to the unresolved F_3^+ and F_2 bands) are observed. Other minor contributions due to the absorption bands of more complex aggregate defects can be detected. The electron penetration depth in LiF ranges from 0.1 to 3.7 μm for energies between 2 and 20 keV. By using the Smakula formula with a reasonable oscillator strength of 0.5 for the F_3^+ and F_2 centres, and by assuming the defect densities to be constant along the estimated penetration depth of 1.57 μm for 12 keV electrons, a concentration of the order of 5×10^{18} cm^{-3} is obtained.

The use of a scanning electron microscope (SEM) equipped with a lithographic system allows a direct writing of predefined patterns onto the LiF surface. Several stripes from 2 to 145 μm wide were defined. By exciting these coloured stripes by blue light from an Argon laser, an intense visible photoluminescence is clearly observed by naked eyes. The emission spectra excited by the 458 and 476 nm lines of an Ar laser at RT of a coloured stripe induced by 12 keV electrons at the surface of a LiF film 2.6 μm thick thermally evaporated on a glass substrate are shown in Fig.2. The luminescence is composed of two bands peaking around 530 and 660 nm, which correspond to the known positions of F_3^+ and F_2 centres.

Active Waveguides Realization On LiF Crystals and Films

Besides the efficient defects formation, the electron bombardment of LiF by a beam of energy in the range of few keV induces an increase of the real part of the refractive index of the irradiated layer in the same wavelength interval where the F_3^+ and F_2 emissions are located. The proper choice of the irradiation conditions, primarily dose and dose-rate once fixed the electron energy, and consequently the depth of the irradiated layer, allows to support at least one propagating mode at the emission wavelengths of the defects laser active in the visible.

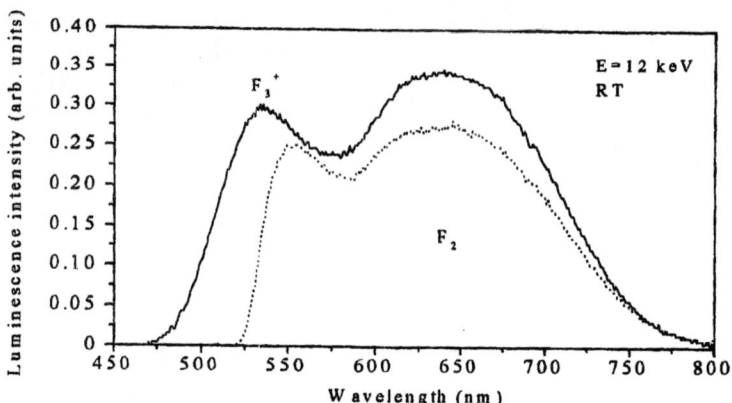

FIGURE 2. RT emission spectra excited by the 458 (solid) and 476 nm (dashed) lines of an Ar laser of a coloured stripe induced by 12 keV electrons on a LiF film thermally evaporated on a glass substrate.

The waveguiding properties of photoluminescent stripes realised by electron beam lithography have been recently demonstrated for the first time in LiF single-crystals [5] and films [6] thermally evaporated on LiF bulk. An increase of the refractive index larger than 5×10^{-3}, suitable for light confinement, was achieved by irradiation with 12 keV electron beams. Losses of the order of 6dB/cm have been measured at 632.8 nm. These values are promising if compared with sizeable optical gain coefficients around 25 dB/cm measured by ASE technique on coloured stripes induced by electron beam lithography on LiF films grown on glass substrates.

Perspectives For LiF Based Active Waveguides on Silicon

The direct-writing lithographic process, that induces at the same time the waveguiding structure and the photoluminescent colour centres, looks like a promising method to design a waveguiding structure on LiF, limited by its low refractive index, ≈ 1.39 at 633 nm, which prevents the use of the most common substrate materials. For this reason the realisation of LiF film based active waveguides on substrates of higher refractive index with respect to silica looks like a more complex task. With the aim of full compatibility with the assessed silicon technologies, coloured stripes have been realised by the same electron lithography technique on LiF films thermally evaporated on a SiO_2 buffer of thickness 11.1 μm grown on a commercial <100> silicon wafer. The RT photoluminescence spectrum is reported in Fig.3. A well resolved interference pattern is superimposed to the expected emission spectrum from the investigated LiF structure and can be ascribed to the presence of the thick SiO_2 optical buffer. A proper choice of its thickness combined with the induced change in the refractive index of the coloured LiF region could allow to design a reliable optically confined structures in the above configuration.

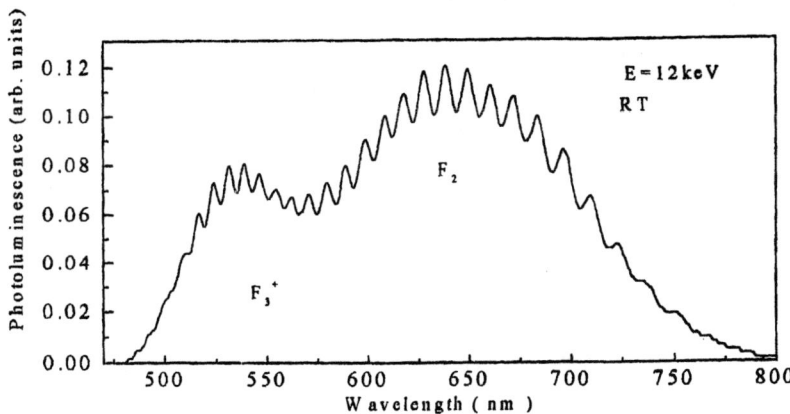

FIGURE 3. RT photoluminescence spectrum excited at 458 nm of a colored stripe induced by 12 keV electrons on a LiF film grown on thick SiO_2 on silicon.

Perspectives For Optical Microcavities Based On Colored LiF Films

In recent times, a lot of attention has been devoted to the investigation of spontaneously emitted radiation modifications inside resonators having at least one size dimension's comparable with the emitted wavelength. Inside a microcavity, the optical confinement in one or more dimensions changes the distribution of the electromagnetic field, thus enabling one to control the emission properties of an emitting materials placed inside such resonators. Recently the fabrication of planar microcavities based on low energy electron beam irradiated thin LiF films evaporated on Bragg reflectors has been achieved. The realization of bidimensional confined structures can be foreseen combining the two approaches, i.e. vertical confinement by passive optical elements and side confinement by electron beam induced patterns.

REFERENCES

1. P.Vincenzini, G.C.Righini, *Innovative Ligth Emitting Materials*, Faenza,Techna Srl, 1999.

2. L.F.Mollenauer, "Color Center Lasers", in *Tunable Lasers*, edited by L.F.Mollenauer and J.C.White, Berlin, Springer Verlag, 1987, pp.225-278.

3. V.V. Ter-Mikirtichev ,T.Tsuboi, *Prog.Quant.Electr.*20,3, 219-268 (1996).

4. R.M.Montereali, G.Baldacchini, L.C.Scavarda do Carmo, *Thin Solid Films* **205**, 106-108 (1991).

5. R.M.Montereali, A.Mancini, G.C.Righini, S.Pelli, *Opt. Comm.* **153**, 223 -225 (1998).

6. L.Fornarini, S.Martelli, A.Mancini, G.C.Righini, S.Pelli, Proceedings of the 9th European Conference on Integrated Optics, ECIO'99, Torino -Italy, April 14-16, 1999, pp.343-346.

On the Possibility of Field-State Reconstruction in Non-Ideal Cavities

H. Moya-Cessa,*† A. Vidiella-Barranco,** P. Tombesi* and J.A. Roversi**

*Dipartimento di Matematica e Fisica, Università di Camerino, and INFM Unità Camerino
I-62032, Camerino (MC), Italy
†INAOE, Coordinación de Optica, Apdo. Postal 51 y 216,
72000 Puebla, Pue., Mexico
**Instituto de Física "Gleb Wataghin", Universidade Estadual de Campinas,
13083-970 Campinas SP Brazil

Abstract. We present a scheme to reconstruct the quantum state of a field prepared inside a lossy cavity at finite temperature. Quantum coherences are normally destroyed by the interaction with an environment, but we show that it is possible to recover complete information about the initial state (before dissipation), making possible to reconstruct its Wigner function.

INTRODUCTION

Recently there have been proposals to reconstruct the quantum state of electromagnetic fields inside cavities. The reconstruction of non-classical states is a central topic in quantum optics and related fields and there have been a number of proposals to achieve it [1–3].

We consider here a single mode high-Q cavity were we suppose that a nonclassical field is prepared. The first step of our method consists in driving the generated state by a coherent pulse. The reconstruction of the field is done after turning-off the driving field, i.e. at a time when the cavity field has interacted with the non-zero environment.

We then show that by measuring the density matrix diagonal elements we can obtain directly the Wigner function. We should remark that to know a state, one has to have information about all the density matrix elements (diagonal and off-diagonal), however, with the method presented here, it is only necessary to have information about the diagonal displaced-decayed matrix elements.

CP513, *Nuclear and Condensed Matter Physics,* edited by A. Messina
© 2000 American Institute of Physics 1-56396-929-7/00/$17.00

THE MASTER EQUATION AND ITS SOLUTION

The master equation in the interaction picture for the reduced density operator $\hat{\rho}$ relative to a driven cavity mode, taking into account cavity losses at non-zero temperature and under the Born-Markov approximation is given by

$$\frac{\partial \hat{\rho}}{\partial t} = (\hat{\mathcal{R}} + \hat{\mathcal{L}})\hat{\rho}, \tag{1}$$

where

$$\hat{\mathcal{L}}\hat{\rho} = (\hat{\mathcal{L}}_1 + \hat{\mathcal{L}}_2)\hat{\rho} \tag{2}$$

with

$$\hat{\mathcal{L}}_1\hat{\rho} = \frac{\gamma(\bar{n}+1)}{2}\left(2\hat{a}\hat{\rho}\hat{a}^\dagger - \hat{a}^\dagger\hat{a}\hat{\rho} - \hat{\rho}\hat{a}^\dagger\hat{a}\right), \quad \hat{\mathcal{L}}_2\hat{\rho} = \frac{\gamma\bar{n}}{2}\left(2\hat{a}^\dagger\hat{\rho}\hat{a} - \hat{a}\hat{a}^\dagger\hat{\rho} - \hat{\rho}\hat{a}\hat{a}^\dagger\right), \tag{3}$$

and

$$\hat{\mathcal{R}}\hat{\rho} = -\frac{i}{\hbar}[\hat{H}, \hat{\rho}] \tag{4}$$

where

$$\hat{H} = i\hbar\left(\alpha^*\hat{a} - \alpha\hat{a}^\dagger\right), \tag{5}$$

where $\hat{a}$ and $\hat{a}^\dagger$ are the annihilation and creation operators, γ the (cavity) decay constant and α the amplitude of the driving field.

It is not difficult to show that

$$[\hat{\mathcal{R}}, \hat{\mathcal{L}}]\hat{\rho} = \frac{\gamma}{2}\hat{\mathcal{R}}\hat{\rho}, \tag{6}$$

and the formal solution of Eq. (1) can then be written as [4]

$$\hat{\rho}(t) = \exp\left[(\hat{\mathcal{R}} + \hat{\mathcal{L}})t\right]\hat{\rho}(0) = \exp(\hat{\mathcal{L}}t)\exp\left[-\frac{2\hat{\mathcal{R}}}{\gamma}(1 - e^{\gamma t/2})\right]\hat{\rho}(0). \tag{7}$$

After driving the initial field during a time t_d, the resulting field density operator will read

$$\hat{\rho}(t_d) = e^{\hat{\mathcal{L}}t_d}\hat{\rho}_\beta(0), \tag{8}$$

where

$$\hat{\rho}_\beta(0) = \hat{D}^\dagger(\beta)\hat{\rho}(0)\hat{D}(\beta), \tag{9}$$

and with

$$\beta = -2\alpha \frac{1 - e^{\gamma t_d/2}}{\gamma}. \tag{10}$$

We now obtain the density matrix (7), by defining

$$\hat{J}_-\hat{\rho} = \hat{a}\hat{\rho}\hat{a}^\dagger, \quad \hat{J}_+\hat{\rho} = \hat{a}^\dagger\hat{\rho}\hat{a}, \quad \hat{J}_3\hat{\rho} = \hat{a}^\dagger\hat{a}\hat{\rho} + \hat{\rho}\hat{a}^\dagger\hat{a} + \hat{\rho}, \tag{11}$$

where the superoperators $\hat{J}_-$, $\hat{J}_+$ nd $\hat{J}_3$ obey the commutation relations $[\hat{J}_-, \hat{J}_+]\hat{\rho} = \hat{J}_3\hat{\rho}$ and $[\hat{J}_3, \hat{J}_\pm]\hat{\rho} = \pm 2\hat{J}_\pm\hat{\rho}$

$$\hat{\rho}(t) = e^{\frac{\gamma t}{2}} e^{\Gamma_{\hat{n}}(t)\hat{J}_+} \left[\frac{e^{-\gamma t/2}}{1 + N_t} \right]^{\hat{J}_3} e^{\Gamma_{\hat{n}+1}(t)\hat{J}_-} \hat{\rho}_\beta(0), \tag{12}$$

where

$$\Gamma_{\hat{n}}(t) = \frac{\bar{n}(1 - e^{-\gamma t})}{1 + N_t}, \quad \Gamma_{\hat{n}+1}(t) = \frac{(\bar{n} + 1)(1 - e^{-\gamma t})}{1 + N_t}, \tag{13}$$

and where we have defined $N_t = \bar{n}(1 - e^{-\gamma t})$.

THE METHOD

Next we calculate the diagonal density matrix elements $< m|\hat{\rho}(t)|m >$ from (8) to obtain

$$\langle m|\hat{\rho}_\beta(t)|m \rangle = \frac{1}{(1 + N_t)} \sum_{k=0,n=0}^{\infty} \binom{k}{n} \binom{m}{n} \frac{[\Gamma_{\hat{n}}(t)]^{m-n}[\Gamma_{\hat{n}+1}(t)]^{k-n} e^{-n\gamma t}}{[1 + N_t]^{2n}} P_k(\beta), \tag{14}$$

where $P_k(\beta) = \langle k|\hat{\rho}_\beta(0)|k \rangle$. Multipliying by χ_s^m, where

$$\chi_s = \frac{\frac{s+1}{s-1} - \Gamma_{\hat{n}+1}(t)}{\frac{e^{-\gamma t}}{[1+N_t]^2} + \Gamma_{\hat{n}}(t) \left(\frac{s+1}{s-1} - \Gamma_{\hat{n}+1}(t) \right)} \tag{15}$$

and adding over m we obtain

$$F(\beta; s) = \sum_{m=0}^{\infty} \chi^m \langle m|\hat{\rho}_\beta(t)|m \rangle = \frac{1}{[1 + N_t][1 - \chi_s\Gamma_{\hat{n}}(t)]} \sum_{k=0}^{\infty} \left(\frac{s+1}{s-1} \right)^k P_k(\beta), \tag{16}$$

Finally, multypliying $F(\beta; s)$ by

$$-\frac{2[1 + N_t][1 - \chi_s\Gamma_{\hat{n}}(t)]}{\pi(s - 1)}, \tag{17}$$

we obtain

184

$$W(\beta; s) = -\frac{2}{\pi(s-1)} \sum_{k=0}^{\infty} \left(\frac{s+1}{s-1}\right)^k \langle k|\hat{\rho}_\beta(0)|k\rangle, \qquad (18)$$

which is the s-parametrized quasiprobability distribution [5].

Therefore, by measuring the diagonal elements of the displaced and interacted density matrix, Eq. (8) one obtains complete information on the initial state (there are many techniques to achieve such information, for one of them see [2]).

CONCLUSIONS

In conclusion, we have presented, a method to reconstruct the Wigner function of an initial nonclassical state at times when the field would have normally lost its quantum coherence. The crucial point of our approach is the driving of the initial field immediately after preparation, that is not only used to cover a region in phase space but also to store quantum coherences in the diagonal elements of the time evolved displaced density matrix, making them robust. In other words, we have shown that the initial displacement transfers the robustness of a coherent state against dissipation to any initial state [6].

REFERENCES

1. Lutterbach,L.G., and Davidovich, L., Phys. Rev. Lett. **78**, 2547 (1997).
2. Moya-Cessa, H., Dutra, S.M., Roversi, J.A., and Vidiella-Barranco, A., J. Mod. Opt. **46**, 555 (1999).
3. Bardroff, P.J., Leichtle, C., Schrade, G., and Schleich, W.P., Phys. Rev. Lett. **77**, 2198 (1996).
4. Arévalo-Aguilar, L.M., and Moya-Cessa, H., Quant. Semiclass. Opt. **10**, 671 (1998).
5. Moya-Cessa, H., and Knight, P.L., Phys. Rev. A **48**, 2479 (1993).
6. Zurek, W.H., Habib, S., and Paz, J.P., Phys. Rev. Lett. **70**, 1187 (1993).

Schrödinger Cat States of Two Bosonic Modes

A. Napoli and A. Messina

INFM and Dipartimento di Scienze Fisiche ed Astronomiche
Via Archirafi 36, 90123 Palermo, ITALY

Abstract. The possibility of generating bosonic bimodal Schrödinger cat states is briefly demonstrated. The experimental scheme here presented relies on the existence of an intrinsically nonclassical effect stemming from the granularity of the quantized electromagnetic field.

Recent years have witnessed a growing interest in studying the paradoxical aspects of quantum mechanics whose understanding remains mysterious up to now. One of the most striking issue in this context is the possibility of drawing the boundary between the microscopic quantum world and the macroscopic classical one. This problem is strictly related to the quantum measurement theory [1] and involves fundamental properties of quantum mechanics itself. Notwithstanding macroscopic objects are made of atoms individually obeying quantum laws, the application of such rules to macroscopic systems gives rise to puzzling behaviours of the nature [2]. As pointed by Einstein [3] the majority of states allowed by quantum mechanics do not exist at the classical level. Indeed, while in the quantum world one frequently comes across coherent superpositions of states, macroscopic quantum superpositions are never observed. This problem was vividly illustrated by E. Schrödinger [4] in 1935 by means of his quite famous cat paradox. With reference to this metaphor, the nonclassical linear combination of macroscopically distinguishable quantum states are often called "Schrödinger cat states". Until few years ago, the possibility of realizing such states in laboratory appeared as an utopian idea only. The great progress made either in laser cooling and trapping of atoms or in the realization of high-Q superconducting microcavities, as well as in the preparation and detection of single Rydberg atoms, offers at least two different physical domains wherein the preparation of nonclassical states, such as Schrödinger cat states, becomes realistic. These two different physical contexts are characterized by an interesting similarity: they both realize simple situations in which an effectively two-level system is coupled to a few bosonic modes. This circumstance is of particular relevance because an hamiltonian model developed in one of the two domains may be successfully used to study correspondent physical situations in the other one. Over the last ten years, there have been many proposals in both contexts aimed at generating nonclassical states having the form of quantum superpositions of classically distinguishable states [5]. At the same time, these proposals provide an useful starting point to progress with the understanding of decoherence phenomena and to learn more about the elusive border between the classical and quantum worlds. In this paper we present the main ingredients of a new theoretical scheme for the preparation of quantum states of two

CP513, *Nuclear and Condensed Matter Physics,* edited by A. Messina
© 2000 American Institute of Physics 1-56396-929-7/00/$17.00

bosonic modes obtained as superpositions of two macroscopically distinguishable bimodal states. We shall, in fact, show that the two contributing terms of the linear combination to be generated may be clearly distinguished by virtue of a marked difference between the expectation values of a physical variable easily interpretable. Following the literature, we have called these highly nonclassical states "bimodal Schrödinger cat states". The scheme we are going to describe requires a single effective two-level Rydberg atom and a degenerate bimodal cavity characterized by a high-Q quality factor. Here "degenerate" means that the resonator supports two independent modes having the same frequency ω but different polarization vectors and/or directions of propagation. Let's indicate by ω_0 the Bohr frequency of the two-level atom and suppose that the two-photon resonance relation, $\omega_0 = 2\omega$, is satisfied. We have demonstrated [6] that, under appropriate conditions concerning both the energy atomic spectrum and the cavity field, when the atom goes through the resonator, a complex atom-field coupling mechanism takes place. The effective Hamiltonian model successfully describing all the different matter-radiation energy exchange channels may be cast in the form [6]:

$$H = \hbar\omega_0 S_z + \hbar\omega \sum_{\mu=1}^{2} a_\mu^\dagger a_\mu + s S_z \sum_{\mu=1}^{2} a_\mu^\dagger a_\mu + \left[\left(r_1 a_1 a_2^\dagger + r_2 a_1 a_2^\dagger S_z \right) + h.c. \right] +$$

$$+ \left[\lambda \left(a_1^2 - a_2^2 \right) S_+ + h.c. \right] \tag{1}$$

The dynamics of the system described by H can be exactly treated thus bringing to the light the existence [6] of an intrinsically quantum phenomenon strictly connected with the granularity of the radiation field. This nonclassical property in the time evolution of the system, christened parity effect, can be successfully exploited to generate entangled bimodal states of the bimodal cavity field having the nature of Schrödinger cats. In order to describe the parity effect, let's suppose to inject a two-level atom, initially prepared in its ground state, into the cavity where one mode is excited in a state exactly containing n photons whereas the other one is left in its vacuum state. It is possible to prove that, after an appropriate n-dependent interval of time, t_n, our system exhibits a peculiar sensitivity to the parity of the initial total number of photons n. Stated another way, the physical system distinguishes between the two different conditions, n odd or n even, showing, after a proper time t_n, macroscopically different quantum behaviours. Suppose now to prepare the cavity leaving, once again, one mode in its vacuum state but now exciting the other one in a coherent state $|\alpha\rangle$.

Such an initial condition can be easily realized in laboratory. The vacuum state is, in fact, reached controlling the temperature at which the experiment is performed. On the other hand, it is possible to excite one mode in a coherent state $|\alpha\rangle$, having a prefixed intensity $|\alpha^2|$ using classical currents. Our experimental protocol starts injecting in the resonator the atom prepared in its ground state. In order to elucidate in which way parity effect previously recalled may be exploited for our goal, it is convenient to cast the initial condition imposed to the system as a sum of the following two contributions:

$$|\psi(0)\rangle = e^{-\frac{|\alpha|^2}{2}} \sum_{n=0}^{\infty} \frac{\alpha^{2n}}{\sqrt{(2n)!}} |2n\rangle|0\rangle|-\rangle + e^{-\frac{|\alpha|^2}{2}} \sum_{n=0}^{\infty} \frac{\alpha^{2n+1}}{\sqrt{(2n+1)!}} |2n+1\rangle|0\rangle|-\rangle \qquad (2)$$

The two terms appearing in the right member of eq. (2), are orthogonal because the first one is such that the initially excited mode is in a linear combination of even number states ($|2n\rangle$) only, whereas the second one contains all the odd number states ($|2n+1\rangle$) appearing in the Fock representation of a coherent state. In view of the linear character of quantum mechanics and as a consequence of the parity effect, the initial state $|\psi(0)\rangle$ evolves toward two macroscopically different states respectively stemming from the two orthogonal contributing terms appearing in eq. (2). We have, in fact, demonstrated that this forecast is correct, analytically evaluating, in particular, the instant of time, $t_{|\alpha|} = \frac{1}{2}\pi|\alpha|^2$, at which the system exhibits in the best way such parity effect-dependent consequences. The knowledge of $t_{|\alpha|}$ represents a key point in the implementation of our experimental scheme for the generation of bimodal Schrödinger cat states. We can, in fact, manipulate the atomic velocity in such a way that the interaction time between the cavity field and the two-level Rydberg atom exactly coincides with $t_{|\alpha|}$. Under this condition, it can be shown that, measuring the internal atomic state immediately after the atom leaves the cavity, the bimodal field is projected into a state having the desired character of a bimodal Schrödinger cat. We have, in particular, proved that, if the internal atomic state detector measures the atom in its ground state immediately after its flight trough the resonator, then the cavity field collapses into a naturalized state having the following form:

$$|\psi_-\rangle = A|\varphi\rangle + B|\chi\rangle \qquad (3)$$

The two orthogonal field states $|\varphi\rangle$ and $|\chi\rangle$ appearing in eq. (3) are macroscopically distinguishable in the sense that in the bimodal state $|\varphi\rangle$ the field energy is essentially concentrated in the initially empty mode whereas in the bimodal state $|\chi\rangle$ all the energy of the field may be found in the other mode.

This behaviour is clearly showed in fig. (1) where we report, as a function of the intensity $|\alpha|^2$, the mean value of the operator $\left(a_1^\dagger a_1 - a_2^\dagger a_2\right)$ in the state $|\varphi\rangle$, fig. 1 (a), and in the state $|\chi\rangle$, fig. 1 (b).

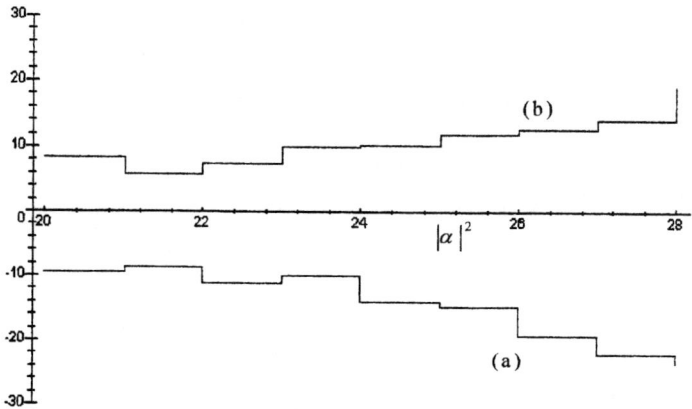

Fig. (1): Mean value of the operator $\left(a_1^\dagger a_1 - a_2^\dagger a_2\right)$ (a) in the state $|\varphi\rangle$ and (b) in the state $|\chi\rangle$.

This operator obviously defines the difference between the photon numbers contained in the initially excited mode (mode 1) and in the other one (mode 2) respectively. We can thus conclude that, exploiting the quite recently brought to the light quantum effect known as "parity effect", we may successfully generate Schrödinger cat states of the bimodal cavity field following all the steps of the procedure briefly reported in this paper.

ACKNOWLEDGEMENTS

The financial support of CRRNSM and MURST 60% are gratefully acknowledged.

REFERENCES

[1] A. Peres, *Quantum Theory: concepts and methods* (Kluwer Academic Publishers, Netherlands 1993);

[2] S. Haroche, *Phys. Today*, July 98, 36;

[3] W.H. Zurek et al, *Phys. Rev. Lett.* **70**, 1187 (1993);

[4] E. Schrödinger, in *Quantum theory and measurement*, J.A. Wheleer, W.H. Zurek eds. (Princeton University Press, 1983), p.152;

[5] X. Maitre et al, *J. Mod. Opt.* **44**, 2023 (1997); C. Monroe et al., *Science* **272**, May 96, 1131;

[6] A. Napoli, A. Messina, *Quantum and Semiclass. Opt.* **9**, 587 (1997).

Quantum Superpositions of Two Equal-Intensity Coherent States

A. Napoli and A. Messina

INFM and Dipartimento di Scienze Fisiche ed Astronomiche
Via Archirafi 36, 90123 Palermo, ITALY

Abstract. We propose an experimental scheme for the generation of quantum superpositions of two single-mode equal-intensity coherent states, characterized by the presence of infinite zeros in their photon number distributions.

Nonclassical properties of the electromagnetic field have received considerable attention in the past twenty years and continue up to now to be an attractive area of research. Intrinsically quantum features as quadrature squeezing, subpoissonian photon statistics, antibunching and so on, have their origin in the interference of the probability amplitudes stemming from one of the most fundamental principle of quantum mechanics: the "superposition principle" [1]. For this reason, the generation of linear combinations of single-mode radiation field states represents an important topic of modern physics providing, at the same time, an effective tool for the investigation of fundamental properties of quantum mechanics itself. Furthermore, from a more applicative point of view, the realization of inherently nonclassical pure states, may represent a significant advance in the modern field of quantum communication and quantum cryptography [2]. Cavity QED experimentalists have quite recently prepared a single-mode radiation field into special linear combinations of two equal-intensity out of phase coherent states [3]. Other groups have performed conceptually related experiments with trapped ions realizing, for example, even and odd coherent states of the quantum mechanically oscillating centre of mass of the ion [4]. Different superpositions of two coherent states, may exhibit different quantum signatures. For example, in the Yurke-Stoler state, the quantum interference between the coherent states $|\alpha\rangle$ and $|-\alpha\rangle$ determines higher order squeezing of the quadrature operators but leaves the photon numbers distribution of this state strictly Poissonian. On the contrary, the photon number distributions of the even and odd coherent states are characterized by oscillations in the sense they contain infinitely many zeros [1]. Generally speaking, the presence of infinite zeros in the photon number distribution of a given field state represents a clear signature of the quantum nature of the radiation field and may be thought of as a direct manifestation of interference between probability amplitudes. It is of relevance to stress that field states characterized by oscillations in their photon number distributions, exhibit quantum features which may be qualitatively and/or quantitatively sensitive to the specific infinite sequence of Fock states which they are orthogonal to. In order to make clear this statement, we may remember, for example, that the even and odd coherent states possesses

CP513, *Nuclear and Condensed Matter Physics,* edited by A. Messina
© 2000 American Institute of Physics 1-56396-929-7/00/$17.00

completely different nonclassical properties being characterized by complementary oscillatory Fock distribution.

At the light of the considerations above, it appears of interest the possibility of generating in laboratory linear superpositions of only two equal intensity coherent states of a single-mode radiation field exhibiting infinitely many zeros in their Fock representations. We have demonstrated [5] that a generic linear combination of two coherent states having the same intensity and characterized by the presence of infinite zeros, in the correspondent photon number distribution, may always be written down in the following form:

$$\left|\varepsilon_\alpha(m,r,n)\right\rangle = N_\alpha(m,r,n)\left(\left|\alpha e^{i\frac{m}{n}\pi}\right\rangle + e^{i\frac{r}{n}\pi}|\alpha\rangle\right) \tag{1}$$

where α is an arbitrary complex number, m is prime to n, r is odd if m is even and, finally, $0 < m < 2n$ and $0 \le r < 2n$. In eq. (1) $N_\alpha(m,r,n)$ is an appropriate normalization constant. The sequence of natural numbers characterizing the Fock states orthogonal to $\left|\varepsilon_\alpha(m,r,n)\right\rangle$ is an arithmetic progression of common difference n or $2n$ depending on the parity of the integer m.

In this paper we briefly propose an experimental scheme aimed at generating a particular subclass of states having the form (1). An aspect making the procedure here presented very interesting from an experimental point of view, is the possibility of varying the target state simply controlling an appropriate parameter having a clear physical meaning. The states we want to generate have the form:

$$\left|\varphi_\beta(p)\right\rangle = M_\beta(p)\left[|\beta\rangle - e^{-i\frac{2\pi}{p}}\left|\beta e^{-i\frac{2\pi}{p}}\right\rangle\right] \tag{2}$$

where $\beta \in C$ and $p > 1$ is an arbitrary integer. It is easy to verify that $\left|\varphi_\beta(p)\right\rangle = \left|\varepsilon_{\beta e^{-i\frac{2\pi}{p}}}\left(\frac{p}{2},1,\frac{p}{2}-1\right)\right\rangle$ if p is even whereas $\left|\varphi_\beta(p)\right\rangle = \left|\varepsilon_{\beta e^{-i\frac{2\pi}{p}}}(p,2,p-2)\right\rangle$ if p is odd, apart from a global phase factor.

The states belonging to the class (2) turn out to be orthogonal to the infinitely many Fock states $|kp-1\rangle$, with $k=1,2,\dots$. In other words the distance between two successive zeros in the photon number distributions correspondent to the states (2) is fixed and equal to p. As we have previously supposed, it is possible to pick out physically interesting features, related to the statistical properties of the states defined by eq. (2), showing a sensitivity to the integer p. For example, the Mandel parameter Q, defined by

$$Q = \frac{\left\langle\varphi_\beta(p)|a^\dagger a^\dagger aa|\varphi_\beta(p)\right\rangle - \left\langle\varphi_\beta(p)|a^\dagger a|\varphi_\beta(p)\right\rangle^2}{\left\langle\varphi_\beta(p)|a^\dagger a|\varphi_\beta(p)\right\rangle} \tag{3}$$

may be explicitly evaluated giving

$$Q = |\beta|^2 \left[\frac{e^{|\beta|^2(1-\cos\vartheta)} - \cos\left(3\vartheta + |\beta|^2 \sin\vartheta\right)}{e^{|\beta|^2(1-\cos\vartheta)} - \cos\left(2\vartheta + |\beta|^2 \sin\vartheta\right)} - \frac{e^{|\beta|^2(1-\cos\vartheta)} - \cos\left(2\vartheta + |\beta|^2 \sin\vartheta\right)}{e^{|\beta|^2(1-\cos\vartheta)} - \cos\left(\vartheta + |\beta|^2 \sin\vartheta\right)} \right] \tag{4}$$

where $\vartheta = \dfrac{2\pi}{p}$. In figure (1) we report Q against the integer p. This plot clearly illustrates the sensitivity of the statistical properties of the states $|\varphi_\beta(p)\rangle$ to the distance p of two successive zeros. In fact, the statistics correspondent to the quantum superposition (2) exhibits a p-dependent poissonian, subpoissonian and superpoissonian character.

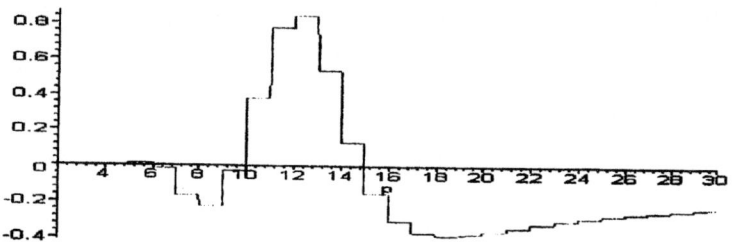

Figure (1): Mandel parameter Q as function of successive zeros p.

Moreover for large enough values of p the Mandel parameter Q remains constant and less than zero thus revealing an asymptotic subpoissonian character of the states $|\varphi_\beta(p)\rangle$. Let's now describe the main ingredients of an experimental scheme for the generation of states having the form (2). Our procedure exploits the two-photon interaction mechanism developed between a single effective two-level Rydberg atom and two modes of a non degenerate bimodal high-Q cavity. Such a two-photon coupling may be effectively described by means the following Hamiltonian model:

$$H = \sum_{\mu=1}^{2} \omega_\mu a_\mu^\dagger a_\mu + \left(\omega_0 + \beta_2 a_2^\dagger a_2 - \beta_1 a_1^\dagger a_1\right)S_z + \lambda\left[a_1 a_2 S_+ + h.c.\right] \tag{5}$$

if the condition $\omega_0 = \omega_1 + \omega_2$ is satisfied. Hereafter we refer to the cavity mode to be manipulated, that of frequency ω_1, as "mode of interest" and to the other one as "auxiliary mode". A Rydberg atom, prepared in its upper state, enters the cavity where the mode of interest is excited in a coherent state $|\alpha\rangle$ and the auxiliary one is left in its vacuum state $|0\rangle$. It is possible to demonstrate appropriately selecting the interaction time τ between the two-level Rydberg atom and the bimodal cavity field, a measurement of the internal atomic state immediately after the atom leaves the resonator, lead to the target field state. If, in fact, the internal atomic state detector finds the atom not excited, then the cavity field is projected in a state well-approximated by a linear superposition of the two equal-intensity coherent states $|\beta\rangle$

and $\left| \beta e^{-i\lambda\tau} \right\rangle$, with $\beta \equiv \alpha\, e^{i\left(\frac{\lambda}{2}-\omega_1\right)\tau}$. Stated another way, indicating by $\left| \varphi_-(\tau) \right\rangle$ the state of the mode of interest after the atomic detection and introducing the state

$$\left| \psi(\tau) \right\rangle = A \left[\left| \beta \right\rangle - e^{-i\lambda\tau} \left| \beta\, e^{-i\lambda\tau} \right\rangle \right] \tag{6}$$

A being an appropriate normalization constant, the scalar product $s \equiv \left\langle \varphi_-(\tau) \middle| \psi(\tau) \right\rangle$ is, practically, coincident with the unity. In particular, it is possible to demonstrate that for $\left| \alpha \right|^2 = 10$, the absolute value of s differs from 1 by a quantity $\varepsilon \approx 10^{-3}$. In this sense we are legitimated to say that, manipulating the atomic velocity in such a way that the condition

$$\lambda\tau = \frac{2\pi}{p} \tag{7}$$

is satisfied, then, following the procedure here only outlined for the sake of brevity, it is possible to prepare the mode of interest in a desired state belonging to the class defined by eq. (2). To claim, however, that our scheme may be effectively thought of as a truly experimental one, we must evaluate its probability of success that is the probability that the atom is detected in its ground state immediately after it leaves the cavity. We have calculated such a quantity proving that, notwithstanding the presence of source of imperfections arising, for example, from the bad control of the atom-field interaction time and/or from the detection of the final state of the atom, the probability of success of our scheme assumes values still of interest from an experimental point of view. In conclusion, we wish to emphasize that the simplicity of our experimental proposal, based on the passage of one atom only through the cavity, enhances the interest toward quantum linear superpositions of two equal-intensity coherent states as good candidates for testing fundamental aspects of quantum mechanics.

ACKNOWLEDGEMENTS

The financial supports of CRRNSM and MURST 60% are gratefully acknowledged.

REFERENCES

[1] V. Buzek, P.L. Knight, in *Progress in Optics* XXXIV (Elsevier Science BV) p.1;
[2] C. M. Caves, P.D. Drummond, *Rev. Mod. Phys.* **66**, 481 (1994);
[3] S. Haroche, *Phil. Trans. R. Soc. Lon. A* **355**, 2367 (1997);
[4] C. Monroe et al., *Science* **272**, May 96, 1131;
[5] A. Napoli, A. Messina, submitted for publication;

Decoherence and Preparation Effects in Mesoscopic Systems

Elisabetta Paladino*,**, Milena Grifoni[†], Ulrich Weiss[††], Giuseppe Falci*, Rosario Fazio*

*Dipartimento di Metodologie Fisiche e Chimiche per l'Ingegneria
and Istituto Nazionale per la Fisica della Materia
Viale A. Doria 6, I-95125 Catania
** Consorzio Ennese Universitario, Cittadella degli Studi, Enna
[†]Dipartimento di Fisica, INFM, Via Dodecaneso 33, I-16146 Genova,
and Institut für Theoretische Physik, Universität Karlsruhe, D-76128 Karlsruhe
[††] Institut für Theoretische Physik, Universität Stuttgart, D-70550 Stuttgart

Abstract. We study the problem of *preparing* and *operating* with a dissipative two-state system in the presence of an external driving field. We derive an exact master equation for the populations and we show that the initial preparation strongly affects the transient dynamics in the underdamped regime and that an appropriately tuned external a.c.-field can slow down decoherence. We finally discuss the connection with the problem of controlling the dynamics of a CJ Qubit.

INTRODUCTION

We study the quantum dynamics of a dissipative two-state system (TSS) [1,2] and show that, by operating with an external driving field, dynamic control of couplings and of decoherence environment-induced can be achieved. An alternative formulation of the problem is to study the effects of preparation on the dynamics determining if they persist for longer times than in the absence of driving.

The dissipative TSS is a fundamental model and the problem we address is relevant for many physical systems. In this contribution we focus on nanofabricated Charge-Josephson (CJ) Qubits [3], which are the elementary unit for a Quantum Computer implemented by CJ devices [4].

In a recent experiment [5] a CJ Qubit has been observed to display coherent evolution of charge states. Motivated by this experiment, we have studied an exact mapping of the dynamics of CJ Qubits in a circuit onto a dissipative TSS [6]. Then our predictions may be cecked using the device of Ref. [5]. It seems that the dynamic suppression of the tunneling coupling has been already observed [9].

CP513, *Nuclear and Condensed Matter Physics,* edited by A. Messina
© 2000 American Institute of Physics 1-56396-929-7/00/$17.00

DYNAMICS OF THE DRIVEN DISSIPATIVE TSS

The dynamics of dissipative TSS has been usually studied considering the particle initially localized in a diagonal state of the reduced density matrix (RDM) [1]. The coupling to the environment results in a reduction of the coherent tunneling [1,2,7], and may lead to localization at zero temperature [1]. Additional time-dependent external forces, partly modify these features and new remarkable effects occur [7].

We study the dynamics of the RDM of the driven dissipative TSS for a general diagonal/off-diagonal initial state [8]. We consider the driven spin-boson model

$$H(t) = -\frac{1}{2}\hbar[\Delta(t)\sigma_x + \varepsilon(t)\sigma_z] + H_B - \frac{1}{2}\sigma_z X . \tag{1}$$

The first term characterizes the TSS in the presence of external fields which modulate the asymmetry energy, $\varepsilon(t)$, and the tunneling amplitude $\Delta(t)$ [7]. The second term describes the bath of harmonic oscillators, $H_B = \sum_i[p_i^2/2m_i + m_i\omega_i^2/2]$, which are bilinearly coupled to the TSS via a collective bath coordinate $X = d\sum_i c_i x_i$ describing the bath polarization energy, $H_{SB} := -\sigma_z X/2$. The bath influence on the TSS is captured by the spectral density $J(\omega) = (\pi/2)\sum_i(c_i^2/m_i\omega_i)\delta(\omega - \omega_i) = (2\pi\hbar/d^2)\alpha_s\tilde{\omega}^{1-s}\omega^s\exp(-\omega/\omega_c)$, α_s is a dimensionless coupling constant, $\tilde{\omega}$ a reference frequency, and d is the distance between the localized states.

For the density matrix of the global system, $W(t)$, we assume the product initial state $W(t_0) = \rho(t_0)W_B$, where W_B is the canonical density matrix of the bath, $W_B = e^{-\beta H_B}/\text{Tr}\{e^{-\beta H_B}\}$ ($\beta = 1/k_B T$), while the TSS has been constrained for times $t_p \leq t \leq t_0$ in the initial RDM state

$$\rho(t_0) = \begin{pmatrix} p_R & a + ib \\ a - ib & p_L \end{pmatrix} . \tag{2}$$

Within the path-integral method, a set of exact equations for the elements of the RDM, $\rho(t) = \text{Tr}_B\{W(t)\}$, is obtained. The population difference of the localized states, $\langle\sigma_z\rangle_t \equiv P(t)$, satisfies a generalized non-Markovian master equation (GME)

$$\frac{d}{dt}P(t) = \int_{t_0}^t dt'[K_A^{(a)}(t,t') - K_A^{(s)}(t,t')P(t')] + 2aK_B^{(a)}(t,t_0) + 2bK_B^{(s)}(t,t_0) . \tag{3}$$

The effects of the dissipative bath and of the driving fields are included into the kernels $K_A^{(s/a)}(t,t')$, defined in Ref. [8]. Effects of the preparation in an off-diagonal state of the RDM are in the inhomogeneous contributions.

From the GME we deduce that (in the undriven case) neither the asymptotic distributions, nor the dephasing and relaxation rates towards the equilibrium state, nor the coherent-incoherent crossover temperature depend on the initiqal state.

Preparation effects are crucial at short times whenever the dynamics exhibits underdamped coherent oscillations. For weak-damping and in the absence of driving from (3) the population difference $P(t)$ is found as ($t_0 = 0$)

$$P(t) = Ne^{-\Gamma_r t} + N_2\cos(\nu t)e^{-\Gamma t} + (2b\Delta/\nu + n_3)\sin(\nu t)e^{-\Gamma t} + P_\infty , \tag{4}$$

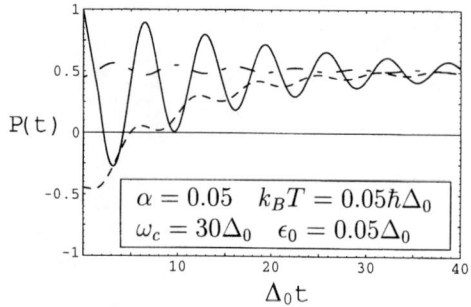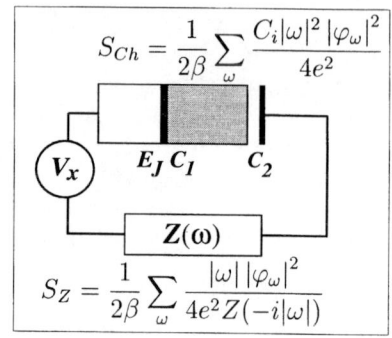

FIGURE 1. (a)Evolution of $P(t)$ for three different initial preparations: particle localized in the right well (solid line), in the ground state (dot-dashed line) and in the excited state (dashed line). (b) Superconducting Box: a superconducting island (shaded) is connected to a voltage source V_x via a Josephson Junction (coupling E_J) and the capacitor C_2; the standard action terms are indicated.

with equilibrium value $P_\infty = (\epsilon_0/\nu)\tanh(\hbar\beta\nu/2)$ and $\nu = (\Delta^2 + \epsilon_0^2)^{1/2}$ (Δ and ϵ_0 denote the adiabatically dressed tunneling coupling, and the bias factor in the undriven case). The incoherent relaxation rate Γ_r and the dephasing rate Γ read

$$\Gamma_r = (\Delta^2/2\nu^2)J(\nu)\coth(\hbar\beta\nu/2) \quad \Gamma = \Gamma_r/2 + 2\pi\alpha\delta_{s,1}(\epsilon_0/\nu)^2 k_B T/\hbar . \quad (5)$$

The second term contributes to Γ only in the Ohmic case ($s = 1$). The amplitudes N_i and n_i are respectively of zero and linear order in the bath coupling. Effects of the particular initial preparation in the transient dynamics are evident in Fig. 1.

In the presence of a high-frequency a.c.-field modulating the bias energy, $\varepsilon(t) = \epsilon_0 + \hat{\epsilon}\cos\Omega t$ ($\Omega \gg \Delta_0, \nu, \Gamma, \Gamma_r$) the dynamics is well described by performing a field-average of the GME [7,8]. For small static bias $\epsilon_0 \ll \Omega$, and $J_0(\hat{\epsilon}/\Omega) \neq 0$ the dynamics behaves like for a static bias, but with effective tunneling matrix element $\Delta_{\text{eff},0} = \Delta_0 J_0(\hat{\epsilon}/\Omega)$. Because $\Gamma_r, \Gamma \propto \Delta_{\text{eff},0}^2 \leq \Delta_0^2$, both the dephasing rate Γ and the relaxation rate Γ_r can be strongly reduced by suitably chosen parameters of the driving field. Analogous features are found when a resonant ($\epsilon_0 = n\Omega$) a.c.-field is applied. In this case, when $J_n(\hat{\epsilon}/\Omega) \neq 0$ the decay rates are proportional to $\Delta_{\text{eff},n}^2 = \Delta_0^2 J_n^2(\hat{\epsilon}/\Omega) \leq \Delta_0^2$. In conclusion, an appropriately tuned external a.c.-field can slow down decoherence and thus allow preparation effects to persist for longer times than in the absence of driving.

APPLICATION TO CJ QUBITS

The dynamic unit of a CJ Qubit is the Superconducting Box (Fig.1.b). Charging effects due to the subfemtofarad capacitances fix the excess charge in the island to $N = 0, 1$ Cooper pair. This effective TSS is subject to (thermal and quantum) fluctuations of the circuit. The imaginary time path-integral generating functional

[4] associated to this system is a proper generalization of the Caldeira-Leggett model [2] and reads

$$\mathcal{Z} = \int \mathcal{D}\varphi_1 \mathcal{D}\varphi_2 \mathcal{D}\varphi_Z \, \delta[\dot{\varphi}_1 + \dot{\varphi}_2 + \dot{\varphi}_Z + \dot{\varphi}_x] \, e^{-S_{ch}[\varphi_1] - S_{ch}[\varphi_2] - S_Z[\varphi_Z] + \int_0^\beta d\tau E_J \cos \varphi_1} . \quad (6)$$

The discreteness of the charge in the island is reflected by the non-trivial boundary conditions $\varphi_1(\beta) = \varphi_1(0) + 2\pi m$ which introduces winding numbers [4]. The last term in the action describes Josephson tunneling [4]. The δ-functional implements the mesh equation for the circuit an allows the elimination of φ_2. We can integrate out φ_Z and we are left with an effective model in the variable φ_1 which we can reexpress in the charge representation in terms of the dual *discrete* variable $\sigma(\tau)$. By considering only $\sigma(\tau) = \pm 1$ (see Ref. [6]) we obtain a driven dissipative TSS, with parameters explicitly given in terms of circuit elements. For instance the imaginary-time interaction and the bare tunneling amplitude are

$$\mathcal{K}(\omega) = \frac{2\pi |\omega| (C_2/C)^2 Z(-i|\omega|)/R_Q}{1 + (C_1 C_2/C) Z(-i|\omega|) |\omega|} \quad ; \quad \Delta_0 = \frac{1}{2} E_J ,$$

where $C = C_1 + C_2$, $R_Q = h/(2e)^2$. For a series resistor, $Z(\omega) = R$, the spectral density $J(\omega)$ corresponding to the interaction $\mathcal{K}(\omega)$ describes an ohmic bath with a Drude cutoff [2] and the static dephasing has been considered in Ref. [3]. For the special case of the last section we obtain the bias energy

$$\epsilon_0 = \frac{2e^2}{C} \left[\frac{CV_{dc}}{e} - 1 \right] ; \quad \hat{\epsilon}(\Omega) = 2e(C_2/C) V_{ac} \, \text{Re}[1 + i(C_1 C_2/C) Z(\Omega) \Omega]^{-1} ,$$

where we considered an external bias $V_x = V_{dc} + V_{ac} \cos(\Omega t)$.

Nakamura has very recently observed [9] that in the setup of Ref. [5] an a.c. field renormalizes the tunneling amplitude but does not affect decoherence. This indicates that decoherence is mainly due to other sources and then an a.c. field may be used for dynamical fine tuning of the time evolution.

REFERENCES

1. Leggett, A. J., Chakravarty, S., Dorsey, A. T., Fisher, M. P. A, Garg, A., and Zwerger, W., *Rev. Mod. Phys.* **59**, 1 (1987); erratum, *Rev. Mod. Phys.* **67**, 725 (1995).
2. Weiss, U. *Quantum Dissipative Systems*, World Scientific, Singapore, 1999.
3. Makhlin, Y., Schön, G., and Shnirman, A. *Nature* **398**, 305, (1999).
4. Schön, G., Zaikin, A., *Phys. Rep.* **198**, 237, (1990).
5. Nakamura, Y., Pashkin, Yu. A., Tsai, J.S., *Nature* **398**, 786, (1999)
6. The Charge Qubit maps *exactly* in a *multistate* system; implications are discussed in J. Siewert, R. Fazio and G.M. Palma, subm. to Phys. Rev. Lett..
7. Grifoni M., and Hänggi, P. *Phys. Rep.* **304**, 229 (1998); Hänggi, P., Talkner, P., and Borkovec, M. *Rev. Mod. Phys.* **62**, 251 (1990).
8. Grifoni, M., Paladino E., and Weiss, U., *Eur. Phys. J. B* **10**, 719, (1999).
9. Y. Nakamura, private communication.

Langevin Approach To Noise Modelling Of Bipolar Microwave Transistors

F.Patti[1], V. Miceli[3] and B.Spagnolo[1,2]

[1]Istituto Nazionale per la Fisica della Materia, Unità di Palermo.
[2]Università di Palermo, D.E.A.F, Viale delle Scienze, Palermo.
[3]ST-Microelectronics-Catania.

Abstract. We present a new approach to study the complete stochastic properties of fluctuations of the output current of microwave transistors. We obtain the π-hybrid model of bipolar microwave transistors with the noise internal sources starting from experimental on-wafer measurements of the scattering and noise parameters. We derive the stochastic differential equations of the Giacoletto model for different loads and source admittances. We give the analytical temporal behaviour of the second moment of the output current, assuming particular given correlation functions between the internal noise sources.

INTRODUCTION

Noise is present in all electronic devices. It is generated by the random motion of electrons in a resistive material, by the random recombination of holes and electrons in a semiconductor, and when holes and electrons diffuse through a potential barrier. The principal noise sources in a bipolar transistors are the thermal noise due to the base spreading resistance, shot noise, flicker noise and burst noise due to the base bias current, and shot noise due to the collector bias current. The study of the internal noise sources is a fundamental aspect of the noise behaviour of microwave transistors to reach high performances device with decreasing dimensions. Miniaturisation process in fact cause an increasing interaction between the internal sources which affect transistor noise behaviour. We derive a generalised Langevin equation, or more exactly a stochastic differential equation, for the Giacoletto model of Bipolar Transistor. The noisy Giacoletto circuit is obtained starting from scattering parameters and noise figure experimental on wafer measurements and using a modelling procedure [1]. The output current is a stochastic process as a consequence of the fluctuations due to the different internal noise sources of the transistor. We consider, as a first step to further investigations, the different noise as a white Gaussian noise. We give the solution of the SDE (stochastic differential equation), obtained for a source impedance purely resistive, as a function of the parameters of the circuit and of the noise sources. We give also the analytical expression of the second moment of the output current as a function of the correlation function between the stochastic processes, which characterise the different noise sources. From the temporal series of

CP513, *Nuclear and Condensed Matter Physics*, edited by A. Messina
© 2000 American Institute of Physics 1-56396-929-7/00/$17.00

the output current we obtain the probability distribution and power spectrum in the presence of internal noise sources (N_s) and without internal noise sources (N_o). Therefore we are able to obtain the noise figure:

$$NF = 10 * \log \frac{N_s}{N_o} \qquad (1)$$

as a function of the operating frequency. We compare our theoretical results with experimental measurements.

CIRCUIT MODEL

The bipolar transistors have been characterised *on wafer* in terms of scattering parameters and noise figure by varying the collector current (Ic) between 0.7 and 10 mA at collector-to-emitter voltage (Vce) value of 1 V over 1-2 GHz frequency range.

Such device is a 2-finger emitter structure and exhibits a current gain β in excess of 100. The test devices and measurements have been furnished by ST-Microelectronics.

As a first step, we extract a broad band π -like circuit model from scattering parameters for two bias condition (Ic=0.7 and 1 mA). After this first optimisation using MMICAD software we add noise generators to the noise free circuit and reset the circuit parameters for the best fitting of the measured noise figure [1].

In figure 1 we report the network obtained with this modelling procedure.

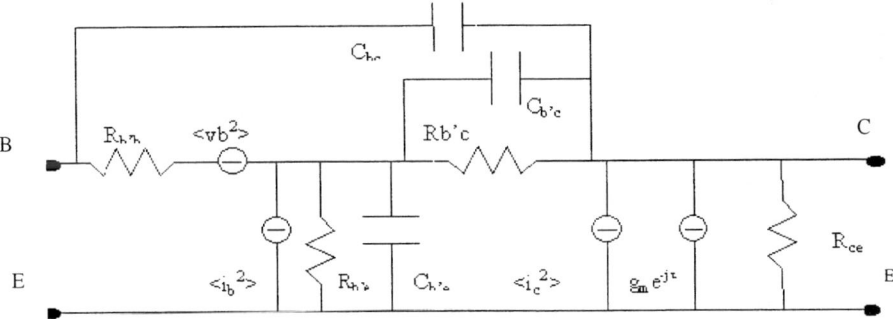

FIGURE 1. Giacoletto model of bipolar transistor over 1 ÷ 2 GHz frequency range.

LANGEVIN APPROACH

To obtain the Langevin equation for the Giacoletto model we derive in the Laplace domain the expression of the output current as a function of the circuit parameters and the noise sources. By antitrasforming this expression we get the dynamical equation with different noise sources [2,3]. For a short circuited output ($R_L=0$) we get:

$$i_{out}(t) = ai_{sn}(t) + bv_{bn}(t) + ci_{bn}(t) + i_{cn}(t) + $$
$$+ \int_0^t [i_{sn}(\tau)a(t-\tau) + v_{bn}(\tau)b(t-\tau) + i_{bn}(\tau)c(t-\tau)]d\tau \qquad (2)$$

where the coefficients a, b and c and the functions a(t-τ), b(t-τ) and c(t-τ) have different expressions depending on the source admittance[3]. For resistive load we obtain:

$$i_{out}(t) = \int_0^t [i_{sn}(\tau)a_1(t-\tau) + v_{bn}(\tau)b_1(t-\tau) + i_{bn}(\tau)c_1(t-\tau)]d\tau \qquad (3)$$

where the functions a_1,b_1 and c_1 are also dependent on the source admittance [3]

From eq.s (2) and (3), by assuming some given correlation functions between the internal noise sources, it is possible to calculate all the moments of the output current.

To calculate the second moment of the stochastic process of the output current we consider all the noise sources expressed in terms of the Wiener process. We assume external noise sources uncorrelated with the internal ones of the BJT, and for cross-correlations between the internal noise sources at different times, Dirac δ-functions:

$$<v_{bn}(\tau)i_{bn}(\tau')> = \sigma_{vb}\sigma_{ib}\,\delta(\tau-\tau')$$
$$<v_{bn}(\tau)i_{cn}(\tau')> = \sigma_{vb}\sigma_{ic}\,\delta(\tau-\tau')$$
$$<i_{cn}(\tau)i_{bn}(\tau')> = \sigma_{ic}\sigma_{ib}\,\delta(\tau-\tau')$$
$$<i_s(\tau)v_{bn}(\tau')> = 0,$$
$$<i_s(\tau)i_{bn}(\tau')> = 0, \quad <i_s(\tau)i_{cn}(\tau')> = 0.$$

By using the statistical properties of the Wiener process we obtain the second moment of the output current for a short circuited load:

$$<i_{out}(t)^2> = \sigma_{is}^2 + b^2\sigma_{vb}^2 + c^2\sigma_{ib}^2 + \sigma_{ic}^2 + $$
$$+ \int_0^t [\sigma_{is}a^2(t-\tau) + \sigma_{vb}b^2(t-\tau) + \sigma_{ib}c^2(t-\tau) + 2\sigma_{vb}\sigma_{ic}b(t-\tau)c(t-\tau)]d\tau +$$
$$+ 2b<v_b(t)i_b(t)> -2b<v_b(t)i_c(t)> -2c<i_b(t)i_c(t)> -+2\sigma_{is}^2a(t) + 2b[\sigma_{vb}^2b(t) +$$
$$\sigma_{vb}\sigma_{ic}c(t)] + 2c[\sigma_{ib}\sigma_{vb}b(t) + \sigma_{vb}\sigma_{ic}c(t)] -+2[\sigma_{vb}\sigma_{ic}b(t) + \sigma_{ic}^2c(t)] \qquad (4)$$

where the cross-correlation functions between the different noise internal sources at the same time are:

$$<i_b i_c> = \sigma_{ib}\sigma_{ic}e^{-t/\tau b} \qquad <v_b i_c> = \sigma_{vb}\sigma_{ic}e^{-t/\tau bc} \qquad <v_b i_b> = \sigma_{vb}\sigma_{ib} \qquad (5)$$

where τ_{bc} is the carriers transit time from base region to collector region.

RESULTS

To obtain the noise figure (eq. (1)) by FFT method we calculate the output current power spectrum with and without internal noise sources. For each frequency value we calculate the output power level, obtaining the noise figure as a function of the operating frequency (see, Figure. 2). In the same figure we report the noise figure behaviour obtained from experimental measurements with the bias points Ic=0.7 and 1 mA

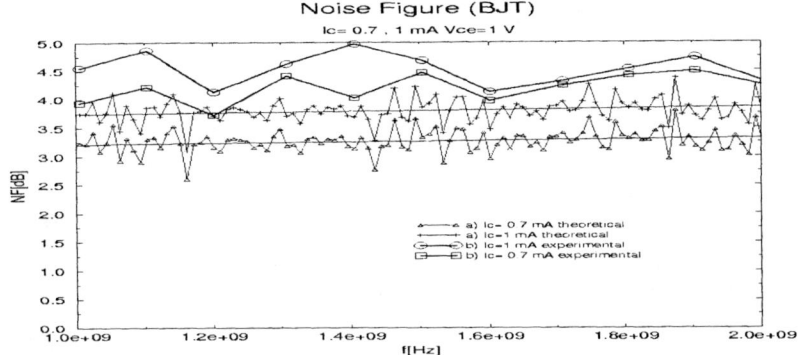

FIGURE 2. Noise figure versus frequency for 2 bias points: a) theoretical b) experimental

We can see that our theoretical results are in good agreement with experimental ones. We can improve our theoretical results of the noise figure in comparison with the experimental ones, by choosing different probability distributions for the noise sources. Finally, we can apply our method to different bipolar devices and for a wider frequency range.

ACKNOWLEDGMENTS

We wish to tank INFM and MURST for financial support.

REFERENCES

1   A. Caddemi, F. Patti, M. Sannino *"Bias Dependence of the Noise performances in Si/SiGe HBT's at Microwave Frequencies"*, IEEE Conference EDMO 98, pp.57-61.

2   V. Miceli, B. Spagnolo, "A Stochastic integral equation for the Giacoletto model", Tech. Rep. PS- ELEN06, University of Palermo, June 1997.

3   V. Miceli, B. Spagnolo, *"Langevin approach to the noise properties of the giacoletto model of the BJT device"*, Tech Rep. PS-ELEN12, University of Palermo, April 1998.

Monte Carlo Simulation of Nonlinear Electron Transport in Semiconductors: Harmonics Generation in GaAs

D. Persano Adorno, M. Zarcone, G. Ferrante

Istituto Nazionale di Fisica della Materia
and Dipartimento di Energetica ed Applicazioni di Fisica,
Viale delle Scienze, 90128 Palermo, Italy

Abstract. The nonlinear response of electrons in a GaAs bulk semiconductor placed in an oscillating electric field with frequency in the far infrared domain is studied using the Monte Carlo method. The drift velocity obtained from the Monte Carlo simulation is used to obtain the efficiency of high order harmonics generation. Harmonics up to the 15th order with an efficiency of 10^{-7} are resolved for a field with amplitude of E=50 kV/cm. It is also reported the dependence of the first four odd harmonics efficiencies on the lattice temperature in the range 80-400 K and on the amplitude of the electric field in the range 5-100 kV/cm.

I INTRODUCTION

The study of the process of high order harmonics generation from a bulk semiconductor subject to intense radiation field is interesting to develop an efficient frequency converter in the THz frequency domain. Moreover harmonics generation in semiconductors is of interest in its own right, being a useful mean for the general understanding of several features of the highly non linear processes of carriers transport in semiconductors. In a previous experiment [1]- [2] the third harmonic generation on the far infrared range has been detected with an efficiency up to 10^{-3} for a n-type silicon at a wavelength of 677 μm for a 2MW incident power radiation. Conversion efficiency for the third and fifth harmonics in Si, GaAs and InP crystals using a Monte Carlo (MC) simulation has also been predicted by Brazis et al [3]. In a doped semiconductor the absorption bands due to photons are in the mid-infrared region so that in the far-infrared region, linear as well nonlinear optical properties are mainly determined by the motion of the free carriers caused by the electric field of the incident waves. In this paper we report the dependence of the first four odd harmonics efficiency on the lattice temperature in the range 80-400 K and on the amplitude of the electric field in the range 5-100 kV/cm.

CP513, *Nuclear and Condensed Matter Physics,* edited by A. Messina
© 2000 American Institute of Physics 1-56396-929-7/00/$17.00

II THE MODEL

The theory of harmonics generation in a semiconductors has been derived in a previous paper [4] and is based on the use of the Maxwell equation for the propagation of an electromagnetic wave in a medium in connection with a Monte Carlo Simulation. Assuming that the medium is transparent to the radiation at frequency ω, i.e. $\omega > \omega_p$, ω_p being the Langmuir frequency, the q-th harmonic intensity normalized to the fundamental one is given by then becomes:

$$\frac{Iq}{I_1} = \frac{1}{q^2} \frac{v_q^2}{v_1^2} \tag{1}$$

The coefficients v_q are given by the Fourier transform of the electrons velocity. The drift velocity of the electrons is obtained making a multiparticle MC simulation of the motion of the electrons in the semiconductor.

The algorithm of MC simulation of the electron motion in the alternating electric field used in this work follows the standard procedure [5]. It involves the non-parabolicity of the band structure and the intervalley and intravalley scattering. For a complete set of n-type GaAs parameters used in our calculations, see Ref [4].

The conduction bands of GaAs are represented by the Γ valley, by four equivalent L-valleys and by three equivalent X-valleys. At thermal equilibrium the electrons are expected to be in the conduction band of the GaAs bulk located in the central Γ valley.

Electrons are expected to exhibit an enhanced nonlinearity in high electric fields due to the equivalent and non equivalent intervalley scattering among the three (Γ, L, X) valleys [4].

III RESULTS AND DISCUSSION

We have performed our calculations for GaAs with an oscillating field at 200 GHz and a free electrons concentration n=10^{19} m^{-3}. The MC simulation has been done for different temperatures in the range 80-400 K and for different field amplitude in the range 5-100 kV/cm. In our model we have considered the following scattering mechanisms: i) the intravalley scattering with acoustic phonons, ionized impurity, acoustic piezoelectric phonons, polar optical phonons and for the L valley also the scattering with optical non polar phonons; ii) the intervalley scattering with the optical non polar phonon among the three valleys.

As shown in Fig.1a the electron drift velocity is modulated at the frequency of the external field ω. The deviation from the sinusoidal form, which cause the harmonic generation, increases with the amplitude of the field [4]. The average value of the electron energy $< \varepsilon >$ and the population of the different valleys presents instead a modulation at a frequency 2ω (Fig. 1b, 1c). In fact, the work per unit time performed by the external electric field on the free electron is given by

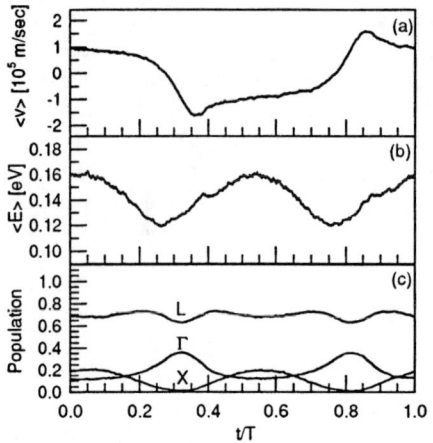

FIGURE 1. (a) carriers drift velocity, (b) average value of the electron energy $\langle\varepsilon\rangle$, and (c) population of the 3 different valleys as a function of the number of the radiation periods [ω=200 GHz, T=300 K, n=10^{19} m^{-3}, E=50 kV/cm].

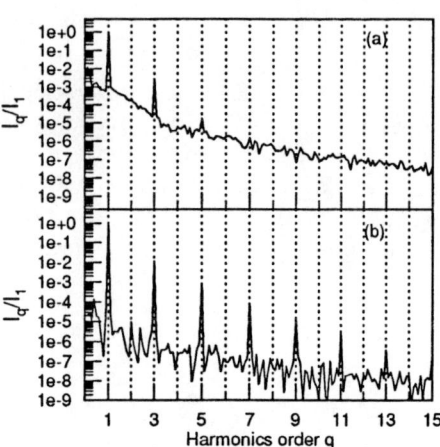

FIGURE 2. Harmonics generation efficiency vs their order [ω=200 GHz, T=300 K, n=10^{19} m^{-3}; (a) E=10 kV/cm, (b) E=50 kV/cm].

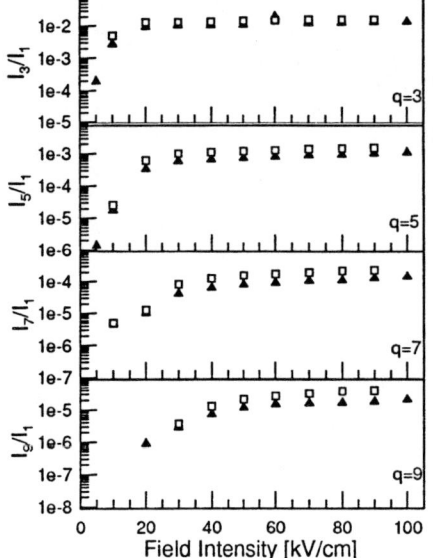

FIGURE 3. Efficiency of the first four odd harmonics as a function of the intensity field for two different values of temperature [□ T=80 K, ▲ T=300 K].

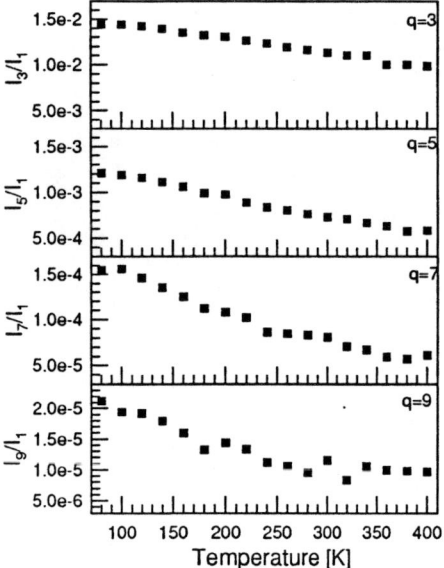

FIGURE 4. Efficiency of the first four odd harmonics as a function of the temperature [E= 50 kV/cm].

$$W = \vec{j} \cdot \vec{E} \qquad (2)$$

Since the velocity $\vec{v}$ and consequently the current density $\vec{j}$ oscillate at the frequency ω of the electric field E, the work W and the average energy $< \varepsilon >$ will oscillate at frequency 2ω. All the transition probabilities for the scattering mechanisms considered in our simulation are strongly dependent on the electron energy. We expect then that the total collision frequency $\nu(\varepsilon)$ will be modulated at frequency 2ω and in turn, the free electrons drift velocity will acquire a component oscillating at frequency 3ω, that will give rise to the third harmonic generation. Further collisions will give rise to a modulations of the drift velocity at higher frequency generating harmonics of higher order. The spectrum for harmonics generation at an oscillating field of amplitude E=10 kV/cm and E=50 kV/cm is given in Fig 2. The efficiency increases with the field. When the field is weak (E=10 kV/cm) only the 3rd and the 5th harmonic are present while for a stronger field (E=50 kV/cm) the efficiency is very high going from 10^{-2} for the 3rd harmonic to 10^{-7} for the 15th one. Fig.3 shows the efficiencies of the first four odd harmonics of the frequency $\omega = 200$ GHz in n-type GaAs as a function of the electric field amplitude at two different lattice temperatures (80 and 300 K). The harmonics efficiency shows an interesting feature: it is nearly independent of the pumping field amplitude above 30 kV/cm. For higher power levels the harmonics generation efficiency exhibits saturation effects.

We have investigated also the temperature dependence of the drift response at the field amplitude of E=50 kV/cm in the range 80-400 K (Fig.4). The lower temperature is representative of a situation where the sample is cooled down to liquid-nitrogen temperature, whereas the 400 K case is a typical value in a room-temperature experiment. The MC simulation shows that an increasing of the temperature from 80 to 400 K makes the harmonics generation efficiency in n-type GaAs decreasing linearly but not significantly (only 2-3 times).

Acknowledgments

This work was supported by the F.E.S.R and by the INFM under the join action "Progetto Sud". Additional financial support from MURST (Italian Ministry of University and Scientific Research) is also acknowledged. One of us (DPA) acknowledges a CNR research fellowship.

REFERENCES

1. Urban M., Nieswand Ch., Siegrist M.R., and Keilmann F., *J. Appl. Phys.* **77**, 981 (1995)
2. Keilmann F., Brazis R., Barkley H., Kasparek W., Thumm M. and Erckmann V., *Europhys. Lett.* **11**, 337 (1990)
3. Brazis R., Ragoutis R., Siegrist M.R., *J.Appl. Phys.* **84**, 3474 (1999)
4. Persano Adorno D., Zarcone M. and Ferrante G., *Laser Physics*, in press (1999)
5. Lebwohl P.A., *J. Appl. Phys.* **44**, 1744 (1979)

Vibrational Mixing and Conformational Heterogeneity In Model-Peptides

Daria Puccia and Maurizio Leone

Department of Physical and Astronomical Sciences and
Istituto Nazionale di Fisica della Materia, University of Palermo (Italy)

Abstract. We report an experimental study on the structural and dynamic properties of model-peptides in solutions. The FTIR spectra of the N-α-acetylmethylglycinamine, a molecular complex owning two peptide units, are measured in different solvents and compared with the ones of the N-Methylacetamide, containing a single peptide group. Both for the complexes, the Amide I and II bands present spectral heterogeneity and appear largely modified by the solvent composition. The comparison between the two model-peptides points out the role played by the interactions between the different peptide units.

INTRODUCTION

The interest in studying water solutions of model peptides derives from the awareness that the functional, structural and dynamic properties of the biological molecules are strictly related to the stereodynamic properties of the solvent. The role played by the peptidic bond in determining the secondary and tertiary structure of the proteins makes simple model peptides as an ideal system to investigate the amide linkage and the proteins structure and dynamics. N-Methylacetamide (NMA) is the simplest molecular model of the peptide linkages in proteins. The IR and Raman spectra of this complex in water are characterized by the Amide bands, that involve CO (COs) and CN (CNs) stretchings and NH (NHi) in-plane bending mode (1). In particular, Amides I and II arise from an out-of-phase combination of the COs with the CNs and with the NHi, respectively (1,2).

In this paper, we compare the IR spectra of NMA with the ones of the N-α-Acetylmethylglycinamine (DNMA). This last complex contains two asymmetric units owning a peptide bond, linked together by a CH_2 group. Aim of this work is to investigate the role played by the presence of different peptide units in modulating the solvent-macromolecule interactions. In this respect, we use the isotopic substitution of

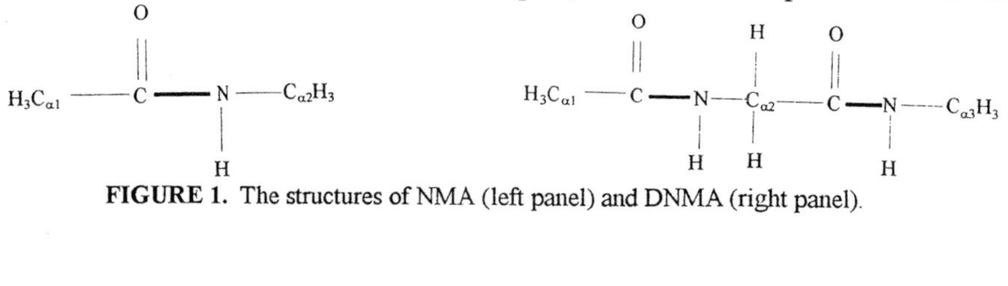

FIGURE 1. The structures of NMA (left panel) and DNMA (right panel).

CP513, *Nuclear and Condensed Matter Physics,* edited by A. Messina
© 2000 American Institute of Physics 1-56396-929-7/00/$17.00

the solvent to distinguish the mechanisms that give rise to the observed spectral heterogeneity.

MATERIALS AND METHODS

NMA and DNMA were purchased from Merck Chemical Co. D_2O (99,8%) was obtained from Aldrich Chemical Co. For each sample, the concentration of NMA and DNMA was adjusted to 0,2 M to avoid the self-aggregation.

The single-beam IR spectra were obtained by using a BIO-RAD FTS-40A spectrometer equipped with a PbS detector, with a spectral resolution of 1 cm^{-1}. All samples were placed in a Specac cell mounting two CaF_2 windows and a 0.025 mm spacer. For each sample, the absorption spectrum was obtained with respect to the empty cell and subtracted by the solvent absorption, after suitable normalization.

RESULTS AND DISCUSSIONS

Fig. 2 shows the FTIR spectra of DNMA in H_2O (a) and in D_2O (b). The spectral region 1500÷1700 cm^{-1} presents, in H_2O, the two Amide structures centered at 1566 (Amide II) and at 1643 cm^{-1} (Amide I), respectively. These bands appear largely modified by the isotopic substitution of the solvent: the Amide II shifts to 1492 cm^{-1} in D_2O, whereas the Amide I remains fixed in peak frequency but shows a higher intensity and a narrower bandwidth. These results are confirmed by the measurements in H_2O/D_2O 50% mixture (data not reported). To analyze in detail the spectral range of

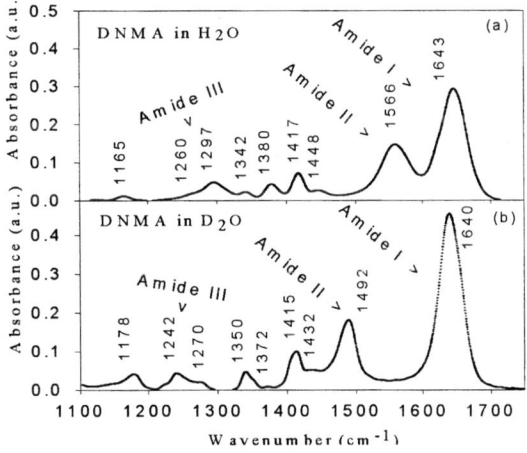

FIGURE 2. FT-IR spectra of DNMA in H_2O (a) and in D_2O (b). The frequency values label the observed peak positions of the IR bands; therefore, they can be somewhat different from the values obtained from the spectral analysis (see Table 1).

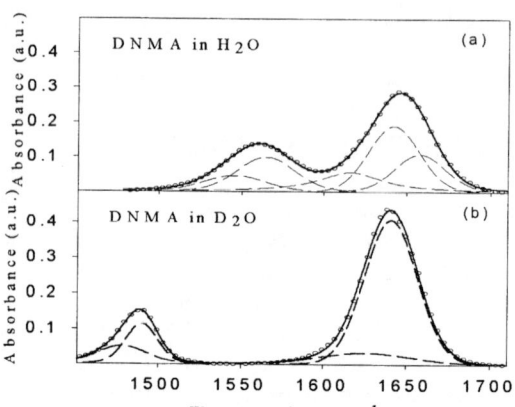

FIGURE 3. Amide I and II spectral region. Circles represent the experimental data reported in Fig. 2, after a linear background correction; the dashed and continuous lines represent each spectral Voigtian component and the overall synthesized profile, respectively. Note that experimental data were reported at 10 point-step, for the sake of clarity.

of Amides I and II, we report in Fig.3 a spectral deconvolution in Voigtian components in the range $1450 \div 1750$ cm^{-1}. Data in Fig.3 and their relative parameters reported in Table 1 together with the same quantities obtained for NMA, reveal a rather complex structure of the spectra. As we can see, for DNMA Amide I in water is resolved into three sub-bands, centered at 1616, 1642 and 1658 cm^{-1}, respectively, whereas Amide I in D$_2$O can be essentially taken into account by a single one, centered at 1641 cm^{-1}, the area of the second component at 1625 cm^{-1} being $\approx$ 10% of total Amide I area.

The comparison between the two model peptides (see Table 1) shows that the DNMA Amide I presents a shift towards higher frequency of the overall band. Moreover, a larger spectral heterogeneity with respect to NMA is found, as revealed by the presence of three sub-bands in water and, both for the solvents, by a larger bandwidth for all the spectral components. Chen et al. (3,4) suggested that, for NMA in water, the mechanism giving rise the decomposition of Amide I is a vibrational mixing between the Amide I mode and water molecules bending mode, at 1640 cm^{-1}. This hypothesis is also confirmed by our measurements in Dimethylsulfoxide (DMSO), an apolar solvent in which this type of vibrational coupling cannot be invoked, that show for NMA essentially a single component of Amide I (5).

This interpretation explains the spectral changes induced by H$_2$O/D$_2$O substitution and suggests that the greater spectral heterogeneity observed for DNMA in H$_2$O with respect to NMA can be ascribed to an increase of the interactions, like hydrogen bonds with the solvent, due to the presence of two peptide groups. Moreover, the Amide I blue-shift suggests that the two peptide planes interact each other inducing an increase of the COs frequency. An inspection of the data reported in Table 1 for Amide II shows that this band presents a decomposition into two spectral components in all studied samples. For DNMA in water, the two sub-bands are red-shifted and more separated than the components of NMA. Both for the model peptides, the H$_2$O/D$_2$O substitution induces a large red shift and a narrowing of the two components. These

TABLE 1. Peak frequencies (v), full-widths at the maximum (FWHM) and intensities (I) of the spectral components detected in the Amides region for DNMA and NMA. Error values: ± 1 cm^{-1} and ± 2 cm^{-1} for v and FWHM, respectively; ± 0.02 for I.

Sample	Mode	v [cm^{-1}]	FWHM [cm^{-1}]	I [a.u.]
DNMA in H₂O	Amide II	1545	44	0,05
		1564	41	0,10
	Amide I	1616	46	0,06
		1642	36	0,11
		1658	36	0,11
DNMA in D₂O	Amide II	1478	37	0,05
		1490	23	0,11
	Amide I	1625	70	0,03
		1640	37	0,41
NMA in H₂O	Amide II	1565	64	0,02
		1581	36	0,06
	Amide I	1623	33	0,14
		1643	36	0,05
NMA in D₂O	Amide II	1492	22	0,07
		1514	15	0,05
	Amide I	1616	54	0,03
		1625	29	0,32

results can be attributed to the contribution of NHi to the Amide II vibrational mode, this bending mode being strongly affected by isotopic substitution of the solvent (1). Also on the basis of our experimental results for the same model peptides in DMSO (data not reported here), that show the same fine structure for this band, the origin of the Amide II splitting seems due to the presence of two different structural conformers, essentially independent from the solvent composition. The clear separation between DNMA Amide I and Amide II peaks suggests that the vibrational modes to which Amide bands correspond are less coupled in DNMA than in NMA.

The above reported results confirm that different mechanisms give origin to the observed spectral heterogeneity in model peptides. In particular, Amide I splitting arises, in water, from a mixing between the Amide I mode and the water bending mode, whereas the Amide II sub-bands, both in H₂O and D₂O, suggest the presence of two different structural conformers. Moreover, Amide I appears very sensitive to the hydrogen bonding between the CO of a peptide group and the NH of the other, so confirming its ability to characterize the secondary structure of polypeptides.

REFERENCES

1. Krimm, S., and Bandekar, J., *Advances in Protein Chemistry* **38**, 181-357 (1986).
2. Seiler, G., and Schweitzer-Stenner, R., *J. Am. Chem. Soc.* **119**, 1720-1726 (1997).
3. Chen, X.G., Krimm, S., Mirkin, N.G., and Asher, S.A., *J. Am. Chem. Soc.* **116**, 11141-11142 (1994).
4. Chen, X.G., Schweitzer-Stenner, R., Asher, S., Mirkin, N.G., and Krimm, S., *J. Phys. Chem.* **99**, 3074-3083 (1995).
5. Puccia, D., Thesis (1999)

Dynamic Properties of Deoxy Hemoglobin Encapsulated in Silica Gels

Magda Rausei and Antonio Cupane

Istituto Nazionale di Fisica della Materia and Dipartimento di Scienze Fisiche ed Astronomiche,
University of Palermo, Via Archirafi, 36- 90123- Palermo, Italy.

Abstract. We have encapsulated native human hemoglobin (Hb) in wet porous silica gels. Suitable treatment of the samples enables to obtain transparent gels in which the Hb molecule is still able to reversibly bind exogenous ligands; however, due to the constraints imposed by the vitreous silica matrix, the Hb tetramer is unable to perform the T-R quaternary transition. Using the gel encapsulation technique we have obtained deoxy Hb samples encapsulated in the "T" or "R" quaternary structure. The study of the temperature dependence of the Soret absorption band in the above samples enables to have informations on the local dynamic properties of the heme pocket and therefore to study the dynamics of the protein in a given ligation state but in different quaternary conformations. The results show that the "T" quaternary conformation is, from a dynamic point of view, intrinsically different from the "R" conformation and is characterized by increased anharmonic motions.

INTRODUCTION

The recently developed technique of protein encapsulation in wet porous silica gels (1) gives the possibility of studying the functional, structural and dynamic properties of immobilized proteins and has therefore attracted the attention of several laboratories, also because of its potential interest in biotechnological applications. For Hb, the gel encapsulation technique enables also to block (or at least to slow down to the timescale of several days) the rate of the T↔R quaternary conformation switch (2), and therefore to investigate the stereodynamic properties of this protein decoupling the effects of quaternary conformation from the often concomitant effects of ligation state (i.e. presence or absence of the exogenous ligand in the active site). We have succeeded in encapsulating deoxy Hb in T or R quaternary conformation in transparent silica gels. Study of the temperature dependence of the Soret band Gaussian linewidth enables to investigate the effect of protein quaternary structure on the dynamic properties of the active site of deoxy Hb.

MATERIALS AND METHODS

Encapsulation of deoxy Hb was performed following the procedure of ref.(2), with some modification. The silica sol was prepared using 60% tetramethylortosilicate, 40% water and $4 \cdot 10^{-2}$ M HCl; this solution was sonicated for 20 minutes. The sonicated sol (2ml) was added to a deoxygenated solution containing 50μl Hb (~7.5%

CP513, *Nuclear and Condensed Matter Physics,* edited by A. Messina
© 2000 American Institute of Physics 1-56396-929-7/00/$17.00

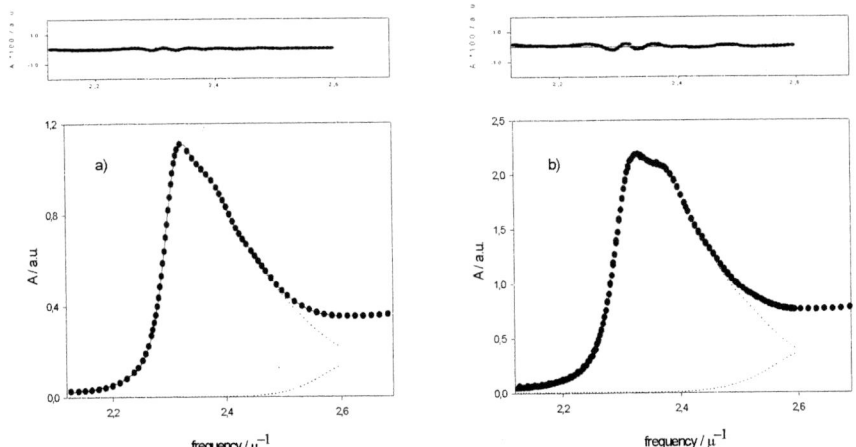

FIGURE 1. Deconvolution of the 30K spectra of deoxy Hb gel (a), and of oxy→deoxy Hb gel (b). Dots represent the experimental points, the continuous lines represent the fittings in terms of Eqs. 1-3. For figure readability not all the experimental points are shown. The residuals are also reported in the upper panels.

concentration), 2ml glycerol, 1ml phosphate buffer (0.1M, pH≅7) and ~3·10^{-4} M sodium dithionite. The mixture was mixed manually for 30 seconds and then laid on a methacrilate slide. The gel (typical thickness ~1mm) formed in about 2 min; it was immediately rinsed several times with a deoxygenated solution containing 65% glycerol, 35% phosphate buffer 0.1 M (pH≅7) in water and ~3·10^{-4} M sodium dithionite; the wet gel was stored at 4 °C for about 12 hrs in an helium atmosphere prior to the measurements. Oxy Hb was encapsulated with the same procedure; however, all steps were performed in air and sodium dithionite was not present. Prior to the spectral measurements oxy Hb gel was deoxygenated by rinsing several times with a deoxygenated solution containing 65% glycerol, 0.1M phosphate buffer pH 7 and ~3·10^{-4} M sodium dithionite, in an helium atmosphere. After about 4 hours, complete deoxygenation of the gel was obtained. We shall call this sample the oxy→deoxy Hb gel. It should be stressed that, at difference from what previously reported in the literature (2), our gels contain 40% glycerol; this modification was necessary to obtain transparent gels even at cryogenic temperature. The final protein concentration was ~5·10^{-5} M in heme. Spectra in the range 500-370 nm were recorded in digitized form at 0.5 nm intervals with a PC-IBM controlled Cary Varian 2300 spectrophotometer. The scan speed was 0.5 nm/s, the integration time 0.5 s and the bandwidth 0.5 nm, corresponding to a spectral resolution of about 25 cm^{-1} at 450 nm. The absorption spectrum of Hb is modeled as a convolution of three terms (3):

$$A(v) = Mv[L(v) \otimes G(v) \otimes P(v)] \tag{1}$$

where M is proportional to the square of the electronic dipole transition moment and v denotes the frequency. L(v) represents all the Lorentzian lines arising from the

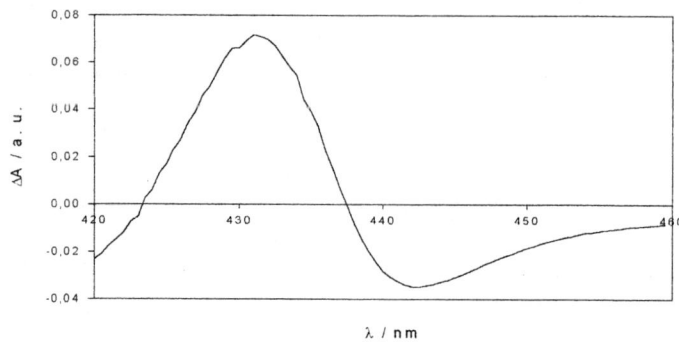

FIGURE 2. Difference spectrum between deoxy Hb gel and oxy→deoxy Hb gel. The two spectra measured at 290K have been subtracted after suitable normalization.

coupling of the electronic transition to the high frequency vibrational modes of the system, according to the Franck-Condon principle. $G(\nu)$ takes into account the coupling of the electronic transition to a bath of low frequency modes of the system. It can be shown that such a coupling causes a Gaussian distribution of the fundamental frequency:

$$G(\nu) = \frac{1}{\sqrt{2\pi}\sigma(T)} \times e^{-\frac{\nu^2}{2\sigma(T)^2}}$$ (2)

where $\sigma(T)$ is the temperature dependent Gaussian halfwidth. $P(\nu)$ takes into account the spectral heterogeneity caused by conformational heterogeneity. For deoxy Hb, the asymmetric distribution first introduced by Champion and coworkers (4) is used. Fits of the 30K spectra of the deoxy Hb and of the oxy→deoxy Hb gels in terms of Eq.1 are shown in Fig1; fits of analogous or better quality are obtained for the higher temperature spectra. If one considers the low frequency bath as a set of N degenerate harmonic oscillators with an average frequency $<\nu>$ and an average coupling constant S (harmonic Einstein approximation), the temperature dependence of the Gaussian width is given by:

$$\sigma^2_{Harm}(T) = NS <\nu>^2 \coth\left[\frac{h<\nu>}{2K_B T}\right] + \sigma^2_{in}$$ (3)

where the term σ^2_{in} reflects inhomogeneous broadening.

RESULTS AND DISCUSSION

The deoxy Hb gel – oxy→deoxy Hb gel difference spectrum at 290K is reported in Fig.2. It is almost identical to the classical "DeoxyT – DeoxyR" difference spectra reported in the literature (5) and shows that in our oxy→deoxy Hb gel the protein, although deoxygenated, retains the quaternary R conformation. Comparison of temperature dependence of spectra in our samples therefore enables to investigate the

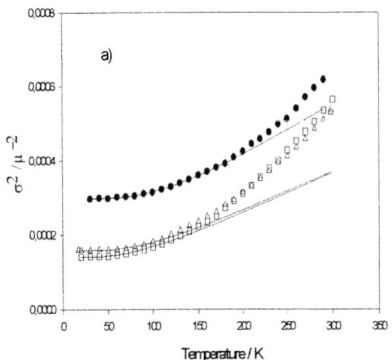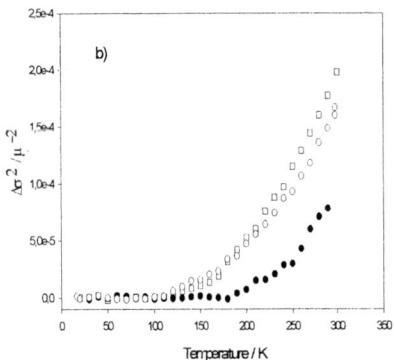

FIGURE 3. a) Temperature dependence of parameter $\sigma^2(T)$ for deoxy Hb in solution (□), for T-state deoxy Hb in gel (o), and for R-state deoxy Hb in gel (●). The continuous lines represent fittings of the data in the harmonic regime in terms of Eq.(3). b) Temperature dependence of $\Delta\sigma^2$ values obtained from the data reported in Fig.3a.

effect of quaternary structure on the dynamic properties of the heme pocket of deoxy Hb. The σ^2 temperature dependence is reported in Fig.3a.The solid lines represent fittings of the low temperature data points in terms of Eq.(3).Deviations from the harmonic behavior become evident at high temperatures and are attributed to the onset of anharmonic motions. They are put in evidence in Fig.3b where $\Delta\sigma^2$ values(i.e. the difference between the σ^2 values actually measured and the predictions of the harmonic model) are reported. Data relative to deoxy Hb in solution are also reported, for comparison, and show that the gelification process alone has only minor effects on the dynamics of deoxy Hb.On the contrary, the behavior of the oxy→deoxy Hb gel (i.e. deoxy Hb in R conformation) is very different from that of the deoxy Hb gel (i.e. deoxy Hb in T conformation). In particular, deoxy Hb in R conformation is characterized by larger conformational/spectral heterogeneity as shown by the higher σ^2 values observed at low temperature, by smaller anharmonic contributions whose onset occurs at higher temperature with respect to deoxy Hb in T conformation (190K as compared to 120K), and by larger average frequency of the soft harmonic modes (240 cm^{-1} as compared to 190 cm^{-1}).The data reported indicate that the protein quaternary conformation, at constant ligation state, has a profound influence on the dynamic properties of the active site. In particular, deoxy Hb in T quaternary conformation appears to be characterized, at room temperature, by much larger anharmonic motions with respect to deoxy Hb in R quaternary structure.

References

1.   Ellerby, L.M., Nishida, C.R., Nishida, F., Yamanaka, S.A., Dunn, B., Valentine, J.S., Zink, J.I., *Science*, **255**, 1113-1115, (1992).
2.   Shibayama, N., Saigo, S., *J .Mol. Biol.*, **251**, 203-209, (1995).
3.   Cupane, A., Leone, M., Vitrano, E., Cordone, L., *Eur.Biophys.J.*, **23**, 385-398, (1995).
4.   Srajer, V., Champion, P.M., *Biochemistry*, **30**, 7390-7401, (1991).
5.   Perutz, M.F., Landner, J.E., Simon, S.R., Ho, C., *Biochemistry*, **13**, 2163-2173, (1974).

Micro- and Mesoscopic Process Interactions in Protein Coagulation

P.L. San Biagio[1,3], V. Martorana[1,3], A. Emanuele[1,2], S.M. Vaiana[1,2]
M.Manno[1,2,3], D. Bulone[1,3], M.B. Palma-Vittorelli[1,2], and M.U. Palma[1,2,3]

[1]Progetto Sud European Union and INFM, Genova, and Crrnsm, Palermo, Italy
[2]Dept. of Phys. and Astr. Sciences, University of Palermo, Via Archirafi 36, I-90123 Palermo
[3]CNR Institute for Interdisciplinary Applications of Physics, Palermo, Italy

Abstract. It has recently been recognized that pathological protein coagulation is responsible for lethal pathologies as diverse as amyloidosis, Alzheimer and TSE. Understanding the coagulation mechanisms is therefore stirring great interest. In previous studies we have shown that on profoundly different systems coagulation is the result of a strong interaction between two processes on different length scales (mesoscopic and microscopic). Here we report experiments on bovine serum albumin (BSA) showing that the overall mechanism is the result of at least 3 distinct and strongly intertwined processes, on both length scales: molecular conformational changes, solution demixing and intermolecular crosslinking. This mechanism involves the statistical mechanics of protein-solvent interaction, its relation to the protein's landscape of configurational free energy and to the solution's thermodynamic stability, and its relation to the topological problem of crosslink-percolation, responsible for coagulation.

INTRODUCTION

Much interest is currently focussed on protein coagulation leading to amyloidoses and neurodegenerative diseases such as Alzheimer and Transmissible Spongiform Encefalopaties. Coagulation involves intermediate protein conformations with increased β-structure content. Bovine serum albumin (BSA) is of interest because, similarly to cases leading to amyloidosis, its coagulation requires a change toward an intermediate conformation, and it can occur even at very low concentration.

It was shown in San Biagio et al. [1] that by incubating 0.1% wt/wt solutions for 20 min at 67 °C a partially unfolded and reasonably stable intermediate form of BSA is obtained. The phase diagram of this intermediate was determined. As shown in Fig.1a, in consequence of the conformational change, the instability region of the solution is shifted downwards to accessible temperatures. In this region the system undergoes a well characterized spinodal demixing [1]. In the resulting high-concentration regions coagulation is promoted [2-9], even at very low overall concentration. The same sequence of demixing and coagulation can occur when solutions of the native form of BSA at a concentration of, say, 1%, are rapidly brought above 65-67 °C . Indeed, as we shall show, at this temperature the protein quickly adopts the intermediate conformation and therefore the solution finds itself in its instability region.

CP513, *Nuclear and Condensed Matter Physics,* edited by A. Messina
© 2000 American Institute of Physics 1-56396-929-7/00/$17.00

Here we show, on the basis of Ref.1 and of new experiments, that demixing of BSA solutions is closely intertwined with molecular cross-linking and with an increase of the β/α structure ratio characterizing the intermediate conformation required for BSA coagulation. This three-process multiple-feedback interaction discloses a richer scenario for understanding micro- and mesoscopic steps as well as thermodynamic drives towards protein coagulation. The same crucial three-process interplay leading (among other things) to local coagulation even at very low overall concentrations, has been recently evidenced in the case of a nonpeptidic polymer [10], and in further ongoing work. Thus it could be a fairly common feature in polymer coagulation.

TECHNICAL DETAILS

Crystallized BSA from fraction V was purchased from Miles Diagnostic (code 81-001, lot 91, purity 99%) and used without further treatments. Solutions of 0.1%wt/wt (0.73% vol/vol) concentration at pH=6.2 (0.1 mol/L phosphate buffer) were used. Fig.1b refers to upward temperature scans (13 °C/hour) on aliquots of the same solution. Tracings in the figure were smoothed by standard procedures.

For circular dichroism a Jasco 715 instrument was used. We measured CD_{197} (solid light grey line) and CD_{208} (solid dark grey line), to monitor secondary structure conformational changes (relative β/α ratio), and CD_{290} (thick black line), to monitor tertiary structure changes (a decrease in this line corresponds to a more compact configuration). Differential scanning calorimetry (DSC) data are taken from Ref.1 (dotted line). Turbidometry (dashed black line) monitored "cloud point" and coagulation. Hydrodynamic radius (thin black line), obtained from the dynamic light scattering equipment of Ref.1), monitored both the actual hydrodynamic radius of Brownian particles in solution and the "characteristic length" of diverging concentration fluctuations (typical of demixing). Viscoelastic modulus, G^* (dashed grey line) was measured using a Rheometrics RFS II Rheomechanical Spectrometer, and monitored cross-linking and consequent gelation.

RESULTS AND CONCLUSIONS

The progressive changes of CD_{197} and CD_{208} up to about $T^*=67$ °C (Fig.1b) reflect a progressive increase of the β/α ratio [11]. The CD_{290} signal indicates, in turn, that the tertiary structure also changes, becoming monotonically more compact up to 60 °C. It shows a trend inversion in the 60°C-T^* interval, consistently with the endothermic DSC signal and with the moderate increase and substantial monodispersity of the hydrodynamic radius R. Note that the conformation at T^* corresponds to the "intermediate" whose phase diagram was shown in Fig.1a, and is now shown to be richer in β structures and more open. Before reaching T^* we see an abrupt increase of turbidity, monitoring demixing and consequent coagulation in higher-concentration regions [10]. The further increase of R thus includes contributions from oligomers and

215

from mesoscopic high concentration regions related to spinodal demixing. A decreased exposure of hydrophobic chromophores to solvent (inversion of the slope of the CD_{290} signal at T^*) and a marked increase of the rate of β-structure formation (break in both the 197 and 208nm CD traces at T^*) are induced by demixing. At higher temperatures, the sharp increase of viscoelastic modulus correlates with a marked increase of the negative slope of the CD_{290} signal, revealing a packing of chromophores that can be ascribed to coagulation and related tertiary structure changes. Summing up: a conformational change towards a β-rich intermediate makes the solution unstable, the consequent demixing amplifies local concentration values in mesoscopic regions, where coagulation is promoted. Further conformational change is promoted by demixing/coagulation, and so on, as visualized in Fig.1c.

In conclusion, three processes have been observed: protein unfolding and misfolding, solution demixing, and protein cross-linking. They are simultaneously present and influence each other (Fig.1c). The same intertwining of processes has also been observed in nonpeptidic polymers [10], as expected from intra- and interpolymer (/protein) interactions *via* the solvent. Thus, it must be taken into serious consideration for a thorough understanding of amyloid deposition . The picture emerging from the present work goes well beyond the simple description in terms of deposition of β-rich intermediates.

REFERENCES

1. P.L. San Biagio, D. Bulone , A. Emanuele and M.U. Palma. *Biophys. J.*, **70**: 494, 1996.

2. P.L. San Biagio, F. Madonia, J. Newman and M.U. Palma. *Biopolymers*, **25**: 2255, 1986.

3. P.L. San Biagio, J. Newman, F. Madonia and M.U. Palma. *Chem. Phys. Lett.*, **154**: 477, 1989.

4. P.L. San Biagio, D. Bulone, A. Emanuele, F. Madonia, L. Di Stefano, D.Giacomazza, M. Trapanese, M.B. Palma-Vittorelli, and M.U. Palma. *Makromol. Chem., Macromol. Symp..*, **40**: 33, 1990.

5. A. Emanuele, L. Di Stefano, D. Giacomazza, M. Trapanese, M.B. Palma-Vittorelli, and M.U. Palma. *Biopolymers*, **31** : 859, 1991.

6. P.L. San Biagio and M.U. Palma. *Biophys. J.*, **60**: 508, 1991.

7. A. Emanuele and M.B. Palma-Vittorelli. *Phys. Rev. Lett..*, **69**: 81, 1992.

8. F. Sciortino, K.U. Prasad, D. W. Urry, M. U. Palma. *Biopolymers*, **33**: 743, 1993.

9. P.L. San Biagio, D.Bulone, M.B. Palma-Vittorelli, M.U. Palma. *Food Hydrocolloids,* **10**: 91, 1996.

10. a) M. Manno and M.U. Palma. *Phys. Rev. Lett..*, **79**: 4286, 1997. b) M. Manno, A. Emanuele, V. Martorana, D.Bulone, P.L.San Biagio, M.B.Palma-Vittorelli, M.U.Palma.*PhysRevE*,**59**,2222, 1999.

11. S.Y. Venyaminov, J.T. Yang, "Determination of Protein Secondary Structure" in *Theory of Circular Dichroism and the Conformational Analysis of Biomolecules,* ed. by G.D. Fasman, N.Y.: Plenum Press, 1996, pp. 69-107.

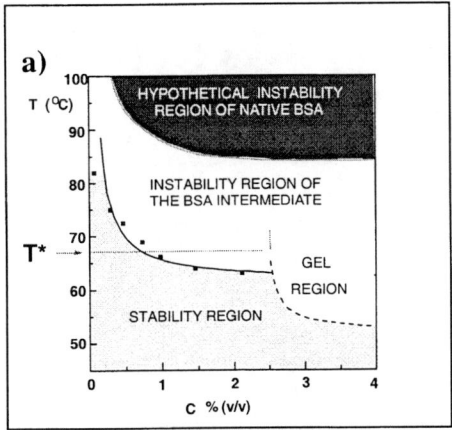

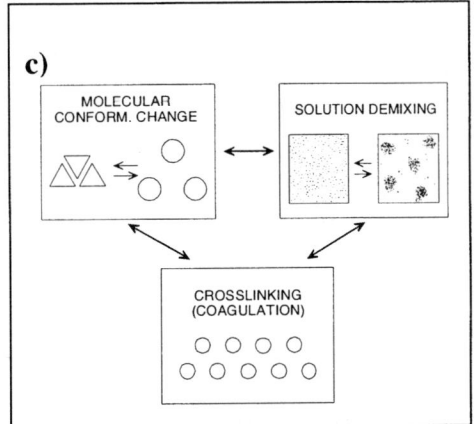

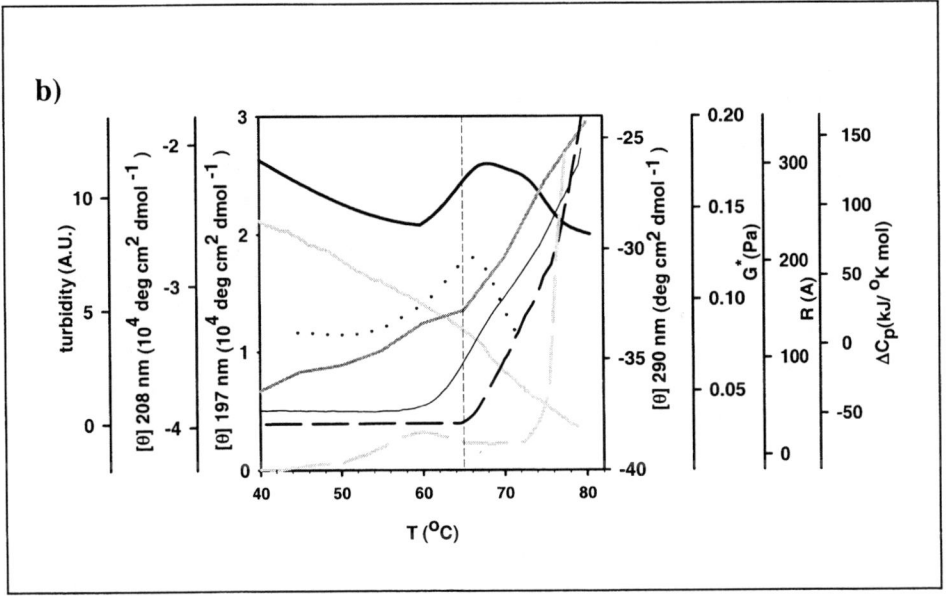

FIGURE 1. a) Phase Diagram (redrawn from Ref.1) of solutions of BSA intermediate, obtained by incubating at 67°C for 20 min. b) Schematic representation of the three processes involved in coagulation of BSA (molecular conformational change, solution demixing, and molecular cross-linking) and of their interactions. c) Results of upward temperature scans (at 13°C/h) on different aliquots of the same BSA solution (0.1% wt/wt, pH=6.2, phosphate buffer 0.1 mol/L): CD_{290} (thick black line), CD_{197} (solid light grey line), CD_{208} (solid dark grey line), differential scanning calorimetry ΔC_p from Ref.1 (dotted line), hydrodynamic radius R (solid thin black line) , turbidity (dashed black line), viscoelastic modulus G^* (dashed grey line).

Structural And Dynamic Properties Of Bulky Ligand Derivatives Of Hemeproteins

Vincenzo Sanfratello*, Alberto Boffi [^], Antonio Cupane*
and Maurizio Leone*

* Department of Physical and Astronomical Sciences and Istituto Nazionale di Fisica della Materia,
University of Palermo, 90123 - Palermo (Italy).
[^]CNR, Center of Molecular Biology, Department of Biochemical Sciences, University Roma La
Sapienza, 00185 - Roma (Italy)

Abstract We report the visible absorption spectra of nicotinate and carbonmonoxy derivatives of soybean leghemoglobin-a and horse myoglobin. The fine structure of the optical bands was taken into account by a deconvolution procedure in terms of vibrational Condon and non-Condon couplings of the heme normal modes to the Q electronic transitions, in the framework of the Herzberg-Teller approximation. The local dynamics of the heme pocket for all the derivatives have been investigated through the temperature dependence, down to cryogenic temperatures, of the spectral line broadening. The results indicate that leghemoglobin-a better accommodate, with respect to horse myoglobin, the structural distortions imposed within the heme pocket by the bulky nicotinate group. The comparative study of the carbonmonoxy derivatives suggests that nicotinate induces smaller heterogeneities and smaller anharmonicities for both proteins.

INTRODUCTION

In understanding the relationships between structure, dynamics and function in hemeproteins, the study of bulky ligand adducts can give precious information in view of the large structural distortions induced within the heme pocket.

In this work, we report a study of the effects of nicotinate binding (-Nic) to two proteins that, although presenting the same globin fold and the same proximal and distal histidines, differ for chemical and functional behavior (1,2): the well known myoglobin (HoMb), that shows a very low affinity for this ligand, and the monomeric soybean leghemoglobin-a (LbHb), that constitutes the natural carrier of nicotinate binding it with high affinity. The experimental approach is based on the analysis as a function of temperature of the optical absorption spectra in the vis-Uv region, whose fine structures are largely dominated, in the case of hemeproteins, by vibronic coupling of the electronic transitions with local normal modes of the macrocycle. In particular,

CP513, *Nuclear and Condensed Matter Physics,* edited by A. Messina
© 2000 American Institute of Physics 1-56396-929-7/00/$17.00

Q bands (450÷600 nm), that originate from an electronically forbidden porphyrin $\pi-\pi^*$ transition according to Gouterman's four orbital model (3), appear widely structured and sensitive to electronic and structural distortions of the heme pocket (4). Moreover, as for the largely studied Soret band (5,6), the broadening of the spectral line as a function of temperature for Q bands is attributed to the coupling of the electronic transition with a bath of soft vibrational modes of the heme-protein complex, thus giving information on the local dynamic properties of heme complex in the surrounding of heme group. The effect of bulky group binding is also put in evidence by the comparison with carbonmonoxy derivatives (-CO) of the same proteins.

MATERIALS AND METHODS

Horse heart myoglobin was obtained from Sigma; leghemoglobin-a is a generous gift of Dr. Jonathan Wittenberg from Albert Einstein College of Medicine. All samples contained 65% (v/v) glycerol/water as solvent and $\approx 10^{-4}$ M of protein. The nicotinate derivatives were obtained by adding at cold 1 M nicotinate buffer, pH 6.2. The carbonmonoxy derivatives were obtained by bubbling CO into the solution.

Optical absorption data were taken by using a Cary 2300 spectrophotometer, with spectral resolution of 0.4 nm; the baseline, measured at room temperature, was subtracted from each spectrum. The analysis of the band profiles was performed by adopting a deconvolution procedure that includes the first order in the electronic dipole moment expansion with respect to the high-frequency vibrational modes of the heme complex (Herzberg-Teller approach)(4,7). The coupling of the electronic transition with the bath of soft vibrational modes of the macrocycle is modeled by a Gaussian distribution of the 0-0 transition frequency, resulting in a temperature dependent broadening (σ^2). In the framework of harmonic approximation and considering the low frequency bath as a set of N harmonic oscillators of average frequency $<\nu>$ and average coupling constant S (Einstein approximation), the temperature dependence of the Gaussian half-width can be expressed as:

$$\sigma_{harm}^2 (T) = N\, S <\nu>^2 \coth\left[\frac{h<\nu>}{2K_B T}\right] + \sigma_{in}^2 \qquad (1)$$

The subscript "harm" indicates that the above expression is valid only in the harmonic regime; the σ_{in}^2 term represents the temperature independent inhomogeneous broadening of the band.

RESULTS AND DISCUSSION

Fig. 1 shows the visible spectra for all the sample studied as a function of temperature. All the spectra consist of two main structures, the low frequency one

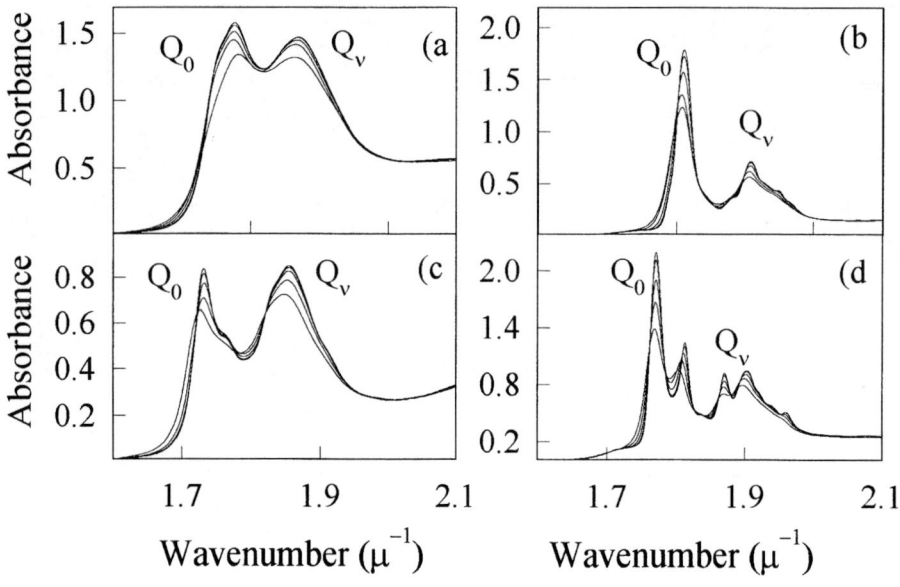

Figure 1. The temperature dependence of Q bands for LbHb-CO (a), LbHb-Nic (b), HoMb-CO (c) and HoMb-Nic (d).

being attributed to the fundamental electronic transition (Q_o) and the other (Q_v) to non-Condon coupling of Q_o with high-frequency vibrational modes of the macrocycle (4). The inspection of the raw data shows that the Nic-CO substitution causes a narrowing of each spectral component and substantial changes in the Q_o/Q_v intensity ratios. Moreover, a clear splitting of the Q_o band is evident for HoMb-Nic; this type of splitting is present in a number of hemeproteins in rigid host matrix and has been ascribed to a symmetry reduction, i.e. a ring deformation, of the heme with respect to the natural D_{4h} symmetry. The comparison with the spectra reported in Fig. 1 (c), where if present the splitting is much less pronounced, suggests that nicotinate induces into heme pocket of myoglobin a noteworthy ring deformation. Figs. 1 (a) and (b) show that for LbHb no Q_o splitting is present, also for nicotinate derivative, so indicating that the structural deformation induced by bulky group can be accommodate in the heme pocket of this protein.

The effects of Nic-CO substitution on the local dynamic properties of heme pocket is sorted out by the results shown in Fig. 2. Data in Fig. 2 obey Eq. 1 only in a limited temperature interval; deviations from the harmonic behavior (continuous line in Fig. 2) are evident at temperatures higher than T=180 K. These deviations are much larger in the case of LbHb-CO and essentially disappear for LbHb-Nic, indicating that in this protein the Nic-CO substitution reduces the iron mobility with respect to the heme plane, that was considered the principal source of anharmonicity (5).

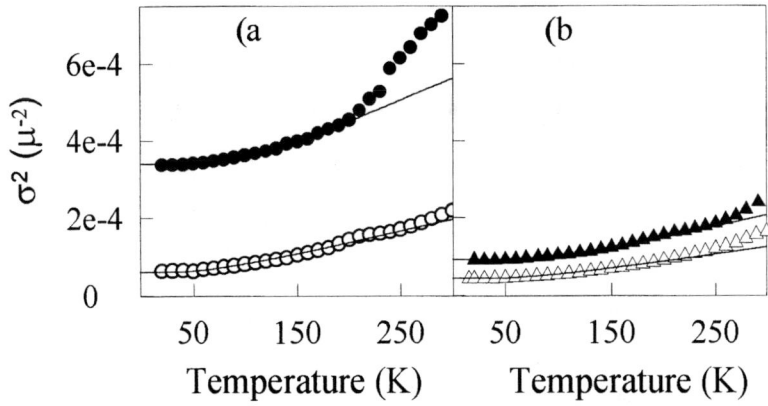

Figure 2. Thermal behavior of Gaussian width: a) LbHb-CO (full circles) and LbHb-Nic (open circles); b) HoMb-CO (full triangles) and HoMb-Nic (open triangles). The continuous lines represent the fitting of Eq. 1 to the experimental data.

In HoMb, the Nic-CO substitution does not induce large dynamic effects, indicating that the heme pocket for this protein is more closely packed with respect to LbHb. A further difference between the two proteins arises from the low temperature σ^2 values (essentially due to the σ_{in} values). The reported results indicate that the Nic-CO substitution causes a less pronounced spectral heterogeneity, this effect being particularly evident for LbHb. Since larger heterogeneities can be due to the presence of conformational substates of the ligand within the heme pocket, the results for LbHb indicate that the heme pocket for this protein is more open with respect to HoMb. For both proteins, however, the presence of nicotinate reduces the accessible conformational substates.

REFERENCES

1. Appleby, C.A., Bradbury, J.H., Morris, R.J., Wittenberg, B.A., Wittenberg J.B., and Wright, P.E., *J. Biol. Chem.*, **258**, 2254-2259 (1983).
2. Hargrove, M.S., Barry, J.K., Bruker, E.A., Berry, M.B., Phillips Jr., G.N., Olson, J.S., Arredondo-Peter, R., Dean, J.M., Klucas, R.V., and Sarath, G., *J. Mol. Biol.*, **266**, 1032-1042 (1997).
3. Gouterman, M., *J. Mol. Spectrosc.*, **6**, 138-163 (1961).
4. Sanfratello, V., Boffi, A., Cupane, A. and Leone, M., submitted.
5. Cupane, A., Leone, M., Vitrano, E., and Cordone, L., *Eur. Biophys. J.*, **23**, 385-398 (1995).
6. Leone, M., Cupane, A., Militello,V., and Cordone, L., *Eur. Biophys. J.*, **23**, 349-352 (1994).
7. Sanfratello, V., *Ruolo delle distorsioni indotte da leganti di grosso ingombro sterico sulle proprietà stereodinamiche del sito attivo di emoproteine*, Thesis (1998).

Modelling Small Angle Neutron Scattering Data from Polymers in Supercritical Fluids

F.Triolo [a,b], A. Triolo [b,c], F. Lo Celso [b], J.S. Johnson, Jr., D.I. Donato [b], and R. Triolo [b]

[a]*Mount Sinai School of Medicine, New York, NY, USA*
[b]*Department of Physical Chemistry, University of Palermo, Palermo, Italy*
[c]*Heriot-Watt University, Edinburgh, UK*

Abstract. In this paper we report a SANS investigation of micelle formation by fluorocarbon-hydrocarbon block copolymers in supercritical CO_2 (scCO$_2$) at 313K. A sharp unimer-micelle transition is obtained due to the tuning of the solvating ability of scCO$_2$ by profiling pressure. At high pressure the copolymer is in a monomeric state with a random coil structure. By lowering the pressure aggregates are formed with the *hydrocarbon* segments forming the core and the *fluorocarbon* segments forming the corona of spherical aggregates. This aggregate-unimer transition is driven by the gradual penetration of CO_2 molecules toward the core of the aggregate and is critically related to the density of the solvent, thus suggesting the definition of a critical micellisation density (CMD).

INTRODUCTION

Super- and near-critical fluids (SCF) have a high compressibility which facilitates large changes in solvent density with pressure. Since solvation of solutes in a SCF increases dramatically with density, this feature can be exploited to control solubilities through both temperature and pressure variation. CO_2 has been most widely researched as a SCF solvent medium being cheap and offering added safety benefits of non-toxicity and non-flammability. However, solubilities of highly polar solutes or apolar macromolecules are often low in supercritical CO_2 (scCO$_2$). Related solubility studies of polymer systems[1] have also demonstrated a higher compatibility of fluorocarbon chains with CO_2, particularly flourinated octyl acrylates (PFOA) . We have reported[2] a SANS study of a block copolymer (PS-b-PFOA) in scCO$_2$ composed of "CO$_2$-phobic" polystyrene (PS) and "CO$_2$-philic" fluorinated octyl acrylate (PFOA). We classified these polymers using their average molecular weights as $<M_n>_S$-b-$<M_n>_{FOA}$. Owing to the difference in solvation by CO_2 of the PS and PFOA blocks the polymers were found to self-assemble into "micelles" with the "CO$_2$-phobic" hydrocarbon blocks forming a central core surrounded by a shell of "CO$_2$-philic" fluorocarbon

CP513, *Nuclear and Condensed Matter Physics*, edited by A. Messina
© 2000 American Institute of Physics 1-56396-929-7/00/$17.00

blocks extending into and solvated by the CO_2. Support for this structure of the aggregates was provided by the SANS data which could be fitted using a simple core-shell model to quantify the aggregation number (N_{agg}) and dimensions of the core (R_{core}) and micelle (R_{total}). The state of aggregation in this and in similar systems[3] depends strongly on the density of the SCF suggesting a shift towards micelle breakdown with increasing "solvation" by the CO_2 and proposing the existence of a critical micellisation density (CMD) for these systems[4].

EXPERIMENTAL

We have investigated the small-angle scattering behavior of a block copolymer in scCO$_2$ containing a "CO$_2$-philic" block (43.1K Dalton PFOA) and a "CO$_2$-phobic" moiety (Poly vinyl acetate, 10.1K Dalton PVAc) and another polymer in which a methyl group of the PVAc moiety was partially deuterated (7.6K Dalton D3PVAc). Experiments were conducted over a range of polymer concentration (6% to 12% w/V) as a function of pressure. Experiments were performed in UK (Rutherford Appleton Laboratory) and in USA (Oak Ridge National Laboratory). SANS measurements were made at 40 °C in the pressure range 190 to 480 bar. The magnitude of the momentum transfer vector (Q=4πλ^{-1} sin(θ), where λ is the incident wavelength and 2θ is the scattering angle) varies between 0.01 and 0.22 Å^{-1} for the experiments performed in UK and between 0.006 and 0.06 Å^{-1} for the experiments performed in USA.

The data were corrected for transmission, incoherent background scattering and normalised to absolute scattering probabilities using standard procedures[3,4]. SANS measurements were performed on the PVA-b-PFOA at three different concentrations (6%, 8% and 12% w/v). All solutions were examined in SCF CO$_2$ at 40 °C over a pressure range of 180 to 450 bar.

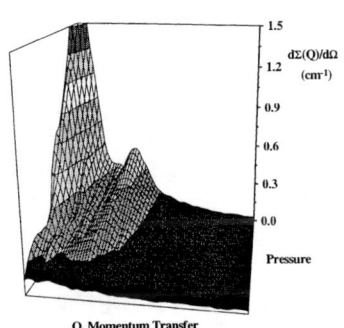

Figure 1. Pressure scan of 8% PVAc-b-PFOA (10.1K/43.1K) in scCO2 at 40 °C.

RESULTS/DISCUSSION

Figure 1 shows the effect of pressure on the scattering function of the 8% (w/v) solution of the PVA-b-PFOA polymer (10.1k-b-43.1k) at 40 °C. Similar behaviour was observed for both the 6% and the 12% solutions. The appearance of a peak accompanied by a great increase in intensity on decreasing the CO_2 pressure is a clear indication of formation of aggregates. All the solutions began to appear opalescent below about 240 bar while the lower pressure phase boundary (LPPB) was found by visual inspection close to 180 bar. This behaviour is common to all SCF solutions for which the solute becomes increasingly less solvated as the pressure is decreased until a solubility boundary is reached at a given pressure. In this regime of incipient instability the micellar aggregates experience stronger interactions which lead to critical fluctuations. The net result is a Lorentzian contribution in the scattering function. Alternatively, when the polymer is not monodisperse, the SCF solvent may be an overall good solvent, although being a poor solvent for the high molecular weight fractions of the polymer[5].Also in this case a Lorentzian contribution in the scattering function is possible[5]. In the regime of low density the neutron scattering curves are consistent with the presence of core/shell micellar aggregates[3,4] (see figure 2). When the pressure is raised, the solubility of the hydrocarbon blocks increases and the CO_2 "invades" the core of the micelles so that ultimately micelles are no longer stable and they break at high pressure. In this high density regime the SANS experiments are then consistent with the presence of random coils of monomeric polymer. This transition from micelles to monomer is reasonably sharp, falling somewhere between 310 and 340 bar and involves all the polymer present. We have used the model developed in a previous paper[3,4] to obtain structural parameters for the copolymer solutions examined. To validate the model we have also performed experiments on the D3PVAc-b-PFOA (7.6K-b-43.1) by making use of the isotopic substitution method and we found good agreement with the results obtained with the H polymer. Similar results are obtained when the temperature is changed. Although concentration and temperature have a small effect on the transition pressure, the critical micellisation density is mainly determined by the solvent density. In fact, the unimer-aggregate transition at 65 °C (Figure 3) happens at higher pressure than at 40 °C, although the density of the solvent is about the same.

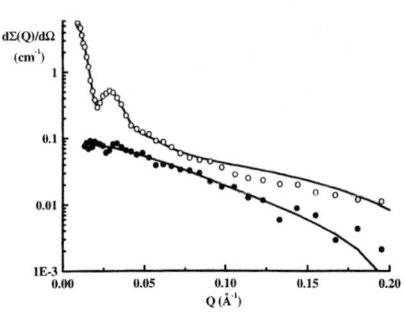

Figure 2. The low density regime (aggregated polymer), open symbol, and the high density regime (random coil), solid symbol. See text for details.

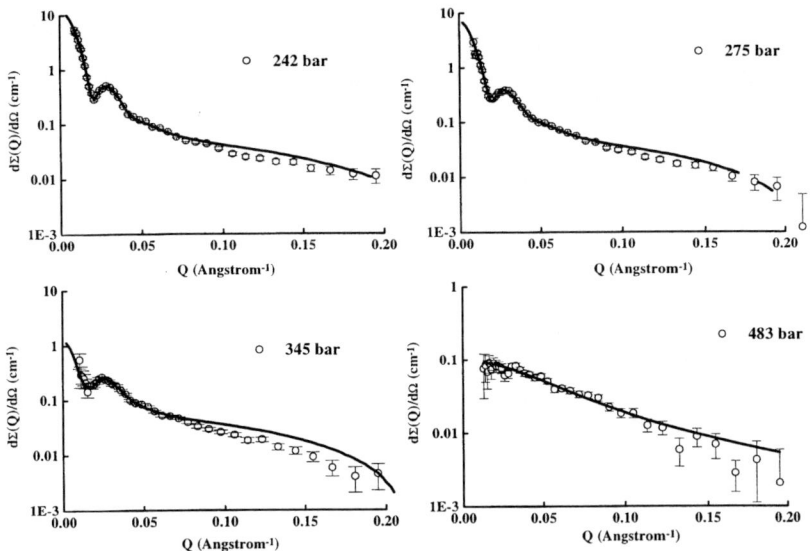

Figure 3. SANS curves at 65 °C for various pressures. At 483 bar the polymer has a random coil structure, while below about 390 bar the polymer aggregates to form core-shell micellar structures.

The fact that at the higher pressures studied (> 340 bar at 40 °C) the polymer has a random coil configuration and upon reaching the CMD begins to aggregate with the hydrocarbon core significantly solvated by the CO_2, makes this class of material interesting for extraction work and in general to perform synthesis of polymers without using enviromentally dangerous solvents. In addition these materials can be used as carriers of chemicals to protect woods and other porous materials. Industrial application of the CMD involves garment dry cleaning and precision cleaning of high-tech components.

REFERENCES

1) DeSimone, J.M. et al., *Macromolecules*, submitted.
2) McClain, J. B. et al., *Science*, **274**, 2049-2052 (1996)
3) Triolo, F. et al., *Langmuir*, **16** (2), 416-421 (2000)
4) Triolo, F. et al., *Phys. Rev. E*, in press.
5) Shimizu, S., Kimura, M., Maruyama, T. & Kurita, K., *J. Appl. Cryst.*, **30**, 712-716 (1997).

Phase Separation Process In Commercial Al-Li Alloys

A. Triolo[a], F. Lo Celso[b], R. Triolo[b]

[a] Department of Chemistry, Heriot Watt University, Riccarton Edinburgh EH14 4AS United Kingdom
[b] Dipartimento di Chimica Fisica, Universita' di Palermo, V.le delle Scienze, Parco d'Orleans II, I-90128 Palermo (Italy)

Abstract. Real-Time Small Angle X-ray Scattering measurements are reported on a commercial aluminum-lithium alloy. This material shows a miscibility gap as soon as it is quenched from high to low temperature, with segregation of a Li-rich phase. The SAXS technique confirms as a valid probe to detect structural features in these alloys. Moreover a detailed analysis is presented for the whole SAXS data set, leading to the time dependence of structural parameters.

INTRODUCTION

A great deal of investigation has been carried on in the last decades on the nature of morphological properties of aluminum-lithium (Al-Li) alloys, due to their relevance in the aircraft industry and their possible application for high-performance secondary batteries.

Al-Li alloys are a suitable substitute to pure aluminum, due to their reduced density and high strength. Most of the mechanic properties of such alloys are strictly related to phase separation phenomena, as it is well known that in the low temperature regime (i.e. T lower than 250 C), precipitation of the coherent δ' phase occurs [1-6]. This phase has a composition nearly close to Al_3Li, and its crystalline parameters are quite close to the ones of the matrix, so that no relevant residual stresses develop during the phase separation process and almost spherical clusters of the minority phase develop [7]. In the present communication, we report on the kinetics of phase separation occurring in commercial Al-Li alloys at low temperature as seen by means of the Small Angle X-ray Scattering technique.

EXPERIMENTAL

SAXS measurements were performed on the 10 m SAXS camera at the Oak Ridge National Laboratory (USA) [8]. Data were collected at the Cu Kα radiation. The sample-to-detector distance ranged between 1 and 5 m. Corrections were made for detector background and sensitivity and the isotropic area detector data were converted to radial averages. The scattering cross sections, $d\Sigma(Q)/d\Omega$, were obtained by calibration with secondary standards provided at the instrument [9]. The alloy we studied is a commercial Al-Li sample (nominally 2090 T8E41), whose Li content is

CP513, *Nuclear and Condensed Matter Physics,* edited by A. Messina
© 2000 American Institute of Physics 1-56396-929-7/00/$17.00

8.00 at. % (for a detailed table containing information on the alloy composition we refer the reader to ref. 7). For details on the experimental procedure we refer to previous work [7].

RESULTS AND DISCUSSION

It is very well known that quench of an Al-Li alloy from a high temperature state (e.g. above 500 C) to a low temperature state (e.g. lower than 200 C) induces a phase separation process, with a δ' phase separating from the homogeneous matrix. The composition difference between δ' phase and the matrix is such that sufficiently high electron density contrast occurs to lead to an appreciable SAXS pattern. It has been shown in the past that it is possible to model the SAXS profiles in terms of a quite simple model, describing the phase separated δ'

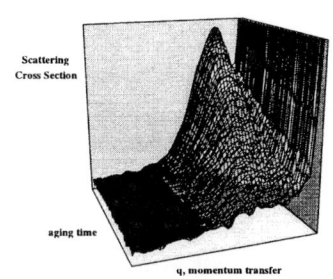

FIGURE 1. Time dependence of the SAXS profile for a 2090 Al-Li alloy, quenched at 75 C.

domains as almost spherical clusters, whose sizes evolve with time according to well defined power laws.

In figure 1 the time dependence of the SAXS profile for an aging process occurring at 75 C. It is evident that in the early stages of the process no interference peak is present, thus indicating the lack of correlation between the clusters, while as soon as the aging time increase a progressive development of an interference peak occurs.

This observation can be confirmed by quantity, which gives an average information of the SAXS profiles: the SAXS invariant [10]. This is defined as:

$$Q_o = \int_0^\infty Q^2 d\Sigma(Q)/d\Omega dQ$$

and contains information on the total amount of the minority phase, as it is: $Q_o = \phi(1-\phi)\Delta\eta^2$, where ϕ is the volume fraction of the minority phase and $\Delta\eta^2$ is the contrast between minority phase and the matrix. The time dependence for $\log(-\log(1-(Q_o/Q_{o,inf})))$ [11] during the aging at 75 C is reported in figure 2. It is evident the occurrence of a well-defined transition between two linear regimes.

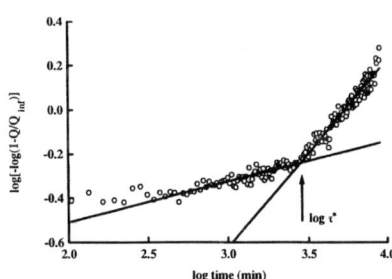

FIGURE 2. Time dependence of the invariant, Q_o, during aging at 75 C. The existence of a definite transition between two regimes is stressed.

In order to further explore the morphological evolution of this system during the phase separation, we modeled the SAXS profiles by means of the following model:

$d\Sigma(Q)/d\Omega = Bkg + C/Q^4 + \rho P^2(Q)S(Q)$, where $P(Q)$ is the Form Factor, $S(Q)$ is the Structure Factor, C/Q^4 is a Porod-like term to describe the low Q behavior, Bkg is a flat incoherent background and ρ is the particle number density. At the low temperature we are exploring, it is well known that the cluster morphology can be described by means of an almost monodisperse distribution of spherical objects whose size evolve with aging time. In these conditions, we have: $P(Q) = \Delta\eta^2[\sin(\kappa) - \kappa\cos(\kappa)]/\kappa^3$, where κ equals $QR_{core}/2$, R_{core} being the size of the spherical scatterers, and $\Delta\eta^2$ is the contrast. Moreover the Structure Factor is described with a hard sphere model in the Percus-Yevick approximation and depends on the volume fraction of the segregated phase, η, and from the inter-cluster distance, R_{str} (we refer the reader to a more detailed description of the model in ref. 7). This modeling then involves the use of three fitting parameters, namely, R_{core}, R_{str}, and η, while the Porod Term, the flat background and the contrast were fixed to their average values

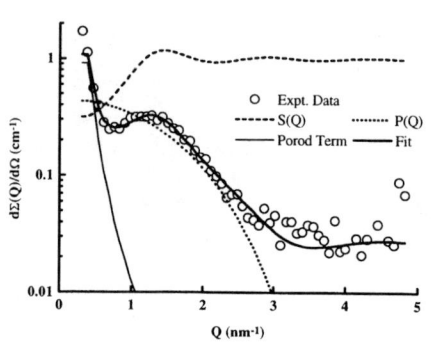

FIGURE 3. Representative fit of SAXS data collected from an Al-Li alloy, aged at 75 C. The different contributes to the whole profile are reported.

during the whole time evolution as the only showed random evolution. In figure 3, a representative example of the fitting result is reported and an overall agreement of the model with the experimental data set can be observed. Figure 4 reports the structural parameters obtained from the fit. It can be seen that R_{core} shows a strong discontinuity in its temporal evolution in the same time range where the invariant changes itself. On the other hand, the R_{str} parameter does not show an appreciable change in slope in the same time regime, although a more careful analysis needs to be done in order to understand the nature of the strong noise in these data. In figure 5, a comparison between the invariant and the volume fraction (as obtained from the fitting procedure) time dependence is reported. It can be seen that these two parameters show an almost parallel behavior, this being an indication of the goodness of the fitting model, as they were obtained by means of independent

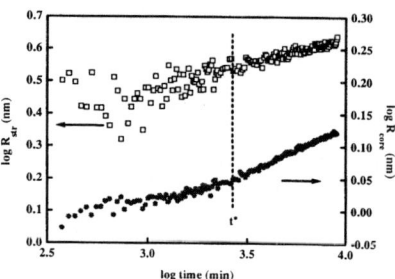

FIGURE 4. Fitting parameters as obtained from analysis of the SAXS profiles during aging of an Al-Li alloy at 75 C.

approaches. Once more a discontinuity is seen in the same time regime. The very same kind of discontinuity has also been observed by simply plotting the position and intensity of the SAXS patterns in the maximum position. A strong evolution of Q_{max} with aging time has also been observed in the time regime where all the other parameters show a discontinuity.

Analysis is still in progress to understand the nature of the observed discontinuity. Nevertheless results so far collected seem to indicate that the morphological changes can be related to the triggering of the correlation between distant clusters.

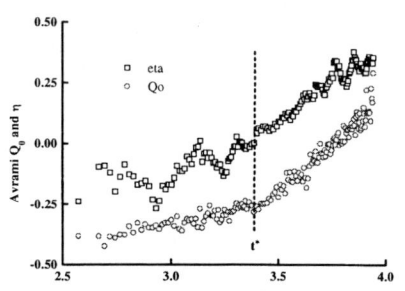

FIGURE 5. Comparison between temporal evolutions of η and Q_o, see the text for details.

CONCLUSION

SAXS measurements have been reported for describing the temporal evolution of phase separation process occurring in a commercial Al-Li alloy. The existence of two different time regimes in the early stages of demixing has been observed. This result was confirmed by application of different analysis approaches aiming to gain structural information on the separating phase. Work is still in progress to try to gain a better understanding of the processes responsible for the observed behavior.

ACKNOWLEDGMENTS

The research was supported in part by the Italian Ministero dell'Universita' e della Ricerca Scientifica (MURST) and the Italian National Research Council (CNR).

REFERENCES

1) B. Noble and G.E. Thompson, Metal Sci. J., 5, 114 (1971)
2) K. Mahalingham, B.P. Gu, G.L. Liedl and T.H. Sander, Acta Metall. 35, 483 (1987)
3) S. Spooner, Materials Research Soc. Symp. Proc., Vol. 41 (Pittsburgh, Pennsylvania: Mat. Res. Soc.), pg. 89 (1985)
4) F. Livet and D. Bloch, Scripta Metall., 10, 1147 (1985)
5) R. Triolo, E. Caponetti, S. Spooner and F. Boschetti, Phil. Mag. A60, 401 (1989)
6) R. Triolo, E. Caponetti and S. Spooner, Phys. Rev. B39, 4588 (1989)
7) E. Caponetti, E. M. D'Aguanno, R. Triolo and S. Spooner, Phil. Mag. B63, 1201 (1991)
8) R.W. Hendricks, J. Applied Phys., 11, 15 (1978)
9) T.P. Russell, J.S. Lin, S. Spooner and G.D. Wignall, J. Appl. Cryst., 21, 638 (1988)
10) Gerold, V. and Kostorz, G., J. Appl. Cryst., 11, 376 (1978)
11) Avrami, M., J. Chem. Phys. 7, 1103 (1939); ibidem 8, 212 (1940); ibidem 9, 177 (1941)

Solid State Polymer Electrolytes: A Structural Characterization.

A. Triolo[a], F. Lo Celso[b], R. Triolo[b]

[a] Department of Chemistry, Heriot Watt University, Riccarton Edinburgh EH14 4AS United Kingdom
[b] Dipartimento di Chimica Fisica, Universita' di Palermo, V.le delle Scienze, Parco d'Orleans II, I-90128 Palermo (Italy)

Abstract. Small Angle X-ray Scattering (SAXS) measurements are reported for Poly Ethylene Oxide (PEO) - Sodium Thiocyanate (NaSCN) mixtures, as a function of composition and temperature. This system is known to possess a complex phase behavior. Accordingly its morphology reflects this complexity in the mesoscopic length scale. SAXS data have been analyzed by means of different approaches leading to a detailed picture of the material morphology.

INTRODUCTION

PolyEthylene Oxide (PEO) is known to form crystalline complexes with several inorganic salts (1). These systems have long attracted attention both in the academic and industrial fields due to their utilization as solid electrolytes for battery (2). Since the early work of Lee and Crist (3), it was well known that PEO-NaSCN mixtures show a complex phase diagram due to the coexistence of different crystalline and amorphous phases in well-defined composition and temperature regimes. In particular, mixtures with salt content as low as $X_{NaSCN}=0.023$ (with $X_{NaSCN}=$ [NaSCN]/([NaSCN]+[-CH_2CH_2O-])), show only two coexisting amorphous (A-PEO) and crystalline (C-PEO) phases, as would be expected for a semicrystalline material like PEO. These phases alternate each other in a lamellar-like way, the added salt being completely dissolved in the amorphous portion. At higher salt content, the existence of a crystalline complex (CC) between PEO and the salt has been characterised (3). Accordingly, in this composition regime, three different phases can simultaneously coexist (i.e. A-PEO, C-PEO and CC).

There are various characterizations (3,4), mainly via SEM technique, concerning the morphology of PEO-NaSCN mixtures. They all converge to a lamellar description of the mesoscopic structure for these systems. Even at temperatures higher that the melting point for C-PEO (i.e. ca. 65°C), a lamellar structure persists and is due to the alternating CC and A-PEO layers. On a higher length scale, a spherulitic morphology has been reported by means of Optical Microscopy. Moreover NMR measurements (5) have been conducted on this system, detecting the coexistence of C-PEO and CC in the composition range $0.05<X_{NaSCN}<0.25$. However, information is still lacking concerning the mutual relationship between the two crystalline phases. Due to the strong relationship between conductivity and microscopic morphology, it is of outstanding value to achieve a detailed description of structural properties for these systems. Almost strangely, no previous SAXS characterization has ever been reported

CP513, *Nuclear and Condensed Matter Physics,* edited by A. Messina
© 2000 American Institute of Physics 1-56396-929-7/00/$17.00

for these PEO-salt mixtures, although this experimental technique can lead to a deep understanding of the microscopic structure.

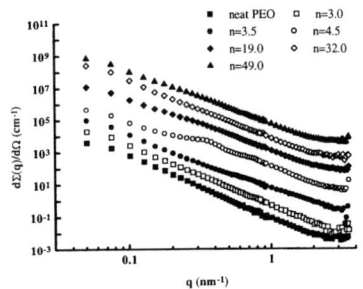

FIGURE 1. SAXS patterns at room temperature for different $(PEO)_n$-NaSCN mixtures.

EXPERIMENTAL.

$(PEO)_n$-NaSCN mixtures were prepared dissolving appropriate amounts of PEO (Aldrich product, MW=600,000) and NaSCN (Aldrich product, reagent grade) in CH_3OH. The solutions were stirred for 12 hours and the solvent was then removed by evaporation at 40°C and by further drying under vacuum at 40°C for 24 h. Samples with different NaSCN content were prepared. SAXS measurements were conducted at room temperature samples with X_{NaSCN}=0.02, 0.05, 0.18, 0.22, 0.25.

SAXS measurements were conducted at the 10-m SAXS camera at the Oak Ridge National Laboratory (USA). Further details can be found elsewhere (6). The momentum transfer, $Q=4\pi\sin\theta/\lambda$, ranges from 0.033 to 0.9 nm^{-1}. The experimental SAXS frames, $d\Sigma(Q)/d\Omega$, were extrapolated to large Q values by means of a Porod-like expression: $d\Sigma(Q)/d\Omega=aQ^b+C$, where b=3.8 and C is a flat incoherent background which was determined for all the samples at room temperature, by collecting the SAXS data on a wider Q range.

RESULTS AND DISCUSSION

In figure 1, the room temperature SAXS patterns for all the samples are reported. An almost linear behavior in the log-log scale can be observed whose slope is close to -4. Only a few samples show interference peaks at room temperature, which are related to the lamellar-like ordering of the morphology. However this

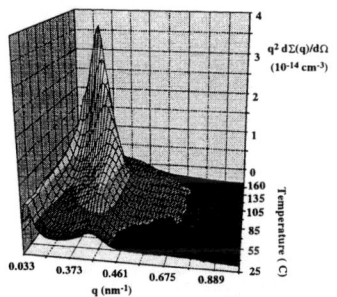

FIGURE 2. Temperature dependence of SAXS patterns for $(PEO)_{4.5}$-NaSCN.

seems to be an effect of a structural inertia, as at slightly higher temperatures all the samples develop a well defined lamellar morphology. Only the X_{NaSCN}=0.18 sample was further studied as a function of the temperature. In figure 2, we report the temperature dependence of the SAXS pattern for this mixture. It can be observed that a low temperature regime exists, where no particular structural evolution is observed. Then, at temperatures higher than 70°C a slight shift in the peak position is observed, while a dramatic evolution of the SAXS pattern is observed in temperature range between 120° and 165°C. This behavior is reflected in the so-called invariant: this is a well known quantity in SAXS investigations and it is defined as:

$$1. \quad Q_0 = \int_0^{\infty} Q^2 \frac{d\Sigma(Q)}{d\Omega} dQ$$

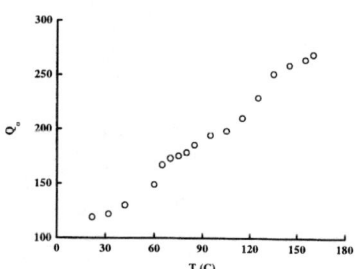

Its relevance in material science is related to the fact that it is related to the total amount of the minority phase dispersed in the matrix, according to: $Q_0 = \phi(1-\phi)\eta^2$, where ϕ is the volume fraction of the minority phase and η^2 is the contrast between the two phases. In figure 3 we report the temperature dependence of the contrast for the

FIGURE 3. Temperature dependence of the SAXS invariant, Q_0, for $(PEO)_{4.5}$-NaSCN.

$X_{NaSCN}=0.18$ mixture. It can be seen that a dual step of the total amount of minority phase occurs. The first one, occurring at ca. 65°C, can be related to the melting of C-PEO phase, while the second step occurs at a temperatures which is definitely lower than the melting point for CC (i.e. ca. 180°C).

To analyze our data we have used the electron density self correlation function approach (7) consisting in Fourier transforming the corrected SAXS data. This function contains average information on the structure of the system and structural parameters for the lamellar morphology can be derived (8). In figure 4, the temperature dependence of K(x) is reported. In the inset a comparison between K(x) at room temperature and at 75°C is shown. At 75°C, the C-PEO phase is melt and thus only the CC and A-PEO phases still coexist. On the other hand at 22°C, the K(x) reflects the coexistence of C-PEO, A-PEO and CC. The slight skewness of the low temperature K(x) reflects then the coexistence of two different lamellar domains, with different composition of the crystalline layer and different structural parameters. In figure 5, we report the structural parameters

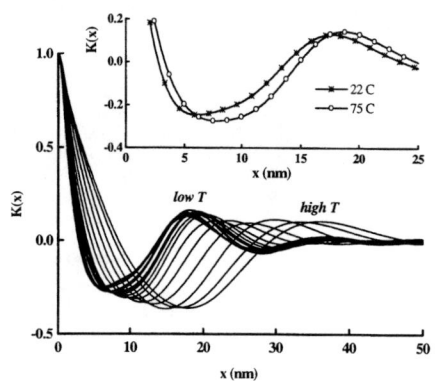

FIGURE 4. Temperature dependence of K(x)'s for $(PEO)_{4.5}$-NaSCN, as obtained by Fourier Transform of SAXS data. In the inset data at 22° and 75°C are reported in a larger scale.

obtained from analysis of curves in figure 4, assuming a pseudo-two phase model (7). In particular, the crystalline and amorphous layers thickness, R_c, R_a, and their sum, the Long Period, LP, are reported. It can be seen that the crystalline layer thickness does not change appreciably in the whole temperature range, while the amorphous layer thickness is profoundly affected by the temperature change. This result seems to be a consequence of a strong temperature dependence of the specific volume of the amorphous phase.

CONCLUSION

SAXS technique has proved to be a suitable probe to detect the structural details of PEO-NaSCN mixtures. The application of different analysis approaches led to a detailed description of the morphological details of this system. The results indicate a complex thermal behavior at temperatures

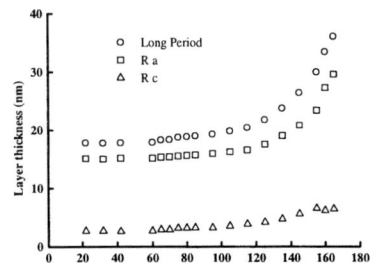

FIGURE 5. Temperature dependence of the crystalline, amorphous and long period layers for a $(PEO)_{4.5}$-NaSCN mixture.

higher than the melting point for C-PEO. A wider exploration in both temperature and composition range is however required to obtain conclusive description of the phase diagram and its implication on morphology.

ACKNOWLEDGMENTS

The research was supported in part by the Italian Ministero dell'Universita' e della Ricerca Scientifica (MURST) and the Italian National Research Council (CNR) and in part by the Division of Material Sciences, U. S. Department of Energy under contract number DE-AC05-96OR22464 with Lockheed Martin Marietta Energy Research Corporation. We thank Dr. G. Visalli and Dr. A. Bartolotta (University of Messina, Italy) for fruitful discussion on data interpretation.

REFERENCES

1) Wright, P.V., "Structure, Morphology and Thermal Properties of Crystalline Complexes of PEO and Alkali Salts" in Polymer Electrolyte Reviews, vol. 2, edited by J. R. MacCallum and C.A. Vincent, London: Elsevier Applied Science, 1989, pp. 61-119.
2) Bruce, P. G., and Vincent, C.A., J. Chem. Soc., Faraday Trans. 89, 3187 (1993).
3) Lee, Y. L., and Crist, B., J. Appl. Phys. 60, 2683-2689 (1986)
4) Robitaille, C., Marques, S., Boils, D. and Prud'homme, J., Macromolecules, 20, 3023-3034 (1987)
5) Bartolotta, A., Forte, C., Geppi, M., Minniti, D. and Visalli, G., Solid State NMR, 8, 231-239 (1997)
6) Triolo, A. et al., Macromolecular Chemistry and Physics, submitted.
7) Strobl, G. R., Schneider, M.; J. Polim. Sci.: Phys. Ed., 18, 1343 (1980)
8) Triolo, A., Silvestre, C., Cimmino, S., Martuscelli, E., Caponetti, E., Triolo, R.; Polymer, 39, 1697 (1998); Triolo, A., Lin, J.S. and Triolo, R., Physica A 249, 362 (1998); Triolo, A., Lin, J.S., Wignall, G.D., Triolo, R., Polymer, 00, 0000 (1999)

233

QENS from Polymeric Micelles in Supercritical CO₂

R. Triolo[1], V. Arrighi[2], A. Triolo[2], P. Migliardo[3], S. Magazù[3], J. B. McClain[4], D. Betts[4], J.M. DeSimone[4], H.D. Middendorf[5]

[1] Dipartimento di Chimica Fisica, Università di Palermo, Italy.
[2] Dept. of Chemistry, Heriot-Watt University, Edinburgh, UK.
[3] Dipartimento di Fisica and INFM, Università di Messina, Italy.
[4] Department of Chemistry, University of N. Carolina, Chapel Hill, USA.
[5] Clarendon Laboratory, University of Oxford, UK.

Abstract. We report QENS measurements from PS-b-PFOA aggregates in supercritical CO₂. These consist of dense cores of CO₂-insoluble polystyrene surrounded by a `corona' of PFOA surfactant molecules whose CO₂-philic groups interface with supercritical CO₂. Lineshapes are dominated by localised diffusive modes and segmental dynamics of the anchored, finite-length PFOA chains. For Q~0.6 Å⁻¹, we obtain effective diffusion coefficients of $\approx 0.8 \times 10^{-6}$ cm²/sec. At higher Q, a single component is not sufficient as shown by excess intensity on the flanks. For Q>1.5 Å⁻¹, the wings reflect contributions due to a distribution of faster, more localised chain modes.

INTRODUCTION

In parallel with small-angle scattering (SANS) studies of the structure of polymer aggregates in supercritical CO₂ (scCO₂) [1-3], we have begun to characterise their molecular dynamics by quasielastic neutron scattering (QENS). Apart from viscosity data providing crude estimates of the time scale of overall Brownian motion, the molecular dynamics of colloidal dispersions in CO₂ has not been investigated. Here we present first results from a medium-resolution QENS study.

EXPERIMENTAL SECTION

We used the pulsed-source spectrometer IRIS [4] at ISIS (RAL, Chilton, UK) to measure QENS from PS-b-PFOA aggregates in CO₂ at pressures of 200 and 350 bar and temperatures between 293 and 313 K [PS=polystyrene; PFOA=poly(1,1-dihydroperfluorooctylacrylate)]. These micelle-like aggregates consist of dense, globular cores of CO₂-insoluble material surrounded by a `corona' of PFOA surfactant chains whose CO₂-philic groups interface with the supercritical solvent. For this first experiment, to ensure a good signal/noise ratio from samples contained in annular-geometry pressure cells, a relatively high polymer micelle concentration of 30% was chosen. QENS from scCO₂ alone was measured under identical conditions,

CP513, *Nuclear and Condensed Matter Physics*, edited by A. Messina
© 2000 American Institute of Physics 1-56396-929-7/00/$17.00

and subtracted from the polymer+CO_2 spectra with appropriate correction factors [5]. Because of the large proton cross-section, the resulting difference spectra are essentially proportional to the incoherent dynamic structure factor, $S_{inc}(Q,\omega)$, for single-particle scattering from the various proton populations in the micelles.

RESULTS AND DISCUSSION

In a number of SAXS and SANS studies [2-3], radii of gyration, core radii, thickness of surfactant shells, polydispersities and other parameters have been determined. The results show that PS-b-PFOA micelles can be modelled as core-shell structures with core radii $R_1 \approx 25$-30 Å and outer radii $R_2 = 70$-90 Å, depending on copolymer size and concentration. At higher Q up to 0.3 Å^{-1}, where scattering from smaller structures becomes dominant, SAXS curves were interpreted by scattering from rod-like segments of the PFOA backbone [2]. The structural data suggest distinguishing between two time scale regions: (i) a relatively slow one quantifying the Brownian dynamics of micelles as a whole, i.e. their rotational and translational motions, and (ii) a faster one relating to localised diffusive modes and segmental dynamics of the anchored, finite-length PFOA chains in the 'corona' region.

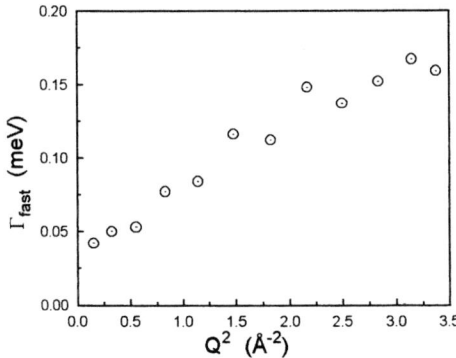

FIGURE 1. HWHM of the fast contribution vs Q^2

In constructing $S_{inc}(Q,\omega)$ models it is essential to distinguish between protons that are immobile or effectively immobile (relative to the longest time scale probed), and protons in translationally and/or rotationally mobile groups contributing to the quasielastic broadenings observed. For IRIS, with data of excellent statistics, width changes down to a few percent are observable, corresponding to times τ_{res} of the order of 1 ns. All contributions to $S_{inc}(Q, \omega)$ carry Debye-Waller factors $\exp(-Q^2 \langle u_p^2 \rangle)$ [6,7]. Here $\langle u_p^2 \rangle$ is an average cross-section weighted m.-sq. vibrational displacement. Measured spectra were corrected with respect to detector efficiencies and detailed-balance factor; the resolution function was determined from spectra measured for vanadium. In this brief communication, convolution operations and Debye-Waller factors are not discussed explicitly. Difference spectra $\Delta S_{inc}(Q, \omega)$ from 30% solutions of PS-*b*-PFOA in scCO$_2$ at 20°C and 40°C and pressures of 200 to 350 bar were analysed in stages as follows. First, inspection of spectra revealed four lineshape contributions: (i) a slight broadening of the central elastic peak, increasing with Q from a few μeV at low Q to values comparable with the resolution width at high Q; (ii) strongly Q-dependent wing broadenings with widths of the order of 100 μeV; (iii) small but noticeable intensity increases with Q on the lower flanks of the q.e. peaks, at energy transfers intermediate between (i) and (ii); (iv) a very broad, essentially flat and background-like underlying component with a Q-dependence that could (at least partially) be due to inelastic processes. For systems of the kind treated here, a useful starting point is to fit scattering laws of the form

$$S_{inc}(Q,\omega) = A_0(Q)\delta(\omega) + (1 - A_0(Q))S_{qe}(Q,\omega) \qquad (1)$$

where the first term accounts for elastic or effectively elastic scattering (i.e. scattering from motions with $\tau > \tau_{res}$), A_o is a partitioning factor, and S_{qe} the scattering law describing quasielastic broadening. A_o can be identified with the elastic incoherent structure factor, or EISF [6]. In solution scattering experiments aimed at deriving $\Delta S_{inc}(Q,\omega)$, however, and especially when pressure cells have to be used, it is difficult to avoid residual elastic scattering due to imperfect subtraction of the normalised spectra of one run (pure solvent + cell) from another (polymer sample + solvent + cell).

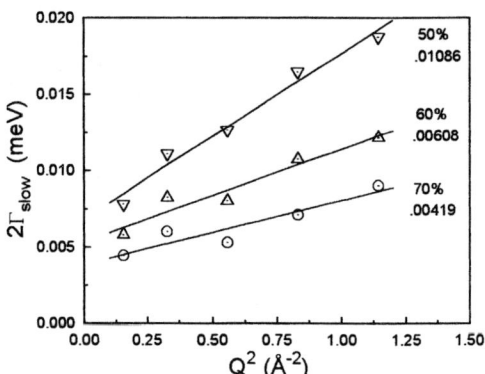

FIGURE 2. FWHM vs Q^2 (see text for details)

Fits of Eq. (1) to measured $\Delta S(Q, \omega)$ reveal the following: (a) outside the central peak and lower flank regions, the spectra are fairly well represented by Eq. (1) if S_{qe} is taken to be a Lorentzian (for restricted translational motions) convoluted with a rotational contribution (Sears model) [6,7]; (b) apart from the inability of Eq. (1) to account for the small but distinct central peak broadenings, the lower flanks <25 μeV give significant deviations (compare Fig. 2) and cannot be modelled satisfactorily in this way. The next step is a simple *ad hoc* extension of Eq.(1) to account for central peak broadenings by allowing the δ-function to broaden and thus to model slow translational motions, as described by a plain Lorentzian of width Γ_{slow} (HWHM). Thus the δ-function is replaced by $A_oL_{slow}(Q,\omega)$. The resulting $\Gamma_{slow}(Q)$ at low to intermediate Q are plotted in Fig. 1. To check the consistency of assuming a Lorentzian, the 60% and 70% level widths are shown along with the 50% widths Γ_{slow} (HWHM). From the slope of Γ_m as a function of Q^2 we obtain an effective diffusion coefficient $D_{eff} \approx 0.8 \times 10^{-6}$ cm^2/sec. This could be interpreted as a measure of the overall micelle mobility; however, the errors in this analysis are large (note in particular the non-zero Q=0 intercept) and μeV-resolution data down to at least 0.1 Å^{-1} are clearly needed. Apart from this, it is difficult to give precise meaning to the qualification `effective' without MD simulations of the dynamics of such core-shell structures. More sophisticated analytical expressions for $S_{inc}(Q,\omega)$ must be able to describe a distribution of relaxation times, instead of one or more averaged times as in composite Lorentzian models [7]. Scattering laws of this kind are *a priori* more appropriate to the morphology inferred from structural studies, i.e. they allow for the fact that we are dealing with surfactant chains that are not only anchored but also polydisperse with a broad length distribution. The main feature of more realistic $S_{inc}(Q,\omega)$ will be a distribution of relaxation times for the rod-like groups along the chains. Our aim in work in progress is to extract from QENS data a width parameter for this distribution, in addition to a time scale. Assuming that this distribution arises from a superposition of relaxation processes that individually display exponential behaviour, the Q,t-dependent changes may be interpreted *via* empirical time correlation functions of the `stretched exponential' kind (KWW function) [8]. We look forward to comparing more extensive QENS results with future MD simulations.

REFERENCES

1. DeSimone, J.M., et al., *Science* **265**, 356 (1994).

2. Fulton, J.L., et al., *Langmuir* **11**, 4241 (1995); McClain, J.B., et al., *Science* **274**, 2049 (1996).

3. McClain, J.B., et al., *J. Am. Chem. Soc.* **118**, 917 (1996); Chillura-Martino, D., et al., *J. Molec. Struct.* **383**, 3 (1996); Londono, J.D., et al., J. Appl. Cryst. **30**, 690 (1997).

4. Carlile, C.J., and Adams, M.A., Physica B **182**, 431 (1992).

5. Windsor, C., *Pulsed Neutron Scattering*, Taylor & Francis, London (1981).

6. Beé, M., *Quasielastic Neutron Scattering*, Adam Hilger, Bristol (1988).

7. Middendorf, H.D., et al., Biol. Chem. **53**, 145 (1994).

8. Arrighi, V. & Higgins, J.S., Physica B **226**, 1 (1996).

Configurational Landscape and Hydration Reconfiguration of a Multi-Element Model Solute in Explicit Water

A.C. Vaiana and M.B. Palma-Vittorelli

INFM and CRRNSM, at the DPAS, University of Palermo, Via Archirafi 36, I-90123 Palermo

Abstract. In this Molecular Dynamics study we use a supersimplified multi-element model solute in explicit solvent. Its configurational space is two-dimensional and its potential energy landscape is flat in vacuo. We study how the landscape is affected by solvent induced interactions, when accounted for to all orders, with no use of Kirkwood's approximation or similar ones. A landscape of potential of mean force (PMF) possessing distinctive protein like features of realistic size is generated by the solvent. Building on a previous study (San Biagio et al., Biophys. J. 1999, in press), effects of different symmetries, geometries and charges on the PMF and effects due to the dynamical solute-solvent configurational coupling are evidenced.

INTRODUCTION

Important properties of proteins are conveniently referred to their configurational energy landscape [1,2]. Interaction with solvent reshapes the potential energy (U) landscape into that of the potential of mean force (PMF) [3], having the character of a free energy (ΔG). In the case of a multi-element solute this ΔG contains individual, pair-wise and higher-order terms: $\Delta G = \Sigma_i \Delta G_{sw}(i) + \Sigma_{i,j} \Delta G_{sw}(i,j) + \Sigma_{i,j,k} \Delta G_{sw}(i,j,k) + \cdots \cdots + \Delta G(1,2,...n)$. The latter are configuration-dependent and often comparable in size with lower-order ones [4-7]. We study the ΔG landscape of multielement super-simplified solutes, modeled as two parallel layers of fixed apolar Lennard-Jones spheres (Fig.1.a). One of the elements bears a negative charge fixed in its center, and a positive charge free to move on a spherical surface (Fig.1.b). Obviously, in the dipole ϑ, φ configurational space, in absence of solvent, the U (potential energy) landscape is flat, since apolar solutes do not interact directly with electric charges. We use Molecular Dynamics simulations in explicit solvent and we compute free energy from the distribution of occupancy probability. This takes into account all interaction terms, with no use of Kirkwood's superposition approximation [3] or equivalent ones. The use of supersimplified solutes allows eliciting unambiguously the important role of explicit solvent. This approach was successfully followed by San Biagio et al. [7] in the simple, yet significant case of a planar array of six Lennard-Jones spheres. In that case, the sole solute-solvent interactions generate two free energy wells, symmetrically located with respect to the solute plane. Here we have studied the effects of different symmetries, geometries and charges on ΔG_{sw} and on the configurational landscape, using four variations of the basic model solute.

CP513, *Nuclear and Condensed Matter Physics*, edited by A. Messina
© 2000 American Institute of Physics 1-56396-929-7/00/$17.00

TECHNICAL DETAILS

The TIP3P potential [8] was used for water-water interactions. The model solutes were depicted using the Lennard-Jones parameters of the TIP3P oxygen. The value of the positive charge was $q^+ = 0.47\,a.u.$ and its (constant) R coordinate was $0.95\,\mathring{A}$. The four model solutes used in this work differ for the values of the negative charge, which was $q^- = q^+$ for solute A, $q^- = 0$ for solute B, $q^- = 2q^+$ for solutes C and D. The center to center distance betweeen the nearest neighbor spheres was $4.6\,\mathring{A}$ for A, B, C and $7\,\mathring{A}$ for D. Simulations were performed using the AMBER (version 5.0) package [9], with 1222 water molecules, *NPT* ensemble, $T = 300\ °K$, $P = 1\ atm$. MD trajectories of 2 ns, after $100\ ps$ thermalisation. We computed the angular distribution function $g(\vartheta,\varphi) = (N\Delta\omega)^{-1} n(\vartheta,\varphi)$, with $n(\vartheta,\varphi)$ frequency of occupancy of a box $\Delta\omega = (4\pi)^{-1} \sin\vartheta\,\Delta\vartheta\,\Delta\varphi$ and $\Delta\vartheta = \Delta\varphi = 10°$. The free energy landscape is computed as $\Delta G(\vartheta,\varphi) = -kT \ln g(\vartheta,\varphi)$. Hydration was computed so as to elicit the solute-solvent configurational coupling suggested by the long dipole switching times in Fig.1, f and g. To this purpose we extracted, from the entire trajectory two different sets of water configurations corresponding respectively, to suitable ϑ, φ "boxes" around the two observed preferred dipole orientations. Hydrations were then computed separately for each of the two sets of configurations. This allowed eliciting a close correspondence between hydration and dipole orientation, implying their dynamic coupling. Actual hydration patterns were drawn using SciAn graphics [10]. Isoprobability surfaces shown in the figure are 2.25 (solute A) and 6.00 (solute C) with respect to bulk water.

RESULTS AND CONCLUSIONS

Space limits allow us to present only two sets of data, relative to solutes A and C, and to discuss briefly other cases. We consistently find that interaction with the solvent, averaged over all solvent configurations, is capable to transform the flat energy landscape of the solute *in vacuo* into a complex free-energy landscape with one or two minima. In the two cases shown in the figure, the related stable configurations correspond to maximum hydration of the negative charge, at the expenses of that of the mobile positive charge. In the case of solute A, the long switching times between minima (10-$100\ ps$) together with the hydration patterns, separately computed for each set of configurations, evidence very clearly a dynamical coupling between solute and solvent reconfiguration. In the case of solute C the effect of the double negative charge is overwhelming and it freezes the hydration pattern, preventing any major solvent reconfiguration. In turn, this freezes the solute configurational dynamics, again (if paradoxically) indicating a strong reconfi-gurational coupling. That is, the coupling between dipole and water reconfiguration is so strong that if the latter is impeded, the former is also impeded. There is only one, very deep free energy minimum. The preferred dipole orientation is distinctly pointing towards internal regions of the solute, so as to shield the positive charge from the solvent. In the case of solute B (not shown), in absence of the negative charge, the region of two shallow minima covers

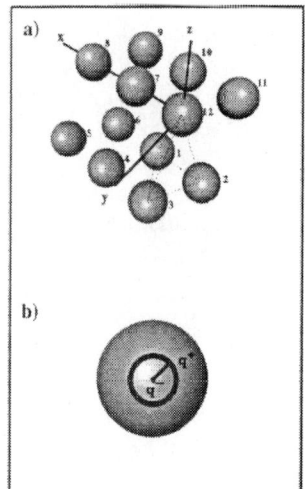

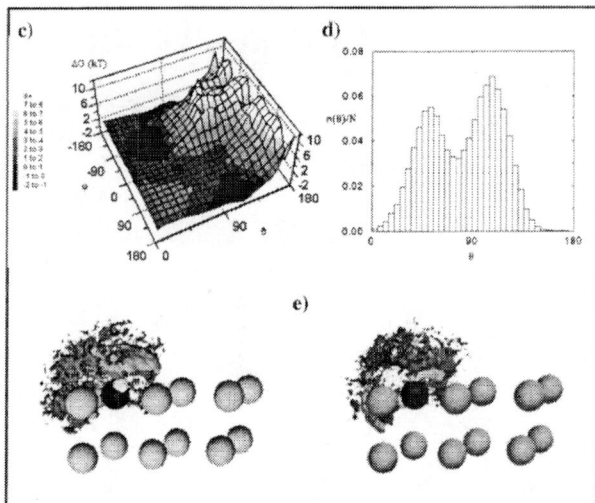

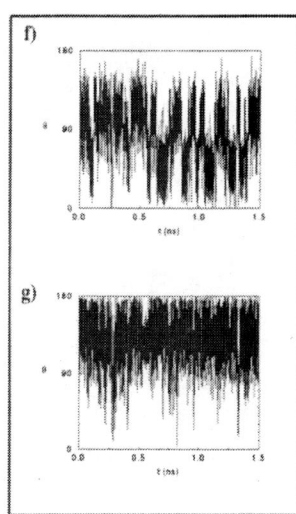

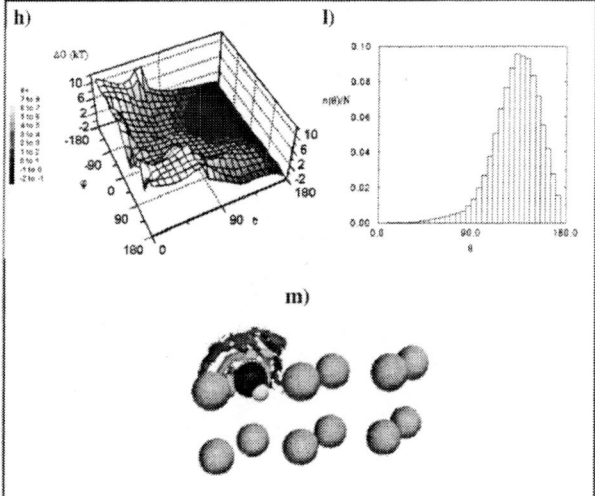

FIGURE 1. Results for solutes A ($q^- = q^+$) and C ($q^- = 2q^+$). Center to center distance *4.6 Å*; a) basic model solute. b) location of charges on element 12. Panels c, d, e and f refer to solute A; panels g, h, l, and m refer to solute C. c) and h) landscape of solute-solvent free energy contributions, in *kT* units (*1kT=2.48 KJ/mol⁻¹*). d) and l) distribution of the ϑ coordinate. e) and m) hydration (in e the two figures refer to the dipole pointing within one or the other of the two wells in fig. c); f) and g) time evolution of the ϑ coordinate.

all dipole orientations corresponding to a high exposure to solvent of the positive charge. This notwithstanding, dynamical coupling between solute and solvent reconfiguration is again evidenced by the two separately coupled hydration patterns. The dipole switching and water reconfiguration dynamics is now faster, as a consequence of the shallow minima. Finally solute D (not shown) differs from solute

C only for a much larger intra-element distance. In this case the minima in the landscape are shallower, all dipole orientations corresponding to a high exposure to solvent of the positive charge.but not significantly modified. This suggests that in the opposite, more realistic case of shorter distances, the landscape will be considerably more rugged and the effect more relevant. This is indeed confirmed by further ongoing results.

The following conclusive remarks are in order: i) Use of supersimplified model solutes in realistic, explicit solvent allows eliciting the role of solvent induced interactions (accounted for to all orders) in generating (not only in modifying) distinctive free energy landscapes. ii) The size and features of the observed effects makes them exquisitely relevant to protein conformation and folding. iii) The dynamical coupling between solute and solvent configurations is also unambiguously confirmed and extended. iv) The observed effects (of the order of several KJ/mol^{-1}) can be viewed as an (indirect, solvent-induced) interaction among apolar and charged elements. These interactions, not existing in vacuo, are strongly dependent upon the charge sign and solute context, in agreement with the strong many body character of solvent induced interactions [4-7]. They are, of course, of great relevance e.g. to chemio-mechanical energy conversion.

REFERENCES

1. H. Frauenfelder and P.G. Wolynes. *Physics Today*, **Feb.**: 58, 1994.

2. H.S. Chan and K.A. Dill. *Proteins*, **30**: 2, 1998.

3. T.L. Hill. Statistical Mechanics, McGraw-Hill, New York, 1956.

4. P.L. San Biagio, D. Bulone, V. Martorana, M.B. Palma-Vittorelli, and M.U. Palma. *Eur. Biophys. J.* **27**: 183-196, 1998.

5. V. Martorana, D. Bulone, P.L. San Biagio, M.B. Palma-Vittorelli, and M.U. Palma. *Biophys. J.*, **73** : 31, 1997.

6. V. Martorana, G. Corongiu, and M.U. Palma. *Proteins Short Comm.*, **32**: 129-135.

7. P.L. San Biagio, V. Martorana, D. Bulone, M.B. Palma-Vittorelli, and M.U. Palma. *Biophys. J.,* Nov issue, **77**, 1999.

8. W. L. Jorgensen, J. Chandrasekhar, J. D. Madura, R. W. Impey and M. L. Klein. *J. Chem. Phys.* **79**: 926-935, 1983.

9. D.A. Case, D. A. Pearlmen, J.W. Caldwell, T.E. Cheatham III, W.S. Ross, C.L. Simmerling, T.A. Darden, K.M. Mertz, R.V. Stanton, A.L. Cheng, J.J. Vincent, M. Crowley, D.M. Ferguson, R.J. Radmer, G.L. Seibel, U.C. Singh, P.K. Weiner, and P.A. Kollmann, 1997. AMBER 5, University of California, San Francisco.

10. E. Pepke and J. Lions. 1993. SciAn user's manual. Supercomputer Computations Research Institutte, Florida State University, Tallahassee, FL, USA.

Self-Dressing Dynamics Of Slow Electrons in Covalent Crystals

D. Valenti, G. Compagno

Istituto Nazionale di Fisica della Materia and Dipartimento di Scienze Fisiche ed Astronomiche dell'Università di Palermo

Abstract. We consider a slow monoelectronic excitation in a covalent crystal at T=0. The interaction with the zero-point longitudinal acoustic phonons leads to the formation of a dressed state, made by the electron accompanied by a deformation field, which has energy lower than the corresponding bare monoelectronic state. To describe the dynamics of the self-dressing process we study the time development of the permanence probability amplitude, $A_e(t)$, in the initial bare monoelectronic state. To follow its time development for long enough times it is necessary to use nonperturbative techniques. In order to be able to follow the decrease of the electron's energy we use a modified van Hove resolvent theory. We show that $A_e(t)$ decays exponentially with a meanlife time τ_e long compared with the inverse Debye frequency ω_D^{-1}.

THEORY

In Quantum Field Theory interaction between sources and fields occurs also for vanishing mean value of the field[1,2]. This interaction causes the creation of a "dressed" state where the bare source is surrounded by a cloud of virtual field quanta[3]. The initial phase of the virtual cloud formation process has been studied both for atoms in Quantum Electrodynamics (QED)[4] and for excitons in Solid State Physics[5]. However the self-dressing process produces an energy lowering of the dressed electron with respect to the corresponding bare one, this energy lowering leading to the completely dressed electron state. This process has been studied in QED for an atom interacting with the electromagnetic field in vacuum state[6]. Here we consider an initial bare monoelectronic excitation in a covalent crystal with the phonon field taken in its vacuum state. The interaction with the longitudinal acoustic modes will then be taken into account and we shall follow the electron's evolution towards the lower energy dressed state. The electron interacting with the acoustic longitudinal modes of the phonon field is described by the Hamiltonian[7]

$$H = H_0 + H_I \tag{1}$$

where

$$H_0 = E(k) a_k^+ a_k + \sum_q \hbar\omega(q)\left(b_q^+ b_q + \frac{1}{2}\right), \tag{2}$$

CP513, *Nuclear and Condensed Matter Physics,* edited by A. Messina

$$H_I = \frac{1}{\sqrt{N}} \sum_{k,q} F_a(\boldsymbol{q}) a_{k+q}^+ a_k (b_q - b_{-q}^+) \qquad (3)$$

represent respectively the imperturbed and interaction Hamiltonians. $F_a(\boldsymbol{q}) = -i(2/3) E_F \sqrt{\hbar |\boldsymbol{q}|/(2Mc_a)}$; $\omega(\boldsymbol{q}) = c_a q$ is the phonon's frequency; c_a is the velocity of the acoustic phonons; M is the total mass of an elementary cell and N is the total number of elementary cells in the crystal; b_q (b_q^+) are the phonon annihilation (creation) operator satisfying the Bose-Einstein commutation rules $[b_q, b_{q'}^+] = \delta_{qq'}$; a_k (a_k^+) are the electronic annihilation (creation) operator satisfying the Fermi-Dirac commutation rule $\{a_k, a_{k'}^+\} = a_k a_{k'}$. The initial monoelectronic state in absence of phonon is described by the state vector

$$|\psi_e^0> = |\boldsymbol{k}; \{0_q\}> = |\boldsymbol{k}> \otimes |\{0_q\}> \qquad (4)$$

where $|\{0_q\}>$ and $|\boldsymbol{k}>$ represent respectively the phonon field vacuum state and the electron bare state having energy $E(\boldsymbol{k}) = \hbar^2 k^2/(2m^*)$ with $\boldsymbol{k}$ wavevector associated to the electron and m^* the effective electron mass. In order to describe the dynamics of the self-dressing process we study the time evolution of the permanence probability amplitude, $A_e(t)$, in the bare state $|\psi_e^0>$. To describe this time evolution for sufficiently long times we must adopt nonperturbative techniques. For this purpose we introduce the van Hove's Theory[3] based on the use of the resolvent operator. This technique has already been used in QED to describe the self-dressing dynamics of an atom interacting with the electromagnetic field[4]. The expression of $A_e(t)$, for $t > 0$, is given by

$$A_e(t) = <\psi_e^0| e^{-i\frac{H}{\hbar}t} |\psi_e^0> = \frac{1}{2\pi i} \int_{+\infty}^{-\infty} G_e(E+i\eta) e^{-i\frac{E}{\hbar}t} \, dE \qquad (5)$$

where η is a positive infinitesimal quantity and $G_e(E+i\eta)$ is the matrix element of the resolvent operator that may be expressed as

$$G_e(E+i\eta) = \frac{1}{E+i\eta - E(\boldsymbol{k}) - R_e(E+i\eta)}, \qquad (6)$$

where

$$R_e(E+i\eta) = <\psi_e^0|H_I|\psi_e^0> + <\psi_e^0|H_I \frac{Q}{E+i\eta - H_0} H_I|\psi_e^0> + \dots . \qquad (7)$$

is the matrix element of the level-shift operator. Truncating expansion (7) up to 2nd order results however in a nonperturbative approximation for $G_e(E+i\eta)$[8]. Within this approximation and substituting eqs. (2) and (3) in (7) we obtain as

$$R_e(E+i\eta) = \frac{4}{9} E_F^2 \frac{\hbar}{2MNc_a} \sum_q \frac{q}{E+i\eta - E(k-q) - \hbar\omega(q)} \qquad (8)$$

Applying the continuum limit $1/N \sum_q \rightarrow 4\pi \int_0^{q_m} q^2 dq$, where $q_M = \pi/(2a)$ (a the lattice constant) is the wave number delimiting the first Brillouin zone, performing the integration in (8) and using the Wigner-Weisskopf like approximation $E = E(k)$ we see that the imaginary part of $R(E(k))$ is zero. This leads to an oscillation like behavior of $A_e(t)$ and the initial bare state doesn't reduce asimptotically to the dressed state. We may improve the van Hove scheme by substituting in (8) in place of $E(k)$ the expression

$$E^*(k-q) = \frac{\hbar^2(k-q)^2}{2\widetilde{m}} - \Delta E \qquad \left(\Delta E = \frac{\pi}{72} \frac{\hbar c_a q_m}{M c_a^2} E_F \right) \qquad (9)$$

representing the dressed energy of the intermediate electronic state between the emission and the reabsorption of virtual phonons. ΔE and $\widetilde{m}$ and are respectively the 2nd order energy correction and the dressed electron mass obtained, in the limit $k << 2m^*c_a/\hbar$, by the ordinary time independent perturbation theory[9]. With this procedure eq. (8) becomes

$$R_e(E+i\eta) = \frac{4}{9} E_F^2 \frac{\hbar}{2MNc_a} \sum_q \frac{q}{E+i\eta - E^*(k-q) - \hbar\omega(q)} \qquad (10)$$

Using again the continuum limit, performing the integral and applying the Wigner-Weisskopf like approximation we obtain

$$R[E(k)] = -\Delta_e + i\frac{\Gamma_e}{2} \qquad (11)$$

where

$$\Delta_e \approx 10^{-1} \alpha E_F, \qquad \Gamma_e \approx 10^{-2} \alpha^2 E_F. \qquad (12)$$

with $\alpha = \hbar c_a q_m/(M c_a^2)$. By using (6) and (10) in (5) and performing the integration in E, for probability amplitude $A_e(t)$ we obtain

$$A_e(t) = e^{-i\frac{E(k)-\Delta_e}{\hbar}t} e^{-\frac{\Gamma_e}{2\hbar}t}. \qquad (13)$$

From (13) we see that Δ_e is the energy shift due to the interaction of the electron with the longitudinal acoustic phonons, while Γ_e is a line width representing the fact that the bare monoelectronic state of wavevector k isn't stationary. $\tau_e = (\Gamma_e / \hbar)^{-1}$ is the lifetime of the bare electronic state. Using $E_F /(2\pi\hbar) \approx 10^{16} s^{-1}$, $\alpha \approx 10^{-2}$ one obtains $\tau_e \approx 10^{-10} s$. The lattice deformation that corresponds to the motion of a virtual phonon cloud occurs with a time scale $T_D \approx \omega_D^{-1} \approx 10^{-13} s$, with ω_D Debye frequency. Successively the electron surrounded by the virtual phonons emits the excess energy and it reaches the polaronic state which represents the new stationary state. This part of the dressing process occurs with a time scale $\tau_e >> \omega_D^{-1}$. In conclusion in our theory the complete process of the electron's self-dressing due to the interaction with the longitudinal acoustic phonons occurs in times ($\sim 10^{-10}$ s) that may be amenable to experimental verification.

ACKNOWLEDGMENTS

The authors acknowledge partial financial support by Ministero dell'Università e della Ricerca Scientifica e Tecnologica, Cofinanziamento MURST, Istituto Nazionale di Fisica della Materia and Assessorato BB.CC.AA. Regione Siciliana.

REFERENCES

1. Wick, G. C., *Rev. Mod. Phys.* **27**, 339-362 (1955).

2. Feinberg, E. L., *Sov. Phys. Usp.* **23**, 629-649 (1980).

3. Van Hove, L., *Physica* **18**, 145-159 (1952); Van Hove, L., *Physica* **21**, 901-923 (1955); Van Hove, L., *Physica* **22**, 343-354 (1956).

4. Compagno, G., Passante R., and Persico, F. *Il Nuovo Cimento* **15 D**, 355-363 (1993).

5. Brown, D. W., Lindenberg, K., and West, B. J., *J. Chem. Phys.* **84**, 1574-1582 (1986); Brown, D. W., Lindenberg, K., West, B. J., Cina, J. A., and Silbey, R., *J. Chem. Phys.* **87**, 6700-6705 (1987).

6. Compagno, G., and Valenti, D., *J. Phys. B: At. Mol. Opt. Phys.*, **32**, 19, 4705-4717 (1999).

7. Davydov, A. S., *Teoria del Solido*, Mosca: Edizioni Mir, 1984, pp. 227-239.

8. Cohen-Tannoudji, C., Dupont-Roc, J., and Grynberg, G., *Atom-Photon Interactions*, New York: John Wiley & Sons, Inc., 1992, pp. 165-197.

9. Cohen-Tannoudji, C., Diu, B., and Laloë, F., *Quantum Mechanics*, New York: John Wiley & Sons, Inc.,1977, pp. 1102-1108.

Structural And Dynamic Effects H-Bond Induced in Monomer-Polymer Solutions

V. Venuti, V. Crupi, A. Faraone, G. Maisano, D. Majolino, P. Migliardo and V. Villari

Dipartimento di Fisica dell'Università & INFM di Messina, C.da Papardo, S.ta Sperone 31, P.O. BOX 55, 98166 S. Agata, Messina, ITALY

Abstract. A study of configurational topology of active groups for H-bond has been performed together with an analysis of vibrational, diffusional and transport properties of Ethylene Glycol (EG: $H-(O-CH_2- CH_2)-OH$), EG Methyl Ether (EGmE: $CH_3-(O-CH_2- CH_2)-OH$) and EG Dimethyl Ether (EGdE: $CH_3-(O-CH_2- CH_2)OCH_3$) and their mixtures with polymers. Infrared (IR) absorption data in the O-H stretching region show various types of H-bond and their evolution as a function of T. The same systems, mixed with Poly(ethylene Oxide) (PEO: $H-(O-CH_2- CH_2)_n-OH$) of different molecular weight, give rise to polymer-monomer interaction effects investigated by viscometry and Photon Correlation Spectroscopy (PCS).

INTRODUCTION

The scientific interest applied to the EG molecule is due to its capability to show inter- and intramolecular H-bonded interactions and to the fact that it represents the monomeric entity of PEO polymer.[1] As far as structure is concerned, many conformation can be created rotating the two CH_2OH groups around the C-C axis, and a computer simulation showed that a *gauche* form stabilized by an intramolecular H-bond is the most energetically favoured. In addition, the OH groups promote, via H-bond, a set of transient crosslinks between neighbouring molecules, giving relevant transient structures. With the aim of clarify the role of H-bond in the structural environments of liquids having identical chemical structure except for the number of the OH-end groups (*two* for EG, *one* for EGmE and *zero* for EGdE), we analysed their vibrational dynamics, vs. T, by IR absorption. In fact, the investigation of the O-H stretching region (3000-3800 cm^{-1}), allows to identify the various, H-bond imposed, inter- and intramolecular conformations. In order to study the behaviour of these polymers in solutions, we report viscometry and PCS results on dilute solutions of PEO, in a wide range of M.W., from 8000 Da to 900000 Da, dissolved in EG, EGmE and EGdE. Viscometry affords a powerful mean of investigating polymer behaviour in solutions, thanks to the possibility to extract the polymer intrinsic viscosity [η] from the concentration dependence of measured shear viscosities. The quality of the solvents has been checked by verifying the Mark-Houwink-Sakurada (MHS) scaling law. Finally, from the PCS data we extrapolated the diffusion coefficient at infinite dilution D_0 and, therefore, the hydrodynamic radius R_H, that gives information on the swelling or collapsing process of a polymer coil in solution.

CP513, *Nuclear and Condensed Matter Physics*, edited by A. Messina
© 2000 American Institute of Physics 1-56396-929-7/00/$17.00

RESULTS AND DISCUSSION

H-bonded liquids are constituted by self-associated systems in which the H-bond promotes inter- and intramolecular arrangements. Considering all the contributions to the H-bond potential energy surface, we expect a correlation between the degree of association and the relative population of these local *transient* structures, generating, in turn, a different dynamic response. So we can explain the spectral variations of the O-H stretching vibration, that spreads out over a large ω-range, undergoes a *red-shift*, and changes dramatically in shape and intensity with respect to the original O-H band centered at ~ 3630 cm^{-1}. This behaviour reflects the existence of an electrical *anharmonicity* in the dipole moment function, other than a mechanical one, in the potential energy evaluation.[2] As a consequence, the relationship:

$$I = k \left(\frac{\partial \mu}{\partial q_i} \right)^2 \qquad (1)$$

that furnishes the intensity of an IR band due to the normal mode q_i, in which k gives information on the number n_i of oscillators involved in it, can be retained valid only for weak H- bonded systems, as ours. According to the formalism introduced by Laubereau et al., successfully previously applied, we define ω_α the O-H stretching mode of free and/or end groups (monomers), ω_β (ω_γ) the one of proton-acceptor (donor) end groups (dimers), ω_δ the O-H vibration of fully bonded OH-end groups (trimers) and, finally, ω_ε as the one of OH groups involved in intramolecular H-bond. The results of the O-H stretching deconvolution (Voigt profiles), in the case of EG and EGmE, are reported in Fig.1 and summarized in Tab.I, where all the assigned centre frequencies ω_i, line widths Γ_i and percentage intensities I_i, are reported.

TABLE 1. IR sub-bands deconvolution fitting parameters for EG and EGmE.

T (°C)	ω_ε (cm^{-1})	Γ_ε (cm^{-1})	I_ε (cm^{-1})	ω_δ (cm^{-1})	Γ_δ (cm^{-1})	I_δ (cm^{-1})	ω_γ (cm^{-1})	Γ_γ (cm^{-1})	I_γ (cm^{-1})
EG									
-10	3210	231.6	62.1	3321	196.1	25.4	3430	197.3	12.5
10	"	"	56.1	"	"	24.9	"	"	19
45	"	"	45.2	"	"	33	"	"	21.8
110	"	"	27.8	"	"	29.5	"	"	42.7
170	"	"	13.9	"	"	26.5	"	"	59.3
EGmE									
0				3297	238	62	3446	202.4	38
25				"	"	60.2	"	"	39.8
45				"	"	48.4	"	"	51.6
65				"	"	44.4	"	"	55.6
85				"	"	32.7	"	"	67.3
100				"	"	30.8	"	"	69.1

For both samples, the absence of ω_α and/or ω_β indicates that there aren't end and/or open monomers. As far as EG is concerned, the existence of intramolecular, H-bond imposed, monomeric structures, and dimeric and trimeric species, has been postulated. The relative intensity of the sub-bands is also strongly T-dependent: this has been related to the presence of a hierarchy of structures whose relative population varies with T by means of microscopic fast dynamic process.

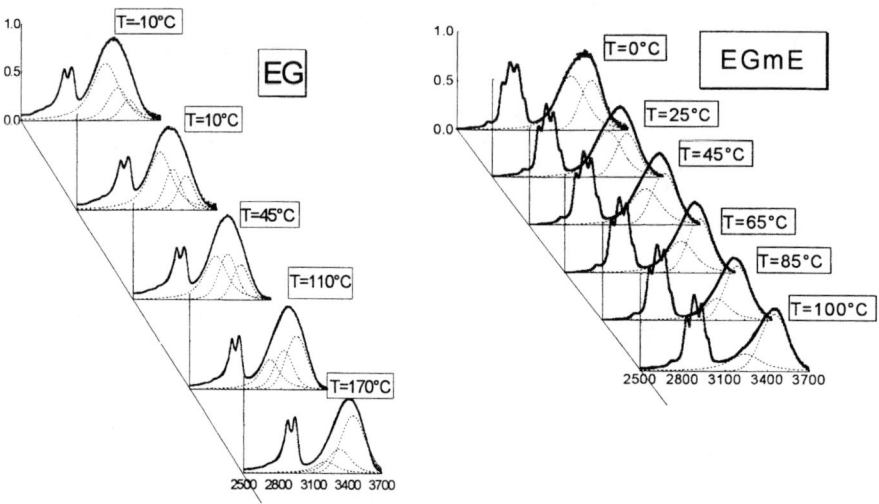

FIGURE 1. IR absorbance vs wavenumber (cm^{-1}) of EG (-10°C≤T≤170°C) and EGmE (0°C≤T≤100°C), with best-fit (continuous line) and components (dashed lines).

In particular, the T-evolution of the intramolecular ω_ε indicates a "kinetic" of transition between *gauche* configurations with or without H-bond. Dimers diminish at high T, and it means that "end-to-end" bonds are the most energetically favoured. Taking into account eqn. (1), because of ω_i, and Γ_i don't change with T, and so also the binding potential, the relative variations for each i-th sub-band can be directly related to n_i. These results revealed an increasing, with T, of trimers at the expenses of intramolecular ε-th structure. In the case of EGmE we observe a dynamics similar to EG, together with the disappearance, as expected, of the intramolecular H- bond. Experimental data suggested an increasing of dimeric species respect to the trimeric fully-bonded structures.

Viscosity measurements results[3] were fitted to a second degree polynomial in concentration c and interpreted in terms of the virial expansion for $\eta(c)$: $\eta(c) = \eta_0(1+[\eta]c+k'[\eta]^2c^2)$ with η_0 viscosity of the solvent, $[\eta]$ the PEO intrinsic viscosity, and k' the Huggin's coefficient. The molecular weight (M_W) dependence of $[\eta]$, reported in Fig.2,

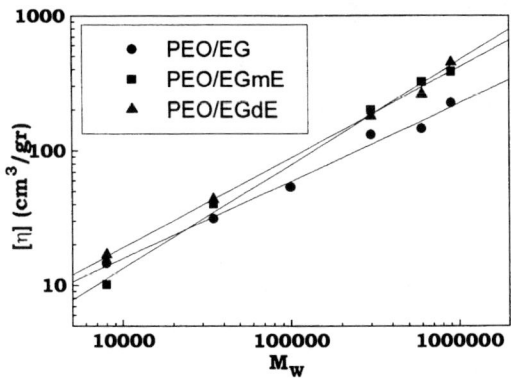

FIGURE 2. Intrinsic viscosity of PEO vs. Molecular Weight.

follows the Mark-Houwink-Sakurada (MHS) power law relation $[\eta]=AM^{\alpha}$ in which α is related to the quality of the solvent. This occurrence suggests that at $T=50°C$ and in the very dilute region the polymer assumes the random-coil conformation in all solvents. From the fit we obtained $\alpha=0.57$ for EG, $\alpha=0.77$ for EGmE and, finally, $\alpha=0.67$ for EGdE. We can conclude that EG is a nearly θ solvent, EGmE is a good one and EGdE has an intermediate quality. The simultaneous knowledge of the diffusion collective coefficient D_c allowed us to get new information on the competitive effects of "polymer-solvent" and "solvent-solvent" interactions. In particular, for EGmE, the "polymer-solvent" interactions" are shown to be stronger than the "solvent-solvent" ones, as results from its *good* solvent behaviour. On the other hand, the tendency of EG in forming inert closed rings triggered by intramolecular H-bond, experimentally observed by FT-IR spectroscopy, gives rise to an extensive hydration layer around the PEO coils. In fact, the R_H values of PEO 35000 in the three solvents, (evaluated through the Stokes-Einstein relation) are respectively: 59.9 Å, 68.8 Å and 54.8 Å. It turns out that, although EGdE is revealed, by viscometry, to be a better solvent than EG, R_H in this case, assumes a greater value.

REFERENCES

1. Crupi, V., Maisano, G., Majolino, D., Migliardo, P., and Venuti, V., *J. Chem. Phys.* **109**, 7394-7404 (1998).

2. Crupi, V., Maisano, G., Majolino, D., Migliardo, P., and Venuti, V., *Anharmonic effects and vibrational dynamics in H-bonded liquids by ATR FT-IR spectroscopy*, submitted to J. Chem. Phys.

3. Crupi, V., Faraone, A., Maisano, G., Majolino, D., Migliardo, P., Venuti, V., and Villari, V., *J. Chem. Phys.*, in press.

Influence of Trehalose on Conformational and Dynamical Properties of Poly(Ethylene Oxide) in Water

V.Villari, C.Branca, A.Faraone, S.Magazù, G.Maisano and P.Migliardo

Dipartimento di Fisica and INFM, Università di Messina, C.da Papardo S.ta Sperone 31, P.O.55, 98166 Messina, Italy

Abstract. This paper reports the results of a study on Poly(Ethylene Oxide)/trehalose/water mixtures performed by Photon Correlation Spectroscopy. The chemical structure of the polymer, simpler than that of proteins and its helical conformation in water, constitute a useful starting point for understanding the more complex protein/trehalose/water interactions. The obtained findings support the «water-replacement» hypothesis, indicating that a direct polymer-trehalose interaction occurs. Furthermore, trehalose is shown to affect the swelling properties of the polymer with temperature, stabilising its conformation.

INTRODUCTION

Trehalose, a diglucose sugar widely distributed in nature, confers to certain plant and animal cells the ability to survive dehydration for decades and to restore activity within minutes of rehydration, a phenomenon known as «anhydrobiosis».[1] The direct interaction with cell membranes and proteins, however, remains a debated subject and a lot of work has been devoted to study the effect of trehalose on proteins' function and dynamics. In a recent paper by Gottfried et al.[2] it has been stressed that in a glassy state trehalose interacts more strongly with the protein than water does. The enhanced interaction could suggest that trehalose directly couples the heme to the complex network-linked solvent molecules that form the glass. Moreover, according to these authors, the absence of substantial conformational effects in a trehalose-embedded protein (hemoglobin) suggested that changes in protein dynamics occur. More recently, by optical absorption and Mössbauer spectroscopy study on carbon-monoxy-myoglobin in a trehalose glass, Cordone et al.[3] found that trehalose coating prevents thermal denaturation by damping large scale fluctuations of protein specific motions that could cause protein unfolding.

The present paper reports results of Photon Correlation Spectroscopy (PCS) measurements on ternary solutions of trehalose in water with the addition of a hydrosoluble polymer, Poly(Ethylene Oxide) (PEO) with average molecular mass 35000 Da. Thanks to the simplicity of its structure, PEO represents a good model system for studying the interaction mechanisms of water and trehalose with hydrophilic surfaces, macromolecules and biological structures.

CP513, *Nuclear and Condensed Matter Physics,* edited by A. Messina
© 2000 American Institute of Physics 1-56396-929-7/00/$17.00

EXPERIMENTAL SECTION

α,α-trehalose and PEO 35000 were furnished by Sigma-Aldrich and Polysciences respectively. PEO/water molar ratio was kept constant at $6.3 \cdot 10^{-6}$, chosen in order that PEO concentration in water was in the dilute regime, but concentrated enough to have a good signal and reproducible correlation functions, whereas the trehalose amount was changed. All the systems were carefully subordinate to a filtering procedure in recirculation with an Amicon Millipore filter 0.2μm of pore size.

Quasielastic Light Scattering measurements were performed by means of PCS technique, using a standard scattering apparatus and a Brookhaven BI-2030 correlator. The detailed description is reported elsewhere.[4] Auxiliary viscosity measurements were performed by means of standard Ubbhelode viscometers of different capillary size in order to minimize the kinetical energy corrections.

RESULTS AND DISCUSSION

The experimental intensity autocorrelation functions clearly reveal the presence of two diffusive decays in the ternary trehalose/PEO/water mixtures for all the concentration values. For this reason the correlation functions have been fitted using a double exponential:

$$g_2(t) = 1 + [A_1 \exp(-\Gamma_1 t) + A_2 \exp(-\Gamma_2 t)]^2$$

The slowest decay can be attributed to the PEO coils and the fastest one to the trehalose molecule diffusion. The amplitude of the slower mode is always greater than that of the fast mode at trehalose concentration values less than c=0.112, because in light scattering each contribution to the scattered intensity correlation function is weighted by the squared mass. At higher sugar concentration values, on the other hand, the trehalose contribution to the scattered light increases and dominates at the highest investigated concentrations.

Fig.1 shows the extracted PEO diffusion coefficient, $[D_2]^{-1} = q^2/\Gamma_2$, in the presence of trehalose as a function of η_{TR}/T (η_{TR} being the viscosity of the trehalose in water mixture[5] and q the exchanged wave-vector). At low concentration it tends to the values corresponding to the trehalose-free system, indicating that increasing the trehalose amount the diffusive behaviour of PEO is slowed down due to the increased viscosity of the solvent (trehalose-water). As a matter of fact, at high η_{TR}/T values, a more and more significant deviation from the Einstein-Stokes (ES) behaviour is observed. Such a result, which indicates a lower diffusion coefficient of the polymer coil than that expected from a mere solvent viscous effect (especially at high η_{TR}/T), supports the hypothesis that the trehalose molecules bind to the polymer coils, so determining a mass increase of the diffusing particles.

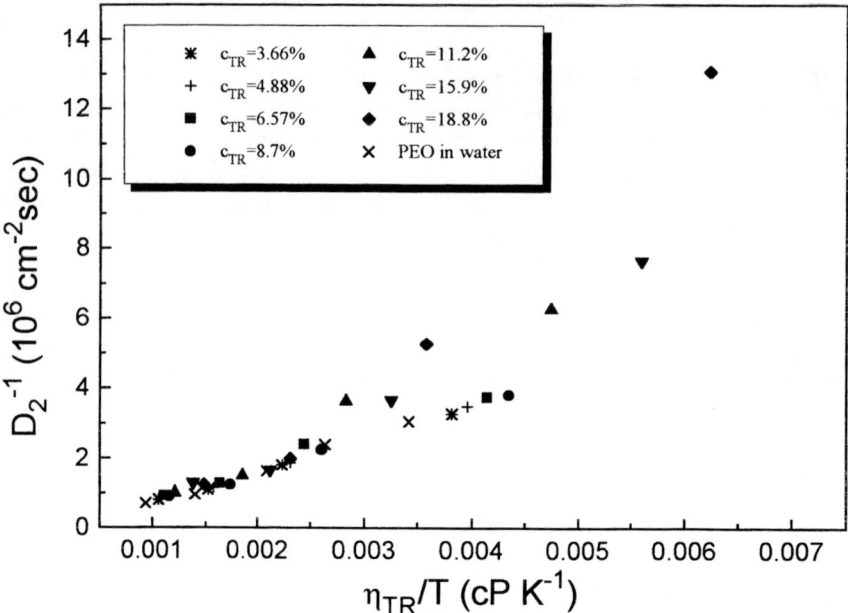

FIGURE 1. D_2^{-1}, obtained from the slow contribution in the correlation functions, as a function of η_{TR}/T at different trehalose amount. The x symbol refers to the PEO diffusion in water without trehalose.

A remarkable feature is that, see fig.2, when the PEO hydrodynamic radius behaviour (open squares), obtained by the ES relation from D_2 at the lowest trehalose amount, is compared with that of PEO coils in water[6], it emerges a stabilising effect of trehalose: the conformational properties of the polymeric coils, due to the presence of trehalose, shows to be less sensitive to temperature changes. These findings suggest that the structural properties of trehalose influences the swelling properties of PEO[4]. In particular, as evidenced in a previous work,[5] the trehalose diffusion coefficient data show, in the 20÷40°C range, a drastic change of the initial slope from negative to positive values and, hence, from attractive to repulsive interactions, which indicate that trehalose molecules tend to a more closed conformation at low temperature. The smaller value of the PEO coils size in the presence of trehalose than that in the binary PEO/water solutions, as shown in fig.2, can be explained considering that trehalose, which directly interacts with the polymeric chains, determines at low temperature a shrinking effect on these latter.

In view of these results we can deduce that trehalose-polymer interactions play a key role in determining the diffusive polymer dynamics. In particular it results that trehalose binds directly to the polymeric coils and that it drives the temperature behaviour of the polymer conformational properties. Therefore, these findings support the «water replacement» hypothesis formulated by Crowe and Crowe[1], indicating, in addition, that the polymer swelling and collapsing effects with temperature are stabilised.

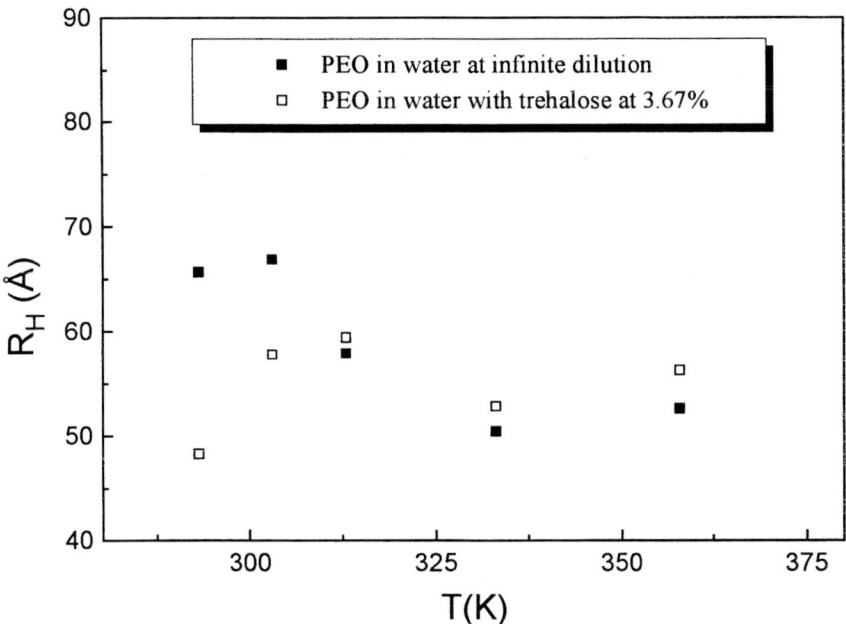

FIGURE 2. Temperature behaviour of the hydrodynamic radius of PEO coils in water[5] and in the water/trehalose medium.

ACKNOWLEDGMENTS

V. Villari acknowledges a grant from Consiglio Nazionale delle Ricerche (CNR).

REFERENCES

1. Crowe, J. H., and Crowe, L. M., *Biological Membranes*, edited by D. Chapman New York, Academic Press, 1984, vol.5, p. 57.

2. Gottfried, S. D., Peterson, E. S., Sheikh, A.G., Wang, J., Yang, M., and Friedman, J. M., *J. Phys. Chem.*, **100**, 12034 (1996).

3. Cordone, L., Galajda, P., Vitrano, E., Gassmann, A., Ostermann, A., and Parak, F., *European Biophys. Journal with Biophys. Lett.*, **27**, 173-176, (1998).

4. Branca, C., Faraone, A., Magazù, S., Maisano, G., Migliardo, P., Triolo, A., Triolo, R., and Villari, V., *J. Phys.: Condens. Matter*, **11**, 6079-6098 (1999).

5. Magazù, S., Maisano, G., Middendorf, H.D., Migliardo, P., and Villari, V., *J. Chem. Phys.*, **109**, 1170-1174 (1998).

6. Magazù, S., Maisano, G., Migliardo, P., and Villari, V., *J. Chem. Phys.*, **111**, (1999) in press.

NUCLEAR PHYSICS

CHIMERA Multidetector at Laboratori Nazionali del Sud

S.Aiello[2], A.Anzalone[1], M.Baldo[2], R.Barnà[5], M.G.Campisi[1], G.Cardella[2],
Sl.Cavallaro[1,3],V.D'Amico[5], E.De Filippo[2],D.DePasquale[5],S.Feminò[5],
E.Geraci[1,3], F.Giustolisi[1,3], P.Guazzoni[4], C.M.Iacono-Manno[1,6],
A.Italiano[5], G.Lanzalone[2,6], G.Lanzanò[2], S.LoNigro[3,6], U.Lombardo[1,3],
G.Manfredi[4], A.Pagano[2], M.Papa[2], S.Pirrone[2], G.Politi[2], F.Porto[1,3],
S.Sambataro[2,3], M.L.Sperduto[1,3], C.M. Sutera[2], L.Zetta[4]

1)INFN Laboratori Nazionali del Sud, Catania
2) INFN, Sezione di Catania
3)Dipartimento di Fisica Università di Catania
4)INFN, Sez. di Milano e Dipartimento di Fisica Università di Milano
5)INFN, Gr. Coll. Messina e Dipartimento di Fisica Università Messina
6)Centro Siciliano di Fisica Nucleare e di Struttura della Materia, Catania

Abstract. The installation of CHIMERA multidetector, designed in order to study central collisions in heavy ion reactions at intermediate energy, is going on at LNS and the first experiment with the forward part (688 telescopes) is running since May 1999. The aim of this contribution is to present the status of the project.

CHIMERA MULTIDETECTOR

The CHIMERA (Charged Heavy Ions Mass and Energy Resolving Array) multi-element detector array was designed to study nuclear multifragmentation in heavy ion collisions at the Fermi energy. This energy domain is now accessible at the heavy ion facility INFN Laboratorio Nazionale del Sud in Catania (LNS) in a broad range of projectile nuclei. The main characteristics of the detector are a systematic measurement of the time of flight, a low multi-hit probability due to the high adopted granularity and a very low energy threshold ($E/A<0.5$ MeV/nucl) for fragments detection [1,2]

The detector is made of 1192 telescopes arranged in 35 rings in a cylindrical geometry around the beam axis. The forward 18 rings are assembled in 9 wheels and they cover the polar laboratory angles between 1° and 30°. The telescopes are placed at variable distances from the nuclear target in the range between 350 cm (at 1°) to 100 cm (at 30°). The remaining 17 rings covering the angular range 30°-176° are assembled in a sphere of 40 cm in radius. The shape and dimensions of CHIMERA make possible to perform, for the first time, a Time Of Flight (TOF) techniques for

CP513, *Nuclear and Condensed Matter Physics,* edited by A. Messina
© 2000 American Institute of Physics 1-56396-929-7/00/$17.00

velocity and mass measurements. Each telescope is made of one silicon detector (300 microns thick) followed by one CsI(Tl) crystal, coupled with a photodiode. The thickness of the crystals varies from 12 cm to 3 cm in order to stop the energetic particles in all the angular range. Considering the beam entrance and outgoing holes, the frames of the detectors and of the target, an overall solid angle of about 94% is obtained. In fig.1 a view of the mechanical structure is shown. An enhanced data acquisition, control and trigger system has been developed to manage the almost 5000 electronic channels of the detectors [3].

Several tests have been done at LNS with heavy ion beams in order to verify the designed performances, namely the time and energy resolutions and the mass and charge identification power of the device[4]. As an example, one wheel made of 64 telescopes has been successfully tested with a ^{58}Ni cyclotron beam during 1998. In fig.2, the charge identification matrix for a typical telescope of CHIMERA is shown: the upper side represents the ΔE full charge to digital conversion dynamics (QDC) and the down side represents 1/8 of the silicon full energy range obtained by an high gain integration. All particles from protons to heavy ions were clearly identified.

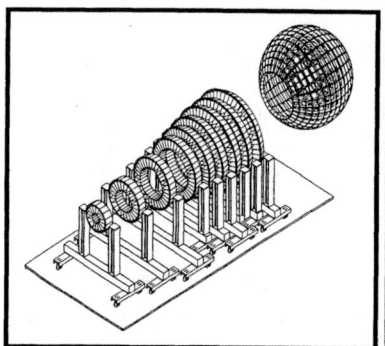

FIGURE 1. Chimera mechanical structure.

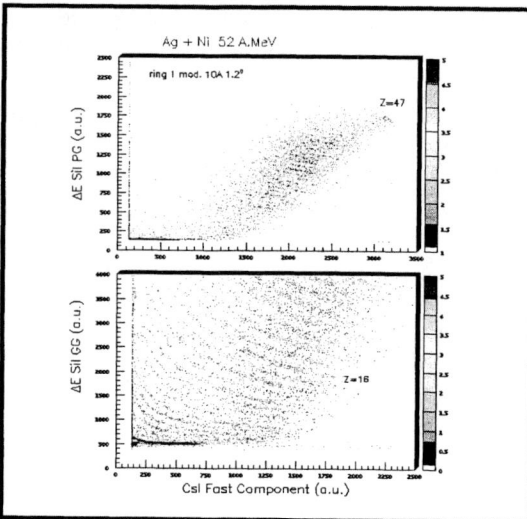

FIGURE 2. Charge identification Matrix.

The signals coming from the photodiode were analyzed by a two-gate integration method allowing light charged particles (LCP) identification[5]. In fig.3 is shown a very good quality LCP identification up to lithium in the ^{58}Ni + ^{27}Al reaction at 30 MeV/A obtained using the pulse shape method.

Another more complete test was performed in GANIL, relatively to INDRA 3rd measurements campaign, where the first wheel of CHIMERA was coupled with INDRA detector. The use of the first wheel of CHIMERA allowed to have a better fragment identification up to 50 charge units and a higher precision on angular correlation measurements, due to the high granularity achieved in the forward region. In fig.4 a ΔE-TOF matrix is shown for the reaction ^{107}Ag+^{58}Ni at 52 MeV/A.

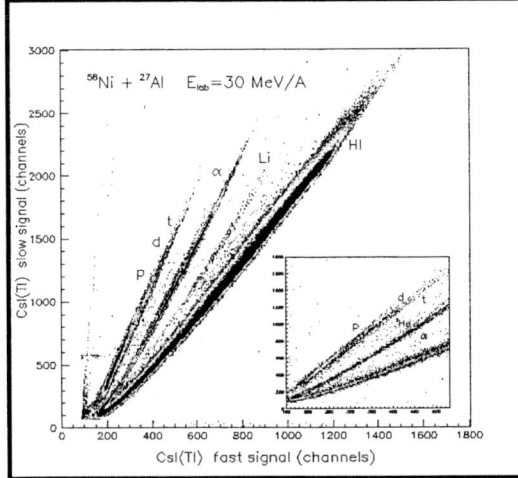

FIGURE 3
LCP identification Matrix.

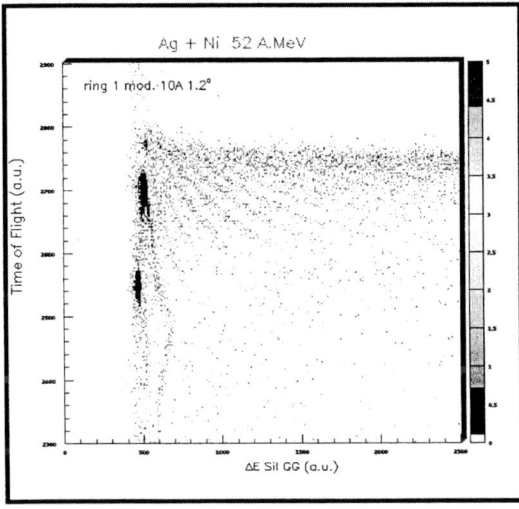

FIGURE 4. ΔE-TOF Matrix.

Reverse experiment

More recently, the forward part of CHIMERA, composed by 688 telescopes covering a polar range from 1° to 30°, has been mounted in the CICLOPE vacuum chamber at LNS and constitutes the experimental apparatus of REVERSE experiment. The experiment, supported by an international collaboration[i], intends to study cluster production and isospin effects of nuclear Equation of State in the nuclear collision ^{112,124}Sn + ^{58,64}Ni, ^{27}Al at 25 and 35 MeV/u.
Several calibration runs have been performed for both silicon and CsI(Tl) detectors. For silicon detectors the energy resolution was quite less than 1% at 15 MeV/A.

For CsI(Tl) detectors an energy resolution less than 2% in the whole detected energetic range was obtained.

ACKNOWLEDGMENTS

We express our gratitude to the Electronic, Mechanical and Lab. Rivelatori staff of INFN sez. Catania and LNS for the great effort done in all phases of the project. Some of us (G. Lanzalone, G. Politi) want to thank the Centro Siciliano di Fisica Nucleare e Struttura della Materia Sez. Catania for financial support.

REFERENCES

1. S. Aiello et al., *Nucl. Phys*, A583(1995), pp 461

2. A.Pagano et al., *Proceedings of the Int. Work. On 4π detectors*, Ed. by Petrovici, Sandulescu, Pelte World Scientific (1996), pp 129

3. E. De Filippo for Chimera Coll., *Proceedings of 11th IEEE NPSS Real Time Conf.*, June 1999, Santa Fe, New Mexico, USA.

4. S. Aiello et al., *Nucl. Instr. And Meth.*, A427, (1999), pp 510.

5. S. Aiello et al., *Nucl. Instr. And Meth.*, A369, (1996), pp 50.

[i]REVERSE Collaboration:

S. Aiello[1], A. Anzalone[2], M. Baldo[2], R. Barnà[3], A. Bonasera[2], B. Borderie[4], A. Botvina[5,6], R. Bougault[7], M. Bruno[7], G. Cardella[1], M. G. Campisi[2], S. Cavallaro[2], M. Colonna[2], M. D'Agostino[5], E. D'Amico[3], R. Dayras[8], N. De Cesare[9], E. De Filippo[1], D. De Pasquale[3], M. Di Toro[2], S. Feminò[3], E. Geraci[2], M. Geraci[1], F. Giustolisi[2], A. Grzeszczuk[10], P. Guazzoni[11], D. Guinet[12], M. Iacono Manno[2], A. Italiano[3], S. Kolwaski[10], G. Lanzanò[1], G. Lanzalone[1],S. Li[17], U. Lombardo[2], S. Lo Nigro[1], C. Maiolino[2], D. Mahboub[2], G. Margagliotti[13], F. Migliardo[3], T. Paduszynski[10], A. Pagano[1], M. Papa[1], M. Petrovici[14], E. Piasecki[15], S. Pirrone[1], G. Politi[1], E. Pollacco[8], F. Porto[2], A. Rapisarda[1], M. F. Rivet[4], E. Rosato[9], S. Sambataro[1], V. Simion[14], M.L. Sperduto[2], J. C. Steckmeyer[7], C. Sutera[1], L. Tassan-Got[4], A. Trifilò[3], M.Trimarchi[3], G. Vannini[13], M. Vigilante[9], J. Wilczynski[16], H. Wu[17], Z. Xiao[17], L. Zetta[11], W. Zipper[10]

1)INFN e dip di fisica Università di Catania 2)LNS e Dipartimento di Fisica Università di Catania 3)INFN e Dipartimento di Fisica Università di Messina 4)IPN,IN2P3-CNRS e Université Paris-sud,France 5)INFN e Dipartimento di FisicaUniversità di Bologna 6)Institute for Nuclear Research, Russian Accademy of Science 7)LPC,ISMRA Université de Caen,France 8)DAPNIA-SPhN,Ce Saclay, France 9)INFN e Dipartimento di Fisica Università di Napoli 10)University of Silesia, Katowice, Poland 11)INFN e Dipartimento di Fisica Università di Milano 12)IPN; IN2P3-CNRS e Université Claude Bernard, France 13)INFN e Dipartimento di Fisica Università di Trieste 14)Institute for Physics and Nuclear Engineering, Bucharest, Romania 15)Institute of Experimental Physics, University of Warsaw, Poland 16)Institute of Nuclear Studies Otwock, Poland 17)Institute of Modern Physics Lanzhou, China

Study of the Quasi-Free Reaction Mechanism in the $^6Li(^{12}C,\alpha^{12}C)^2H$ reaction: Astrophysical Implications.

M. Aliotta[1,2], S. Cherubini[3], L. Gialanella[4], P. Figuera[2], M. Lattuada[2,5], Ð. Miljanic[6], A. Musumarra[3], M.G. Pellegriti[1,2], R.G. Pizzone[1,2], C. Rolfs[4], S. Romano[2], N.Soic[6], C. Spitaleri[1,2], F. Strieder[4], A. Tumino[2,5], S. Typel[7], H.H. Wolter[7]

[1]*Dipartimento di Metodologie Fisiche e Chimiche per l'Ingegneria, Università di Catania, Catania, Italy*
[2]*Laboratori Nazionali del Sud-INFN, Catania, Italy*
[3]*Institut de Physique Nucléaire, Université Catholique de Louvain, Louvain-la-Neuve, Belgium*
[4]*Institut für Experimentalphysik III – Ruhr Universität Bochum, Germany*
[5]*Dipartimento di Fisica, Università di Catania, Catania, Italy*
[6]*Institut Rudjer Boskovic, Zagreb, Croatia*
[7]*Ludwig Maximilians Universität München, Germany*

Abstract. The Trojan-Horse method has been applied to the $^6Li(^{12}C,\alpha^{12}C)^2H$ reaction at 18 MeV in order to study the two-body α-^{12}C interaction around the Coulomb barrier. Coincidence spectra show clear evidence of the quasi-free α-^{12}C scattering, in good agreement with calculations performed in the framework of the Plane Wave Born Approximation. The virtual excitation function of the α-^{12}C elastic scattering has been extracted and compared with the behaviour of the free scattering cross section.

1 INTRODUCTION

The $^{12}C(\alpha,\gamma)^{16}O$ reaction plays a key role in astrophysics both for the nucleosynthesis of elements and the final evolution of massive stars. Together with numerous direct measurements [1] (and references therein), aimed at the determination of the reaction cross section as close as possible to the relative Gamow energy region (E_G~300 keV), several indirect methods have been used to derive information on the nuclear parameters of the two sub-threshold states which dominate the radiative capture process [1] at astrophysical energies.

We report here on measurements of the quasi-free $^6Li(^{12}C,\alpha^{12}C)^2H$ scattering and compare the extracted two-body excitation function with the results of direct α-^{12}C elastic scattering measurements in the energy range 2.5-3.5 MeV. Such a study represents a first step to determine under what conditions the Trojan-Horse Method (THM) [2] can be applied to the $^{12}C(\alpha,\gamma)^{16}O$ reaction.

CP513, *Nuclear and Condensed Matter Physics*, edited by A. Messina
© 2000 American Institute of Physics 1-56396-929-7/00/$17.00

2 PWBA FORMALISM

In the Plane Wave Born Approximation (PWBA) the full scattering wave function in the final channel and the distorted wave function in the initial channel are replaced by the corresponding plane waves with the appropriate momenta.

The hypotheses underlying the quasi-free mechanism imply that the incident particle interacts with only one cluster at a time in the target nucleus, leaving the other one unperturbed. Since the ^{6}Li nucleus can be considered as being strongly clustered into α+d, an off-line selection can be made of those events corresponding to the α-^{12}C interaction, with the deuteron behaving as a spectator. Within these approximations the cross section of the ^{6}Li(^{12}C,α^{12}C)^{2}H quasi-free scattering can be factorized by a term describing the α-d structure of ^{6}Li and a term describing the α-^{12}C interaction. Since the momentum distribution of the deuteron in ^{6}Li is known [3], the study of the three-body reaction can therefore be used to infer information on the α-^{12}C interaction pole. Finally, the three-body cross section can be expressed as

$$\frac{d^3\sigma}{dE_{\alpha C}d\Omega_{\alpha C}d\Omega_{0d}} = KF \ \left|\chi_{Li}(\vec{p}_d)\right|^2 \frac{d\sigma}{d\Omega_{\alpha C}} \qquad (1)$$

where:

a) KF is a kinematic factor which can be calculated according to the experimental conditions;

b) the second term is the deuteron momentum distribution in ^{6}Li, which depends only on the internal wave function and is in principle independent of the experimental conditions;

c) $d\sigma/d\Omega_{\alpha\text{-}C}$ is the two-body differential scattering cross section for the α-^{12}C subsystem.

In order to determine completely the kinematic properties of the final state, the total kinetic energy of the two outgoing particles must be measured in coincidence at specified angles, θ_α-θ_C. Since the deuteron momentum distribution inside ^{6}Li is peaked around zero momentum (the α-d relative motion is in an l=0 state) [4], the chosen angles correspond to the case p_d=0. Being related to the most probable value of the deuteron momentum, these pairs of *quasi-free angles* define the kinematics conditions where the quasi-free process, if present, should be dominant.

Since the momentum distribution of the deuteron in ^{6}Li is known and the kinematic factor can be calculated, the two-body cross section is derived from the measured three-body cross section using equation (1).

3 EQUIPMENT AND SETUP

The first runs on the ^{6}Li(^{12}C,α^{12}C)^{2}H reaction were performed at the 4MV Dynamitron Tandem Laboratorium in Bochum (Germany). Additional data were taken at the SMP Tandem Van de Graaff accelerator of the Laboratori Nazionali del Sud (LNS), Catania. A 145 μg/cm^2 thick LiF target (enriched in ^{6}Li to 95%) was evaporated on a thin carbon backing. It was bombarded with an 18 MeV ^{12}C beam, having a spot size on target of about 1 mm diameter. The reaction products were detected in coincidence by means of two ΔE-E telescopes, each one consisting of an ionization chamber (IC) and a position sensitive detector (PSD). The PSD's (1000 μm thick) were placed behind their corresponding IC's and were centered at $\theta_\alpha=22.5°$ and $\theta_C=12.5°$ on opposite sides of the beam direction, at respective distances of 18.7 cm and 33.2 cm from the target. The in-plane angular ranges covered by the telescopes were $\Delta\theta_\alpha=14.9°$ and $\Delta\theta_C=8.6°$, with solid angles of $\Delta\Omega_\alpha=12.6$ msr and $\Delta\Omega_C=2.2$ msr.

4 DATA ANALYSIS AND RESULTS

After identification of the He and C ions in the ΔE-E matrix, the locus of the events due to the ^{6}Li(^{12}C,α^{12}C)^{2}H reaction in the E_α-E_C plane was defined by the corresponding three-body kinematics. Other three-body reactions occurring in the target were identified in the same way. Since the kinematic loci for different reactions do not overlap with the one of interest, a graphical cut was then used to select the events corresponding to the reaction ^{6}Li(^{12}C,α^{12}C)^{2}H.

Fig.1 shows the coincidence yield as a function of the α-^{12}C relative energy E_{cm}, for different ranges of the deuteron momentum p_d.

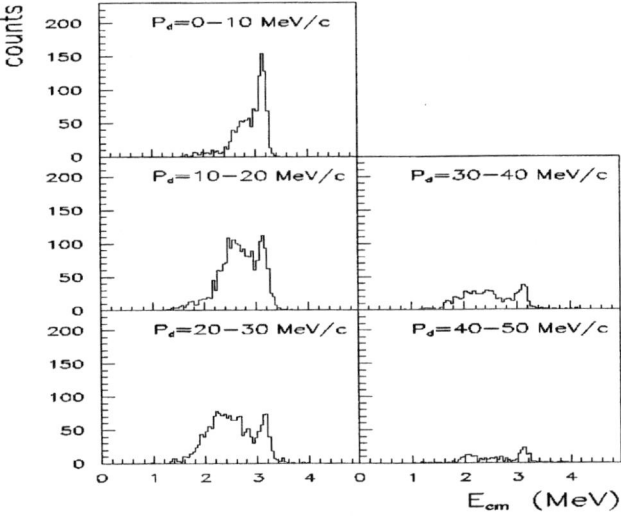

Figure 1 Coincidence yield as a function of the relative $E_{\alpha\text{-}c}=E_{cm}$ energy for various ranges of the deuteron (i.e. spectator) momentum.

A peak at $E_{cm}=3.1$ MeV is most prominent for $p_d < 10$ MeV/c and decreases significantly for higher values of the spectator momentum. This peak has been identified with the $J^\pi=4^+$ state at $E_x=10.35$ MeV in ^{16}O. The experimental evidence of a strong correlation between the magnitude of the observed peak and the value of the deuteron momentum supports strongly the dominant presence of the quasi-free mechanism at zero deuteron momentum.

5 MONTECARLO SIMULATION

The experiment was simulated by means of a Montecarlo calculation under the assumption that the mechanism giving rise to the reaction is purely quasi-free, so that the three-body cross section can be calculated according to eq.(1). The momentum distribution has been obtained from the Fourier transform of the ^{6}Li-ground state wave function, assuming for the α-d interaction a Woods-Saxon potential adjusted so as to reproduce the ^{6}Li binding energy (B=1.475 MeV). A fit to the Fourier transform using the product of a Lorentzian and a Gaussian function gives a FWHM of 73 MeV/c, consistent with observation [6]. The two-body cross section entering eq.(1) was calculated from the energy-dependent phase shifts for partial waves l=0→6. These phase shifts were taken from a multilevel R-matrix parameterization of elastic α–C scattering derived from a phase-shift analysis given in [7]. Finally, the geometrical efficiency of the experimental setup, as well as the detection thresholds of the two telescopes, have been taken into account.

Fig.2 shows both the calculated three-body cross section (histogram) and the experimental data (points) with various conditions on the deuteron momentum and for θ_{cm}=120°±2.5°.

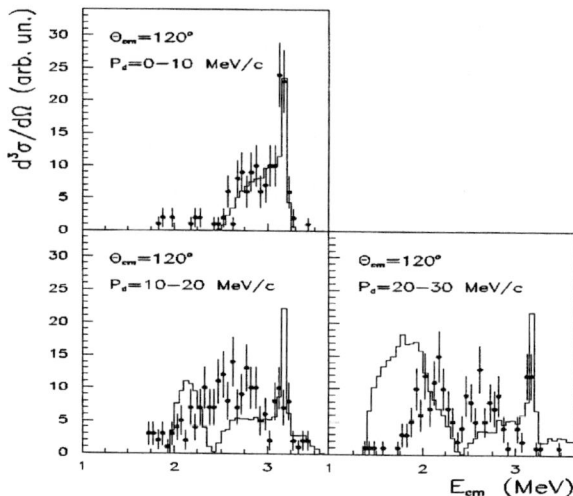

Figure 2 Calculated three-body cross section (histogram) and data for the given experimental conditions.

The normalization factor between the theoretical calculation and the experimental data has been obtained through a χ^2-minimization procedure for the case with $p_d \leq 10$ MeV/c. The comparison shows an excellent agreement at low spectator momenta (i.e. $p_d \leq 10$ MeV/c). The results demonstrate that the quasi-free mechanism is present and the approximations used to describe the cross section are plausible and give results in agreement with the experimental data.

6   THE TWO-BODY CROSS SECTION

According to the results of the simulation (fig.2), only events with $p_d \leq 10$ MeV/c were taken into account for the extraction of the two-body cross section. In order to divide the three-body cross section by the momentum distribution and the kinematic factor, these latter have been calculated as a function of the relative $E_{\alpha\text{-}C}$ energy. Geometrical detection efficiency has also been taken into account. The original statistical errors of the primary data have been preserved throughout the whole procedure. The resulting two-body excitation function is shown in fig.3.

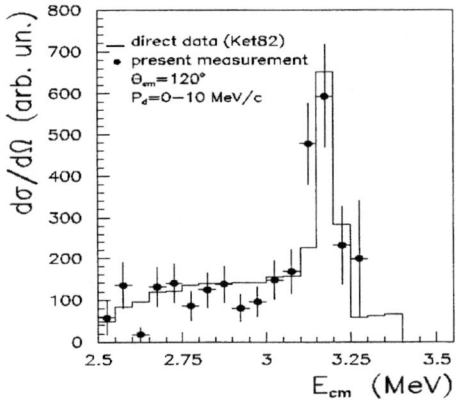

Figure 3 Comparison between direct (solid curve) and indirect excitation function for the two-body scattering α+C at θ_{cm}=120° and E_{cm}=2.5-3.5 MeV.

Shown are also the direct data [5], after having been rebinned in 50 keV steps. Good agreement is noted between the two data sets, thus confirming the validity of the hypotheses used in the derivation of the two-body cross section.

REFERENCES

1.   R.Plaga et al., *Nucl. Phys. A* 465 (1987) 291.

2.   G.Baur, *Phys. Lett. B* 178 (1986) 135.

3.   M.Lattuada et al., *Nuovo Cimento A* 83 (1984) 151.

4.   G.Calvi et al., *Lettere al Nuovo Cimento*, 37 (1983) 279.

5.   K.U.Kettner et al., *Z. Phys. A*, 308 (1982) 73.

6.   S.Barbarino et al., *Nucl. Phys. C*, 21 (1980) 1104.

7.   R.Plaga et al., *Nucl. Phys. A*, 465 (1987) 291.

Excited States of ^{11}Be

F.Cappuzzello[a,b], A.Cunsolo[a,b], S.Fortier[d],
A.Foti[a,c], H.Laurent[d], H.Lenske[e], J.M.Maison[d],
A.L.Melita[a,b], C.Nociforo[a,b], L.Rosier[d], C.Stephan[d],
L.Tassan-Got[d], J.S.Winfield[g], H.H.Wolter[f]

(a) Università di Catania, Italy
(b) I.N.F.N.-L.N.S., Catania, Italy
(c) I.N.F.N.-sez.CT, Catania, Italy
(d) I.P.N., Orsay, France
(e) Inst. Theor. Physik, Univ. of Giessen, Giessen, Germany
(f) Sektion Physik, Univ. of München, München, Germany
(g) University of Surrey, Surrey, England

Abstract. The ^{11}B(^{7}Li,^{7}Be)^{11}Be reaction at 57 MeV incident energy was used to explore the ^{11}Be excitation energy spectrum at forward angles. Angular distributions were extracted for the transitions to the ground and to the states of ^{11}Be at excitation energies of E^*=0.32, 1.78, 2.69, 3.41, 3.89, 3.96, 6.05 MeV combined with the ground and the first excited state of ^{7}Be. Also the SDR [1][2] oscillation mode was observed at E^*=9.5 MeV and FWHM≈9 MeV and a new peak at E^*=6.05 MeV and FWHM≈0.3 MeV was observed. QRPA calculations in the G-matrix representation are in progress in order to describe the continuum structure of ^{11}Be. DWBA calculations have been started to evaluate transferred angular momenta both in the one step and in the two steps dynamical framework.

INTRODUCTION

Investigations of dripline nuclei are of central interest for modern nuclear structure and dynamics. At large charge asymmetry the nucleon-nucleon isovector interactions are strongly repulsive on the excess nucleons and push them towards the continuum threshold. In such a situation a transition from mean-field behaviour to the dominance of residual interactions takes place. A prominent example for a one-neutron halo nucleus is ^{11}Be, for which the static and dynamic properties are strongly determined by the interaction between a single valence neutron and the ^{10}Be core. Moreover, due to isovector repulsion the ^{10}Be core itself is rather soft and the system is easily polarizable. In such conditions the external neutron can gain energy in exciting the core. This effect, known as dynamical core polarization, accounts for about 25% of the spectroscopic amplitude of the ^{11}Be ground state, and explains, to some extent, the experimentally established parity inversion of the ^{11}Be $(1/2)^+$ ground state with respect to the shell model systematics. Theoretically, the structure of ^{11}Be and the dynamics involving such a nucleus is up to now controversial. In the following we report on our approach based on the ^{11}B(^{7}Li,^{7}Be)^{11}Be reaction to excite the ^{11}Be states. The (^{7}Li,^{7}Be$_{g.s.}$) transition has an excellent isospin selectivity and, at low incident energies,

CP513, *Nuclear and Condensed Matter Physics*, edited by A. Messina
© 2000 American Institute of Physics 1-56396-929-7/00/$17.00

mainly proceed via a (l=0, Δs=0) Fermi transition with small admixtures of (l=0, Δs=1) Gamov-Teller and higher spin transfer components. On the contrary, the (^{7}Li,^{7}Be$_{1/2}^-$) transition is mainly determined by the GT strength. Moreover, at this energy the reaction includes both sequential proton-neutron transfer and direct charge exchange. The first transfer contributions are expected at least of second order, therefore the information on nuclear structure is less direct. The direct charge exchange is a one-step process mediated by the exchange of virtual isovector mesons such as the pion or the ρ-δ mesons. In this case the cross sections are directly related to nuclear structure entering into the scattering amplitudes via the nuclear proton-neutron transition densities. The shapes of angular distributions of the one- and two-step contributions are expected to be rather different, even at $E_{lab}\approx10$ MeV per nucleon and it seems possible to separate the two processes at forward angles [5].

EXPERIMENTAL DATA AND DWBA CALCULATIONS

The experiment was performed at the IPN-Orsay Tandem facility using a 57 MeV ^{7}Li^{+++} beam and a 95% enriched ^{11}B self supporting target about 130 μg/cm^2 thick. ^{7}Be ejectiles were detected by means of the IPN-Orsay Split-Pole spectrometer. The solid angle covered was about 0.5 msr and the overall energy resolution was about 50 KeV.

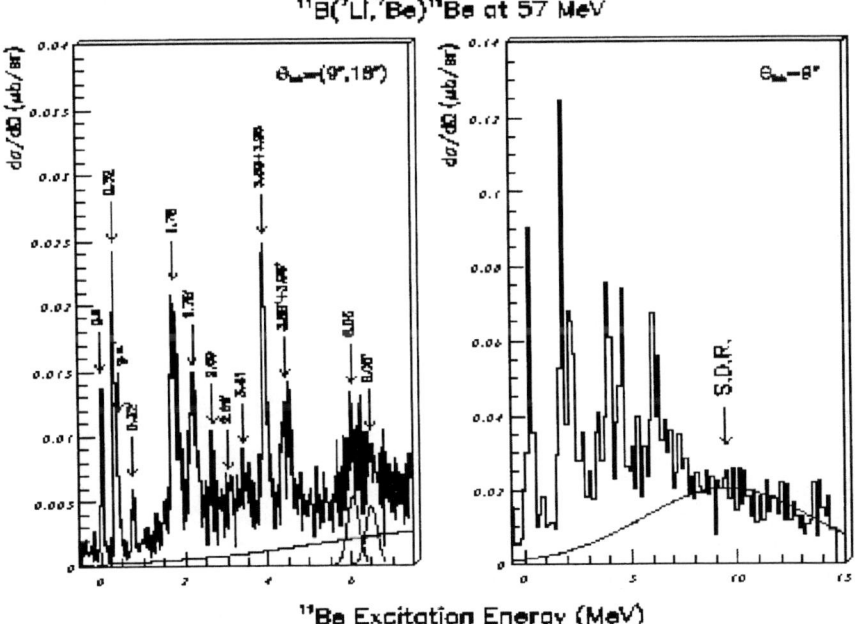

FIGURE 1. Excitation energy spectrum for ^{11}B(^{7}Li,^{7}Be)^{11}Be reaction at 57 MeV. In the left panel the spectra (14 KeV/channel) at $\vartheta_{lab}=9°$, 12°, 15° and 18° were added, while in the right panel the compressed (140 KeV/channel) $\vartheta_{lab}=9°$ spectrum is shown. The solid curves are the result of a peak fitting analysis. See text for detail.

Nine peaks are assigned to transitions in which ^{7}Be remains in its ground state while ^{11}Be is either in the ground state or in the 0.32, 1.78, 2.69, 3.41, 3.89, 3.96, 6.05, 9.5 MeV excited states. Peaks marked by an asterisk refer to transitions in which ^{7}Be is in the 0.43 MeV(1/2$^-$) first excited state. Peaks due to the ^{12}C impurities in the target were subtracted using the spectra measured in separate runs with a ^{12}C target. At forward angles (9$\leq$ 7.1°), there is a strong contribution from the p(^{7}Li,^{7}Be)n reaction which obscured some ^{11}Be peaks at certain angles. As can be inferred from Table 1, the reaction mainly populates single particle states according to the spin-isospin selection rules Δs=0,1; Δt=1. An interesting feature of the spectra is the previously-unobserved structure at 6.05 MeV. In line with the other transitions, this new structure has been supposed to contain contributions both from transitions to the ^{7}Be ground and first excited states.

TABLE 1. States populated in the ^{11}B(^{7}Li,^{7}Be)^{11}Be reaction.

E*(MeV) [4]	Γ(KeV) [4]	E*(MeV)(*)	Γ (MeV)(*)	J$^\pi$	Structure [3][4]
g.s.	bound			1/2$^+$	^{10}Be(g.s.)⊗(s1/2)
0.320±0.002	bound	0.32±0.02		1/2$^-$	^{10}Be(g.s.)⊗(p1/2)
1.748±0.004	104±21	1.77±0.02	100±4	5/2$^-$	^{10}Be(g.s.)⊗(d5/2,d3/2)
2.642±0.009	228±21	2.67±0.02	199±8	3/2$^-$	^{10}Be(g.s.)⊗(p3/2)$^{-1}$
3.398±0.006	104±17	3.40±0.02	128±6	3/2$^-$	^{9}Be(g.s.)⊗(sd)$^2_{0+}$
3.888±0.001	<10 [9]	3.89±0.02	<50	3/2*	^{10}Be(2$^+$)⊗(s1/2)
3.855±0.001	15±5 [9]	3.96±0.02	<50	3/2$^-$	^{9}Be(g.s.)⊗(sd)$^2_{2+}$
		6.05±0.02	322±12		
~10 [1]	~7 (MeV)	9.4±1	9±1	SDR	

(*) present

The two gaussian fit, shown in Fig.1 gives a width of about 0.3 MeV for the ^{11}Be state at 6.05 MeV. The large difference between its excitation energy and the ^{11}Be neutron emission threshold (S$_n$=504 KeV) makes it unlikely that this structure could not due to the large radial overlap, at threshold, between the neutron halo wave function and continuum states. In a similar manner to the g.s. and the 1.78 MeV (5/2$^+$) state, the structure for the 6.05 MeV state could be a ^{10}Be core coupled to a (sd) neutron and would correspond to the predicted 6.70 MeV (3/2$^+$) state [4]. As regards the SDR mode, our results agree with the previously measured [1] values of its centroid and width. The experimental angular distributions are generally structureless due to the many transferred angular momenta involved [6]. One example is shown in Fig.2. The uncertainty in the value of cross section is about 20%. We have started the analysis of the angular distribution in the framework of DWBA calculations by means of the HIDEX code [7] for the direct CEX reaction mechanism and the JUPITER code [7] for the two-step contribution to the total cross section. Both the central and tensorial part of the isovector nucleon-nucleon interaction are taken into account. The optical parameters are taken from [8] while spectroscopic amplitudes are calculated in the QRPA approximation theory, using the G-matrix Dirac-Brueckner coupling constants taken from [7]. In Fig.2 the calculated direct component of the angular distribution for the ^{11}Be ground state transition is drawn as a continuous line, showing a good agreement with data, especially at forward angles. At backward angles a systematic lack of strength seems to indicate a non-negligible contribution from the 2-step processes.

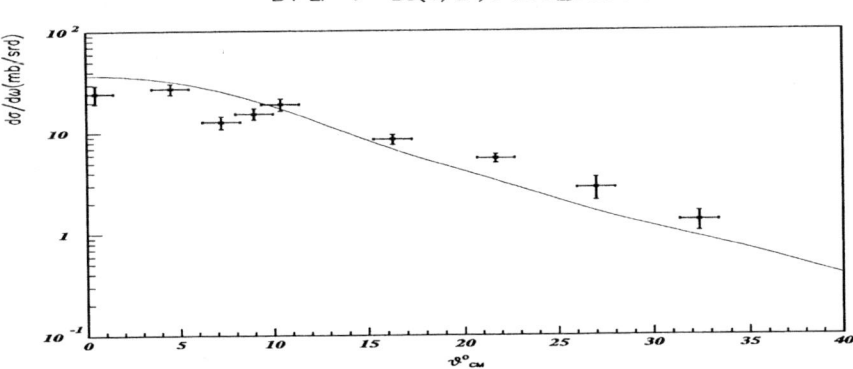

FIGURE 2. Angular distribution for the $^{11}B(^{7}Li,^{7}Be)^{11}Be$(g.s.) transition. The solid line refers to the direct CEX process.

CONCLUSION

The main finding in this work is the observation in the ^{11}Be spectrum of a state lying at 6.05 MeV with a width of about 0.3 MeV. Preliminary QRPA calculation indicates for this state a ^{10}Be core structure coupled to a (sd) neutron. We have started the analysis of the angular distributions of the states populated in the reaction in the DWBA framework, even if the non zero spins of the ^{11}B(g.s.), ^{7}Li and ^{7}Be nuclei hamper a clear determination of the main transferred angular momentum. Further work is in progress.

REFERENCES

1. Sakai H. et al., *Phys. Lett. B*, 7-12 (1993).

2. Yamaya T. et al., *Phys. Rev. C***51**(2), 493-499 (1995).

3. Fortune H.T., Koltenuk D. and Lau C.K., *Phys. Rev. C***51**(6), 3023-3025 (1995).

4. Liu G.B. and Fortune H.T., *Phys. Rev. C***42**(1), 167-173 (1990).

5. Sakuta S.B. et al., *Europhysics Letter* **22**(7), 511-515 (1993).

6. Jänecke J. et al., *Phys. Rev .C***54**(3), 1070-1083 (1996).

7. Lenske H., private comunications and *J.Phys G: Nucl. Part.Phys.* **24**, 1429-1438 (1998).

8. Cook J. et al. *Nucl.Phys* A**466**, 168-188 (1987).

9. Ajzenberg-Selove F. *Nucl.Phys.* A**433**, 1 (1985).

MAGNEX: A Large Acceptance Magnetic Spectrometer for EXCYT

A.Cunsolo[a,b], F.Cappuzzello[a,b], A.V.Belozyorov[d], A.Elanique[b], A.Foti[a,c], A.Lazzaro[a], O.Malishev[d], L.Melita[a,b], W.Mittig[e], C.Nociforo[a,b], P.Roussel-Chomaz[e], V.Shchepunov[b,d], D.Vinciguerra[a,b], A.Yeremin[d], J.S.Winfield[f].

(a) Università di Catania, Catania, Italy
(b) INFN-LNS , Catania, Italy
(c) INFN- Sez.di Catania, Catania, Italy
(d) Flerov Laboratory of JINR, Dubna, Russia
(e) GANIL, Caen, France
(f) Surrey University, Surrey, UK.

Abstract. We describe the large acceptance magnetic spectrometer called MAGNEX. It will be used in nuclear reaction experiments with both stable and radiactive beams at the LNS-INFN laboratory, Catania, Italy.

INTRODUCTION

The advent of the new facilities for the production of Radiactive Ions Beams (RIB) like EXCYT [1], presently under construction at LNS-INFN, Catania, will give rise to a "new physics". This new research field is connected with the study of nuclear structure and reaction dynamics induced by nuclei with particular properties such as large values of isospin, haloes, large deformation, and with the measurement of nuclear observables of astrophysical interest. The driving force is the large number of hitherto unobtainable projectile-target combinations. For this new physics, a new detection philosophy is needed, which takes into account the relatively low intensities of RIBs, their wide energy and mass ranges, and, in the case of EXCYT, the excellent properties of Tandem beams. The large acceptance MAGNEX spectrometer has been conceived as a general purpose device to explore a lot of such a nuclear physics [2] both in inclusive measurements and in coincidence with other detectors [3][4]. On the April 4th 1997, a stimulating workshop called "II Giornata EXCYT" was held at the LNS in order to verify the physics interest of the scientific community on the EXCYT RIBs including the possible use of MAGNEX [3][4].

MAGNEX OVERVIEW

The MAGNEX spectrometer, which is schematically represented in Fig.1, consists of:

CP513, *Nuclear and Condensed Matter Physics,* edited by A. Messina
© 2000 American Institute of Physics 1-56396-929-7/00/$17.00

- an entrance position-sensitive start detector (PSD), based on a microchannel plate which provides the start signal for TOF measurements.

- a quadrupole magnet (Q), vertically focusing the particles.

- a dipole magnet (D), deflecting particles toward the focal plane.

- a focal plane detector (FPD), used to measure ion direction, energy, charge, atomic number and mass.

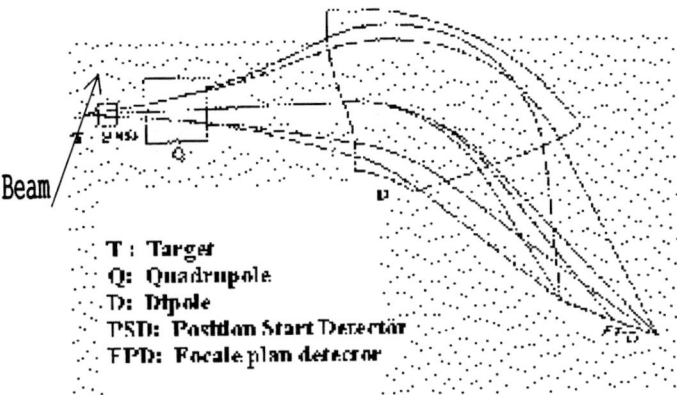

T : Target
Q: Quadrupole
D: Dipole
PSD: Position Start Detector
FPD: Focale plan detector

FIGURE 1. Schematic view of the MAGNEX spectrometer layout.

The main features of MAGNEX are shown in Table 1. The optical properties of the system have been investigated by means of raytracing programs (RAYTRACE and ZGOUBI) to achieve an optical configuration which gives the best compromise between the very large angular and momentum acceptance and the high momentum and mass resolution.

TABLE 1. Main parameters of MAGNEX.

Parameters	Values
Maximum magnetic rigidity	1.8 T·m
Solid angle	53 msr
Momentum acceptance	±10 %
Bending radius	1.6 m
Horizontal angular acceptance	(-90÷110) mr
Vertical angular acceptance	±130 mr
Deflection angle	55°
Focal plane tilt angle	61°
Focal plane length	81 cm
Momentum dispersion in focal plane	4.04 cm/%
1st order momentum resolving power	4200 (1mm beam spot size)
Path length of the central ray	5.92 m
Final momentum resolution*	~ 2000 (1 mm beam spot size)
Final mass resolution	~ 250

*for ^{7}Li at 49 MeV and ^{24}O at 90 MeV (see Table 2)

The aberrations and the kinematical effect of such a system, which would be unacceptably large if not corrected, have been compensated up to the 5th order [4], partly by a 2nd order spline shaping of the dipole entrance and exit effective field boundaries and by the introduction, inside the dipole, of an exapolar surface coil (hardware compensation), and partly by the trajectory reconstruction technique (software correction by COSY INFINITY algorithm [3][4]).

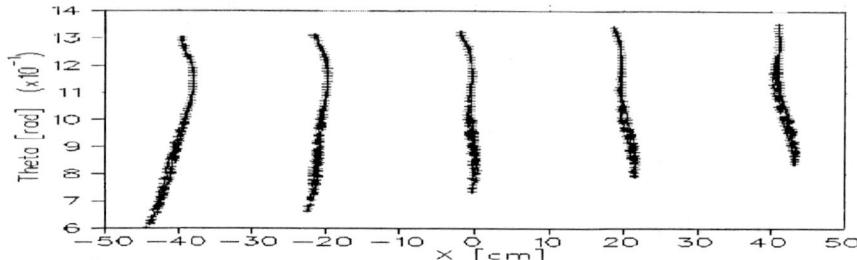

FIGURE 2. (x_f, ϑ_f) scatter plot in the focal plane after correction with the entrance and exit EFBs of the dipole in the total horizontal angular acceptance.

The large focal plane tilt angle, usual for QD systems, forces the positioning of the FPD along the focal plane in order to achieve the desired position resolution [3][4]. Moreover, the length (~ 80 cm) and the height (~30 cm) of the focal plane are rather large and the covering of the whole focal plane with a FPD becomes complicated. The proposed solution is a trapezoidal shaped box containing two position sensitive counters (drift chambers) separated by an ionization chamber. Besides the second position counter, a wall of silicon detectors measures the residual particle energy and provided a good timing signal (Δt ~1 ns).

GEANT SIMULATION

A complete simulation program GEAMAG has been developed for MAGNEX using the GEANT package. The magnetic field map for the quadrupole was constructed analytically while the dipole map was obtained by spline interpolation of the ZGOUBI mesh map. It is proposed to mount the silicon array in columns staggered along the back of the focal plane detectors. GEAMAG simulations have been used to estimate the efficiency of the array. It appears that a useful gain in the efficiency (from 88% to 95% in the case of (5x5) cm^2 detectors) can be made by partially overlapping (20%) the columns. Initial studies with GEAMAG have been performed to calculate the position (and energy) resolution at the focal plane under various conditions. Table 2 shows position resolution at the FPD for various ions (^{7}Li, ^{24}O and ^{40}Ca). In the table, "full" indicates the contribution of the optics, the straggling produced by the target, the PSD emissive foil, the FPD entrance window and gas, the effect of broadening of electron swarm along the drift path and the intrinsic resolution of the counters. On the other hand, the "bare" calculation, besides having no target or PSD detector, has only a skeleton FPD with no gas except in the position counters volumes. Table 2 shows that for ^{7}Li, the 123 µg/cm^2 mylar target and the PSD foil (70 µg/cm^2) have only a small effect on the resolution. As might be expected, for ^{24}0 and ^{40}Ca they give a larger contribution even though the final resolutions are similar.

TABLE 2. Position resolution (in mm) at the focal plane observed in GEAMAG simulations.

Incident ion	^{7}Li (49 MeV)	^{24}0 (90 MeV)	^{40}Ca (160 MeV)
Bare (optics)	0.95	0.95	0.71
PSD (+ optics)	1.41	1.11	2.15
Target (+ optics)	1.42	1.96	1.20
Full	1.90	2.03	2.44

TEST OF DETECTOR PROTOTYPES

Two sessions of beam tests [4] have been performed at the LNS Catania laboratory for the PSD and FPD detector prototypes. In the first session (May 1997) a PSD start detector based on a (7×7 cm^2) microchannel plate [5] and a drift chamber (92×10×20 cm^3) based on stripped anode read-out were tested. A position resolution of about 2.4 mm as FWHM was measured for the PSD both for the horizontal and for the vertical coordinate, while horizontal intrinsic resolution of about 0.2 mm and a vertical one of 1.1 mm were measured for the drift chamber. In the second session (July 1999), the same PSD was tested, this time equipped with a magnetic field for electron focusing. Concurrently, an hybrid gas ionization detector (96×10×37 cm^3) with 3 proportional wires alternate with 3 energy-loss sections and followed by a silicon detector was tested. The position resolution for the improved PSD was about 0.8 mm, while for the FPD prototype, values of 1.2 mm and 1.1 mm respectively for the horizontal and vertical resolutions were achieved. GEAMAG simulations show that the effect of such resolutions in the final energy measurement is not very big (see Table 2).

CONCLUSION

The INFN has approved the MAGNEX project and initial orders for the dipole and quadrupole magnet will soon be placed.

REFERENCE

1. Ciavola, G. et al., *Nucl. Phys.*A**616**, 69c-76c (1997).

2. Cunsolo A. et al., "Giornata EXCYT" LNS-INFN, Proceeding of the Workshop, Catania, 1996 pp. 143-161 and "II Giornata EXCYT" LNS-INFN, Proceeding of the Workshop, Catania, 1997 pp. 71-80.

3. Cunsolo A. et al., "MAGNEX: a large acceptance magnetic spectrometer for EXCYT", Report LNS-INFN, 1998.

4. Cunsolo A. et al., Addendum to Magnex report "Geant simulation, Electronics and Acquisition system", Internal status report LNS-INFN, 1998 and Cunsolo A. et al., Addendum to Magnex report "Test of detector prototypes", Internal status report LNS-INFN, 1999.

5. Odland, O.H., et al., *Nucl.Instr.Met.* A**378**, 149-155 (1996).

Study of a Water-Cooled Convective Divertor Prototype for the DEMO Fusion Reactor

P. Di Maio, E. Oliveri, G. Vella

Department of Nuclear Engineering (DIN) of the University of Palermo
Viale delle Scienze, 90128 Palermo, ITALY

Abstract. The plasma facing components of a fusion power reactor have a large impact on the overall plant design, its performance and availability and on the cost of electricity. The present work concerns a study of feasibility for a water-cooled prototype of the convective divertor component of the DEMO fusion reactor. The study has been carried out in two steps. In the first one thermal-hydraulic and neutronic parametric analyses have been performed to find out the prototype optimized configuration. In the second step thermo-mechanical analyses have been carried out on the obtained configuration to investigate the potential and limits of the proposed prototype, with a particular reference to the maximum heat flux it can undergo without incoming both in critical heat flux and in mechanical stress limits. The results show that the proposed divertor prototype is able to safely withstand peak heat fluxes of 9 MW/m^2.

INTRODUCTION

The potential use of the nuclear fusion reaction for the electric power generation on an industrial scale is being, at the moment, highly considered by the international scientific community which has started an intense Research and Development (R&D) programme aiming to the realization of a Fusion Power Reactor (FPR).

In the framework of the European Union R&D programme it is foreseen the design and the construction of a DEMOnstration reactor as intermediate step before the FPR realization, whose main aim is to demonstrate the commercial feasibility of such a kind of reactor. The DEMO reactor is a TOKAMAK machine with a double-null configuration working with a D-T plasma in a continue mode for 20000 hours (1). In the framework of the DEMO reactor design the so-called plasma facing components play a pivotal role being devoted to withstand the high heat fluxes coming from plasma which may reach peak values of some tens of MW/m^2, during plasma disruption. They have, then, a large impact on the overall reactor design, its performance and availability and on the cost of electricity. In particular, among the plasma facing components, the divertor is the one that undergoes the highest heat fluxes and recently it has been more deeply investigated in view of an improvement of the economic features of the DEMO reactor.

The DIN of the University of Palermo is involved in research activities on the DEMO plasma facing components and, at the present, it is cooperating with the CEA, Saclay (FRANCE) on the topic of the water-cooled divertor potential and limits.

CP513, *Nuclear and Condensed Matter Physics,* edited by A. Messina
© 2000 American Institute of Physics 1-56396-929-7/00/$17.00

THE DEMO WATER-COOLED CONVECTIVE DIVERTOR

The divertor is the high heat flux component of a TOKAMAK fusion reactor devoted to control the boundary plasma characteristics and to limit the interactions between the plasma and its surrounding shielding structure. It diverts the external lines of the confining electro-magnetic field in such a way to attract the exhaust plasma, the He particles and the impurities of the plasma-wall interactions driving them far from the main plasma, which can be kept clean (2). The divertor undergoes, then, an high heat flux and it has to be equipped with an highly efficient cooling system preferably able to extract the deposited power at a sufficient high temperature level to integrate it in a power conversion cycle, improving the overall reactor efficiency (3). The DEMO divertor is composed off two sub-components called top and bottom divertor (Fig.1. a) each one subdivided in 32 Segment Divertors (SD), one for each reactor sector.

The proposed prototype foresees that each SD has a layered structure composed of a supporting plate and a matrix housing the cooling tubes. It has a toroidal width varying from 865 to 1260 mm, a poloidal length of about 3000 mm and a radial depth of 70 mm. The supporting plate has a merely structural function and its reference material is a reduced activation ferritic-martensitic steel. The matrix, which is directly exposed to the plasma, is entirely made of Carbon Fiber Composite (CFC) N112. It is composed of poloidal monoblocks, each one containing a cooling tube (Fig.1.b). Each monoblock, on its plasma facing surface, is characterized by toroidal and poloidal castellations of 6 and 50 mm respectively in order to reduce the thermal stress level. The cooling tubes are U-shaped and their reference material is the TZM Mo-alloy. The coolant flowing into the tubes is subcooled water at pressure of 15.5 MPa having an inlet temperature of 265 and an outlet one of about 325 °C in order to let the divertor

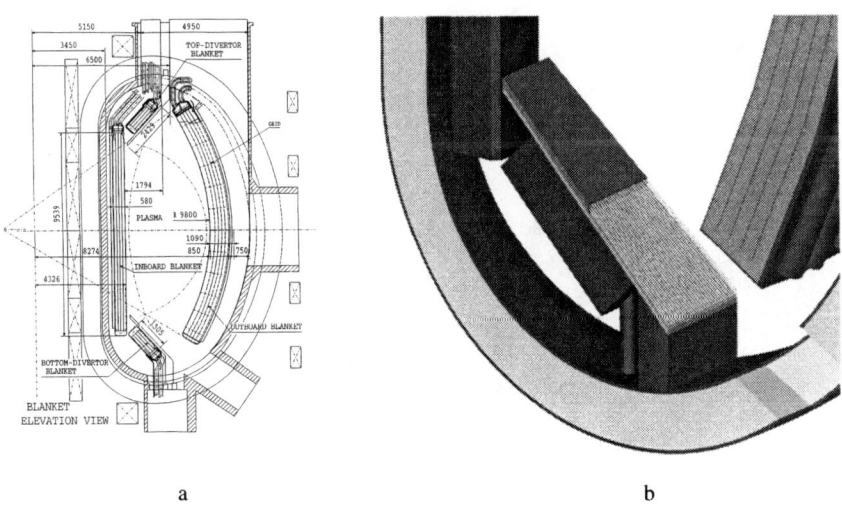

a   b

FIGURE 1. a) Poloido-Radial Section of the DEMO Reactor; b) Isometric Exploded View of a Bottom Segment Divertor of the DEMO Reactor.

275

cooling system to be integrated in the power conversion cycle of the whole reactor (3). Swirl tapes are foreseen into the tubes to enhance the heat transfer and raise the critical heat flux limit.

THE STUDY OF THE DIVERTOR PROTOTYPE

A feasibility study of the above mentioned divertor prototype has been performed to optimize the monoblock lay-out and to investigate its limits and performances. The study has been carried out in two steps. In the first one thermal-hydraulic and neutronic parametric analyses have been performed to find out the prototype optimized configuration. In the second step thermo-mechanical analyses have been carried out on the obtained configuration to investigate the potential and limits of the proposed prototype, with a particular reference to the maximum heat flux it can undergo without incoming both in critical heat flux (4) and in stress limits (2).

Thermal-Hydraulic and Neutronic Parametric Analyses

In order to determine the optimized configuration it has been, conservatively, assumed that the whole length of the divertor is uniformly subjected to a heat flux of 5 MW/m^2 and that the coolant flows at a velocity of 12 m/s. Starting from the mass and energy steady state conservation equations, relevant to the divertor system, the cooling tubes number N has been selected as the parametric variable. A detailed set of thermal-hydraulic and neutronic parametric analyses (the last one using the MonteCarlo method) have been iteratively performed evaluating for each value of N the Tritium Breeding Ratio (TBR), the deposited power density and the cooling tube diameters.

The results show that the best configuration which maximizes the TBR to a value of about 1.102 (5) seems to be the one with 25 cooling tube, 17 mm inner diameter.

Thermo-Mechanical Analyses

The objective of these analyses was the determination of the maximum incident heat flux that the divertor is able to safely withstand, owing to the following constraints:

- CFC matrix temperature < 1500 °C;
- TZM tubes temperature < 700 °C;
- DNBR > 1.1;
- Equivalent Von Mises stress in the structure < $3S_m$ (2).

TABLE 1. Thermo-Mechanical Results.

	Design Heat Flux (5 MW/m²)	Peak Heat Flux (9 MW/m²)
T_{max} (CFC matrix) [°C]	836	1375
T_{max} (TZM tube) [°C]	428	502
DNBR	1.4	1.14
Stress Margin vs $3S_m$ [MPa]	470	330

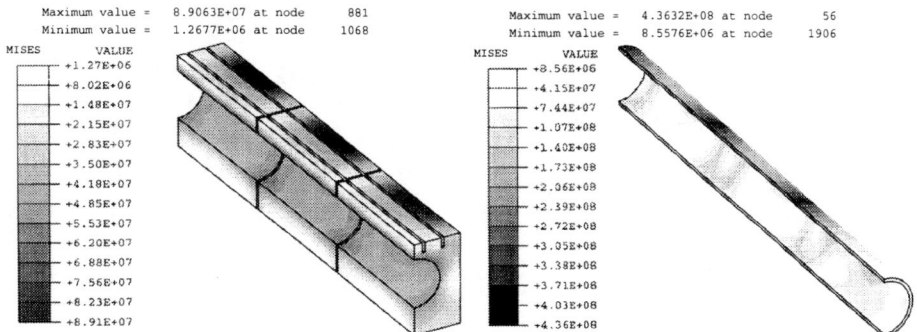

FIGURE 2. Von Mises Stress Field [Pa] of the Central Region of a Monoblock in Peak Flux Case.

For the thermal calculation an heat flux Gaussian distribution with the peak located in the central divertor region has been assumed. For the stress calculation the conservative assumption of a fixed joint between the monoblock and the tube has been made. If a ductile joint could be used the prototype performances would, substantially, increase. Numerical non linear steady state coupled thermo-mechanical analyses on a three dimensional divertor model have been performed using the ABAQUS finite element code. Table 1. lists the principal results and Fig. 2. shows the equivalent Von Mises stress field for the central region of a monoblock in the peak heat flux case.

CONCLUSIONS AND HINTS

The results show that the prototype seems to safely withstand the design heat flux and the peak ones up to values of 9 MW/m^2. Further studies should be performed to investigate the impact of the use of W-alloys or Beryllium as monoblock materials.

REFERENCES

1. Giancarli, L., Dalle Donne, M., and Dietz, W., *Fusion Engineering and Design* **36**, 57-74 (1997).

2. Bruno, L., Castiglia, F., and Vella, G., *Analisi Termica e Meccanica di un Prototipo di Divertore Convettivo. Modelli Numerici e Verifiche Sperimentali*, Quaderni del Dipartimento di Ingegneria Nucleare, Università di Palermo **3**, (1997).

3. Futterer, M. A., Vella, G., et alii, *Potential and Limits of Water-Cooled Pb-17Li Blankets and Divertors for a Fusion Power Plant*, presented at 5th ISFNT 19-24 September, 1999, Rome.

4. Nariai, H., and Inasaka, F., *Fusion Engineering and Design* **9**, 245 (1989).

5. Ruvutuso, G., *Potenzialità e Limiti del Progetto del Reattore Nucleare a Fusione DEMO a LiPb e Refrigerato ad Acqua. Analisi Neutroniche con il Codice MCNP per l'Ottimizzazione del Mantello e del Divertore.*, Tesi di Laurea in Ingegneria Nucleare, Anno Accademico 1998-1999.

Different Aspects of ^{24}Mg Formation and Decay Using a Radioactive ^{13}N Beam.

P.Figuera[a], F.Amorini[a], W.Bradfield-Smith[c], M.Cabibbo[a], G.Cardella[b], T.Davinson[c], A.Di Pietro[c], W.Galster[d], P.Leleux[d], A.Musumarra[d], A.Ninane[d], M.Papa[b], G.Pappalardo[a], F.Rizzo[a], A.C.Shotter[c], C.Sukosd[e], S.Tudisco[a], P.J.Woods[c].

[a]INFN-Laboratori Nazionali del Sud, Via S.Sofia 44, 95123 Catania Italy
[b]INFN-Sezione di Catania, Corso Italia 57, 95129 Catania Italy
[c]Department of Physics and Astronomy University of Edinburgh. Edinburgh U.K.
[d]Institut de Physique Nucleaire Universitè de Louvain. Louvain la Neuve Belgium.
[e]Department of Physics University of Budapest. Budapest Hungary.

Abstract. Different aspects of the formation and decay of ^{24}Mg in the collision ^{13}N+^{11}B have been studied using a large solid angle and highly segmented Silicon strip detector. Results concerning the fusion cross section, the 6 α decay of ^{24}Mg and the GDR gamma ray emission are discussed.

INTRODUCTION AND EXPERIMENTAL SET-UP

Using a post accelerated ^{13}N beam, we have studied different aspects of the ^{24}Mg formation and decay in the reaction ^{13}N +^{11}B at E_{lab}=45.0 and 29.5 MeV. The experiment was performed at the radioactive beam facility of Louvain la Neuve. A large solid angle and highly segmented (224 strips) silicon strip array "Leda+Lampshade", sketched in figure 1, was used to detect charged particles which have been identified with the standard TOF technique. Three BaF$_2$ clusters, covering the backwards angles, were used to detect high energy gamma rays (E$_\gamma$>4 MeV).

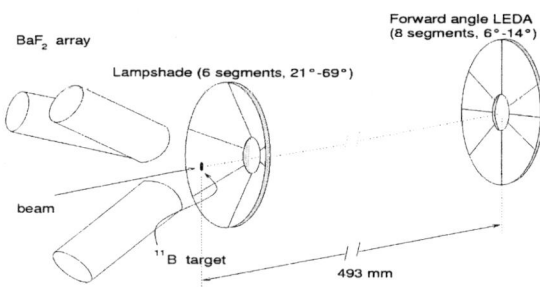

FIGURE 1. Experimental set-up.

EXPERIMENTAL RESULTS AND DISCUSSION

The energy spectra of the evaporation residue are well reproduced by statistical model Montecarlo calculations as shown in figure 2. The fusion cross section, obtained integrating the angular distributions, is in agreement, within the experimental error, with the fusion excitation functions of similar systems ($^{14}N+^{10}B$, $^{12}C+^{12}C$) leading to the same compound nucleus. Therefore we do not observe strong structure effects on the fusion process due to the presence of the weakly bound proton (S_p=1.9 MeV) in ^{13}N.

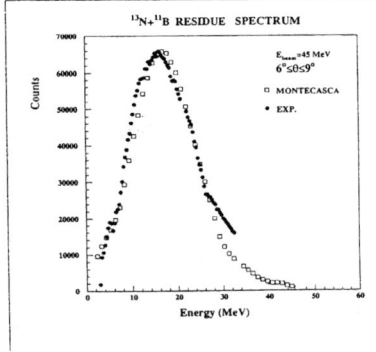

FIGURE 2. Comparison between an experimental evaporation residue spectrum for the reaction $^{13}N+^{11}B$ at 45 MeV and the corresponding calculated one using a Montecarlo statistical code.

The ^{24}Mg at E(^{13}N)=45 MeV is populated with an excitation energy $E^*\approx47$ MeV. In this excitation energy region of ^{24}Mg several resonances have been observed with $^{12}C+^{12}C$ scattering [1,2]. In particular a broad resonance was originally observed in [1] and attributed to the 6α linear chain configuration in ^{24}Mg. We studied [3], for both bombarding energies, the decay of the ^{24}Mg intermediate system into two $^{12}C^*$ each of them decaying into $^8Be+\alpha$ as sketched in figure 3.

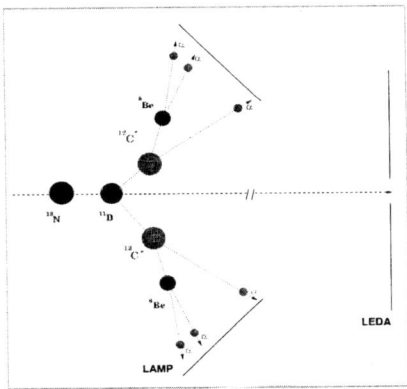

FIGURE 3. Symbolic sketch of the studied reaction channel

Events with 6α in the final channel have been selected with the help of E-TOF identification spectra. In addition we constructed the Q value spectrum $Q=\Sigma(E_i)-E_{inc}$ and put a gate around the value $Q=-500$ KeV corresponding to the $^{13}N+^{11}B\rightarrow6\alpha$ reaction (figure 4 left). The two ^8Be were selected by looking at the relative energy ($E_{rel}\approx90$ KeV) of α particles detected in adjacent strips. By measuring energies and angles of these two ^8Be and α particles, the double excitation energy spectrum for the two primary $^{12}C^*$ nuclei has been reconstructed. The result for the 45 MeV run is shown in figure 4 right. Here the arrows mark different possible combinations of ^{12}C excited states for the $^{12}C^*-^{12}C^*$ pair. To be exact, we remind that in our analysis we may have an ambiguity due to the wrong association between the two ^8Be and the two alphas. However, such miscorrelated events can only contribute in the hatched excitation energy region in figure 4 right. The calculated detection efficiency ε of our experimental setup for different kinds of events is shown in the same figure (dashed line). Although our data set (collected in a reaction induced by a low intensity radioactive beam) is much smaller, there is a reasonable consistency between the present $^{13}N+^{11}B$ results and what observed by other authors in $^{12}C+^{12}C$ inelastic scattering [2] were feeding from the $3_1^- - 3_1^-$ and $0_2^+ - 3_1^-$ states of ^{12}C to the six alpha break-up is observed together with a reduced feeding from $0_2^+ - 0_2^+$. Bearing in mind the single particle structure of the $^{13}N+^{11}B$ channel compared to the α cluster nature of $^{12}C+^{12}C$, this finding seems surprising.

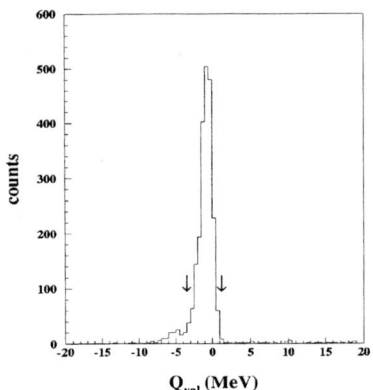 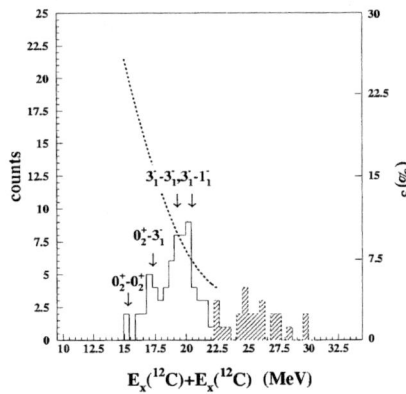

FIGURE 4. Left: Q value spectrum for the selected events with multiplicity 6; the arrows mark the selected Q value window for the reaction $^{13}N+^{11}B\rightarrow6\alpha$. Right: double excitation energy spectrum for the two $^{12}C^*$ in the exit channel.

The GDR gamma decay from the reaction $^{13}N+^{11}B$ (total Isospin T=0,1 in the entrance channel) has been studied and compared with the one for the collision $^{14}N+^{10}B$ (T=0) leading to ^{24}Mg at the same excitation energy $E^*\approx47$ MeV. Due to the E1 decay selection rules, a GDR gamma decay from an initial state T=0 in a self conjugated nucleus will have to populate one of the less numerous T=1 states at lower excitation energy and the GDR yield will be suppressed. However, if a self conjugated nucleus is formed in a collision with a T=0 entrance channel, due to isospin mixing T=1 states can also be populated. Therefore the GDR yield, in the reaction with the T=0 entrance channel, will be sensitive to the degree of isospin purity/mixing in the formed CN. Using radioactive and stable beams we are now forming the same CN ^{24}Mg, at the same excitation energy $E^* \approx 47$ MeV, with two very similar entrance channels characterized by T=0,1 (^{13}N+^{11}B) and T=0 (^{14}N+^{10}B). The calculated (Cascade statistical model calculations -histogram) and experimental (symbols) ratios between the gamma spectra are shown in figure 5. A larger GDR yield for the reaction ^{13}N+^{11}B is clearly seen. As discussed above, this is due to the isospin suppression of the GDR yield in the ^{14}N+^{10}B reaction which is sensitive to the degree of isospin purity/mixing. Within the experimental error, our data have been reproduced by statistical model calculations using an isospin mixing coefficient α^2=0.1. This appears to be coherent with previous studies [4] on the same C.N. performed producing different isotopes of Mg in stable beam induced reactions.

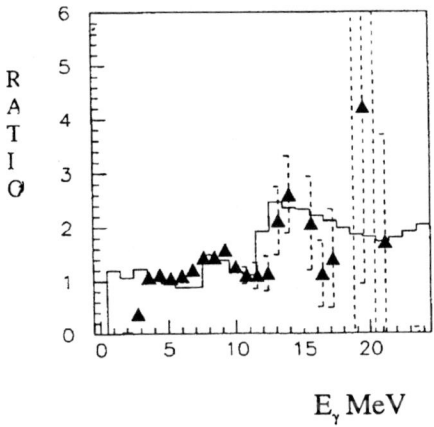

FIGURE 5. Ratio between the gamma spectra $E\gamma$-(^{13}N+^{11}B)/$E\gamma$-(^{14}N+^{10}B). Symbols: experimental data. Histogram: Cascade calculations (E_{GDR}=17.4, Γ_{GDR}=10.8, α^2=0.1)

REFERENCES

1. A.H. Wuosmaa et al., Z.Phys. A349, 249, (1994)

2. R.A. Le Marechal et al., Phys. Rev. C 55, 1881, (1997)

3. A. Di Pietro et al. Phys. Rev. C59,1185,(1999)

4. M.N. Harakeh et al., Phys. Lett. B176, 297, (1986)

Fusion Reaction within a Microcrack in a Crystalline Lattice at Room Temperature

Fulvio Frisone

Department of Physics, University of Catania, Corso Italia 57, 95129 Catania
CSFNSM - Catania -

Abstract. This communication presents the results obtained analysing the reaction of deuteron fusion, catalysed by the plasmons in crystalline lattices with cubic structure, on varying the temperature. The enhancement of the tunnelling effect, a consequence of increasing the temperature and the concentration of impurities, suggests the hypothesis that a sort of chain reaction favoured by the microcracks formed in the structure as a result of lattice deformation, could promote the process. This result could be interpreted considering the trend of the potential which describes the effective interaction between deuterons within the metal: in effect it shows that the coupling of plasmons and deuterons, in the presence of impurities, can not only reduce the thickness but also lower the height of the Coulomb barrier K.

INTRODUCTION

This communication presents the results obtained analysing the influence of temperature on the phenomenon of deuteron fusion within crystalline lattices with CFC structure. In particular, the research considers the possibility that variations in the thermodynamic conditions could produce dislocations in the lattice[5], micro-deformations that may be able to concentrate in their vicinity a relevant fraction of the deuterons present in the metal, catalysing a form of chain reaction which favours the process.

Further, it has been suggested[1] that the deuteron-plasmon coupling could increase the rate of the fusion reaction, effectively acting as an attractive interaction between the deuteron nuclei and thus reducing the distance at which Coulomb repulsion becomes dominant. With this perspective, a study was made of crystalline lattices with more than 10 electrons in the "d" band. Here we will concentrate on Pd because it has been observed that when subjected to thermodynamic stress, this metal yields results which are interesting form both the theoretical and the experimental aspect.

We further suggest that the phenomenon of fusion is also conditioned by the level of impurities present in the metal, correlated with the deformation of the lattice itself. The numerical calculation, conducted on Palladium as a function of the temperature, the total energy and the concentration of impurities, allows the conclusion that the probability of fusion is in effect significantly enhanced by increasing these parameters. The present study will attempt to discover whether, in the three-dimensional isotope case, the formation of microcracks can be caused by lattice deformation, in a similar manner to that hypothesised for variations in temperature. This would offer a further verification of the hypotheses proposed[6].

CP513, *Nuclear and Condensed Matter Physics,* edited by A. Messina
© 2000 American Institute of Physics 1-56396-929-7/00/$17.00

TABLE 1. Adopting the Morse potential, the probability of fusion Γ, normalised to events per minute, was calculated within a microcrack in the presence of deformation on "impure" Pd (J $\approx 0.75\%$) for different values of temperature (varying between 100 K and 300 K) and of energy (varying between 150 eV and 250 eV). The probability generally increases with T and E.

Palladium $\quad$ J $\approx$ O.75% $\quad$ T - range $\approx$ 100 - 300 K $\quad$ $\alpha \approx$ .34 Å $\quad$ $\lambda = 10^{-3}$ eV / min $\quad$ $M_{Pd}/(\mu g)$

T $\approx$ 100 K		T $\approx$ 200 K		T $\approx$ 300 K	
E $\approx$ 150	$\Gamma \approx 10^{-80}$	E $\approx$ 150	$\Gamma \approx 10^{-63}$	E $\approx$ 150	$\Gamma \approx 10^{-60}$
E $\approx$ 160	$\Gamma \approx 10^{-75}$	E $\approx$ 160	$\Gamma \approx 10^{-61}$	E $\approx$ 160	$\Gamma \approx 10^{-58}$
E $\approx$ 170	$\Gamma \approx 10^{-72}$	E $\approx$ 170	$\Gamma \approx 10^{-60}$	E $\approx$ 170	$\Gamma \approx 10^{-56}$
E $\approx$ 180	$\Gamma \approx 10^{-69}$	E $\approx$ 180	$\Gamma \approx 10^{-59}$	E $\approx$ 180	$\Gamma \approx 10^{-55}$
E $\approx$ 190	$\Gamma \approx 10^{-67}$	E $\approx$ 190	$\Gamma \approx 10^{-57}$	E $\approx$ 190	$\Gamma \approx 10^{-50}$
E $\approx$ 200	$\Gamma \approx 10^{-66}$	E $\approx$ 200	$\Gamma \approx 10^{-55}$	E $\approx$ 200	$\Gamma \approx 10^{-48}$
E $\approx$ 210	$\Gamma \approx 10^{-65}$	E $\approx$ 210	$\Gamma \approx 10^{-53}$	E $\approx$ 210	$\Gamma \approx 10^{-47}$
E $\approx$ 220	$\Gamma \approx 10^{-63}$	E $\approx$ 220	$\Gamma \approx 10^{-52}$	E $\approx$ 220	$\Gamma \approx 10^{-46}$
E $\approx$ 230	$\Gamma \approx 10^{-60}$	E $\approx$ 230	$\Gamma \approx 10^{-51}$	E $\approx$ 230	$\Gamma \approx 10^{-45}$
E $\approx$ 240	$\Gamma \approx 10^{-59}$	E $\approx$ 240	$\Gamma \approx 10^{-49}$	E $\approx$ 240	$\Gamma \approx 10^{-43}$
E $\approx$ 250	$\Gamma \approx 10^{-58}$	E $\approx$ 250	$\Gamma \approx 10^{-48}$	E $\approx$ 250	$\Gamma \approx 10^{-41}$

TABLE 2. Adopting the effective potential, the probability of fusion Γ, normalised to events per minute, was calculated within a microcrack for "pure" Pd (J$\approx 0.25\%$), under the same dynamic conditions as those in Table 1. Also here Γ is systematically greater by some orders of magnitude.

Palladium $\quad$ J $\approx$ 0.25% $\quad$ T - range $\approx$ 100 - 300 K $\quad$ $\alpha \approx$ .34 Å $\quad$ $\lambda = 10^{-3}$ eV / min $\quad$ $M_{Pd}/(\mu g)$

T $\approx$ 100 K		T $\approx$ 200 K		T $\approx$ 300 K	
E $\approx$ 150	$\Gamma \approx 10^{-75}$	E $\approx$ 150	$\Gamma \approx 10^{-69}$	E $\approx$ 150	$\Gamma \approx 10^{-65}$
E $\approx$ 160	$\Gamma \approx 10^{-74}$	E $\approx$ 160	$\Gamma \approx 10^{-68}$	E $\approx$ 160	$\Gamma \approx 10^{-63}$
E $\approx$ 170	$\Gamma \approx 10^{-73}$	E $\approx$ 170	$\Gamma \approx 10^{-67}$	E $\approx$ 170	$\Gamma \approx 10^{-60}$
E $\approx$ 180	$\Gamma \approx 10^{-71}$	E $\approx$ 180	$\Gamma \approx 10^{-65}$	E $\approx$ 180	$\Gamma \approx 10^{-58}$
E $\approx$ 190	$\Gamma \approx 10^{-69}$	E $\approx$ 190	$\Gamma \approx 10^{-64}$	E $\approx$ 190	$\Gamma \approx 10^{-57}$
E $\approx$ 200	$\Gamma \approx 10^{-67}$	E $\approx$ 200	$\Gamma \approx 10^{-62}$	E $\approx$ 200	$\Gamma \approx 10^{-56}$
E $\approx$ 210	$\Gamma \approx 10^{-65}$	E $\approx$ 210	$\Gamma \approx 10^{-60}$	E $\approx$ 210	$\Gamma \approx 10^{-55}$
E $\approx$ 220	$\Gamma \approx 10^{-64}$	E $\approx$ 220	$\Gamma \approx 10^{-59}$	E $\approx$ 220	$\Gamma \approx 10^{-54}$
E $\approx$ 230	$\Gamma \approx 10^{-63}$	E $\approx$ 230	$\Gamma \approx 10^{-57}$	E $\approx$ 230	$\Gamma \approx 10^{-50}$
E $\approx$ 240	$\Gamma \approx 10^{-61}$	E $\approx$ 240	$\Gamma \approx 10^{-55}$	E $\approx$ 240	$\Gamma \approx 10^{-49}$
E $\approx$ 250	$\Gamma \approx 10^{-60}$	E $\approx$ 250	$\Gamma \approx 10^{-54}$	E $\approx$ 250	$\Gamma \approx 10^{-45}$

This study has shown that, at room temperature, microcracks and the deformation of the lattice constitute important factors influencing fusion.

In fact, moderately high values were obtained for the probability of interaction within a microcrack in the presence of a deformation on impure metals at room temperature. On the other hand, results much closer to the experimental data have been found within a microcrack see[6]: for example, for $J=0.75\%$, $E=250$ eV, $T=300$ K, $\Gamma=10^{-21}$. From a comparison between the two theoretical calculations, this experimental procedure would therefore appear to reduce the rate of fusion rather than favour it.

To verify, from another viewpoint, the influence of the concentration of impurities on the process, the trend of the potential within the pure metal lattice ($J \approx 0.25\%$) was evaluated and a very high curve was obtained. To cross the barrier, therefore, would require a total energy greater than the potential, as shown in Figure 1 for Pd. If the potential barrier is evaluated under the same conditions with a concentration of impurities $J \approx 0.75\%$, under the same thermodynamic conditions of the system, it is seen that the probability of fusion could be greater than that observed with pure metals, with a total energy less than the potential so that the tunnelling effect is amplified.

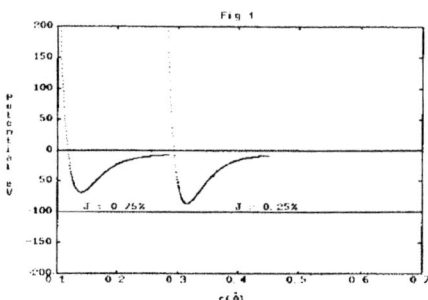

Figure 1. Tunnelling appears amplified for those metals with a concentration of impurities
$J \approx 0.75\%$.The "Morse" potential was calculated at $T = 290$ K. The Coulomb barrier appears high in the case of pure metals with $J \approx 0.25\%$.. The effective potential was calculated at $T = 290$ K.

REFERENCES

1. Frisone F., *Il Nuovo Cimento*,"Can variations in temperature influence deuteron interaction within crystalline lattices ? " Vol.20 D.N.10, 1998, pp. 1567-1580.
2. Mc Kubre, M.C.H. and Tanzella, F.L., "Matrerials Issues of Loading Deuterium into Palladium, and the Association with Excess Heat Production"-*ICCF - 7* - April 19 - 24, 1998,pp. 230 – 235.
3. Horowitz, C.J., *Phys. Rev. C*, "Cold nuclear in metellic hydrogen and normal metals" Num.4 Vol.40, 1989, pp. 1555-1558.
4. Nabarro,F.R.N., Adv. Phisycs a Quarterly supplement of the Philosophical Magazine,Vol.1 N.3 1952, Adv. Phys, pp.271–390.
5. Hirt, J.Price and Lothe J., *"Theory of dislocation"*, McGraw Hill, Z. Phys., 1960, pp. 116-129.
6. Frisone, F., *"Calculation of the probability of interaction between deuterons in crystalline lattices at room temperatures"*, submitted for publication in EPJ-AP

Experimental Study Of Thermal Crisis In Connection With Tokamak Reactor High Heat Flux Components

D. Gallo[1], M. Giardina[1], F. Castiglia[1]
G. P. Celata[2], A. Mariani[2], G. Zummo[2], M. Cumo[3]

[1]Nuclear Engineering Department, University of Palermo
[2]National Institute of Thermal – Fluid Dynamics, ENEA, Rome
[3]Nuclear Engineering Department and Energy Conversion,
University "La Sapienza", Rome

Abstract. The results of an experimental research on high heat flux thermal crisis in forced convective subcooled water flow, under operative conditions of interest to the thermal-hydraulic design of TOKAMAK fusion reactors, are here reported. These experiments, carried out in the framework of a collaboration between the Nuclear Engineering Department of Palermo University and the National Institute of Thermal – Fluid Dynamics of the ENEA – Casaccia (Rome), were performed on the STAF (Scambio Termico Alti Flussi) water loop and consisted, essentially, in a high speed photographic study which enabled focusing several information on bubble characteristics and flow patterns taking place during the burnout phenomenology.

INTRODUCTION

Recently, in the framework of a collaboration between the Nuclear Engineering Department of Palermo University and the National Institute of Thermal – Fluid Dynamics of the ENEA – Casaccia (Rome), a research program has been carried out on high heat flux (about 10 MWm^{-2}) thermal crisis, in forced convective subcooled boiling water flow, under operative conditions of interest to the thermal hydraulic design of TOKAMAK nuclear fusion reactors, where the critical heat flux (CHF) mechanisms are different from those relevant to nuclear fission light water reactors. The experimental campaign, performed on the STAF (Scambio Termico Alti Flussi) water loop in correspondence of a wide variety of test conditions (system pressure, liquid inlet temperature and velocity, heater surface finiture, etc.), consisted in setting up a high speed photographic study (50 f.p.s.) of the thermal crisis phenomenology. The burnout images, analysed by using a PC digital image processing system, enabled focusing several qualitative phenomenological information, such as vapour bubble shape and dimensions, flow pattern under different heat transfer regimes, hot spot appearance and heater burnout. Vapour bubble width and height were also measured and the influence of thermal-hydraulic tested conditions on these parameters was estimated. The next sections report the experimental apparatus, the performed measurements and the obtained results.

CP513, *Nuclear and Condensed Matter Physics*, edited by A. Messina
© 2000 American Institute of Physics 1-56396-929-7/00/$17.00

EXPERIMENTAL APPARATUS AND TEST PROCEDURE

The high speed photographic study of burnout in water subcooled flow boiling was performed using the water loop (already extensively described in [1]), shown in figure 1, and the optical equipment schematically shown in figure 2. The loop, made of type 304 stainless steel and filled with deionized water, consists of main alternative pump, damper, filter, turbine flow meter, heated test section and water cooled tank. The test section is vertically oriented with water flowing upwards. For each experimental run, the procedure consisted in setting the mass flow rate and, when this became steady, thermal power was delivered step by step to the test section up to burnout. The burnout was evidenced by heater destruction and detected by the sharp drop in the electrical power. In order to capture and "freeze" the motion of the vapour bubbles a 150 mW laser and a mechanical shutter was used. To take movies of the burnout occurrence, a CCD video camera, 50 f.p.s. speed, and a VHS video recorder were chosen. The images were analyzed through a PC digital image processing system.

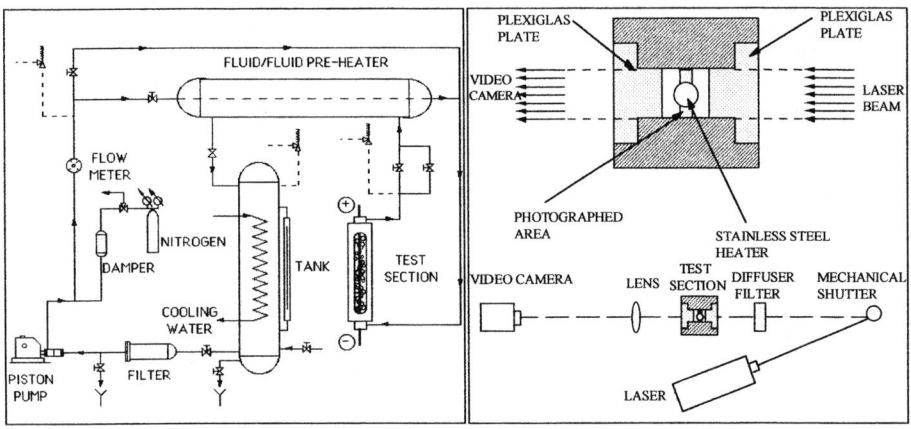

FIGURE 1. Schematic of the STAF.　　　　　**FIGURE 2.** Schematic of the optical bench.

EXPERIMENTAL RESULTS

The summary of test conditions (liquid velocity, system pressure and inlet temperature) for expected experimental activity is presented in Table 1 [2].

TABLE 1. Test conditions (the values reported are the inlet temperatures in °C).

u [ms⁻¹] \ p [MPa]	0.5	1.0	2.0	3.0
3.0	20	20-40-60	20	20
5.0	20	20	20-40-60	20
7.5		20	20	20
10.0			20	20

287

Wall Temperature And Bubble Parameters Measurements

The heater wall temperature was deduced from the thermal elongation of the heater rod measured at each heat flux step. The values of these temperatures were correlated directly to the heat transfer and flow regimes. The measured geometrical parameters, i. e. bubble width L_B and height D_B, were estimated for each heat flux value until the burnout occurred. The information on flow pattern and vapour bubbles shape for different heat fluxes from ONB (Onset of Nucleate Boiling) to the burnout were drawn from the video images analysis. The results about images observations, wall temperature measurements and vapour bubbles parameters for increasing values of heat flux are represented in figure 4 for one of the runs.

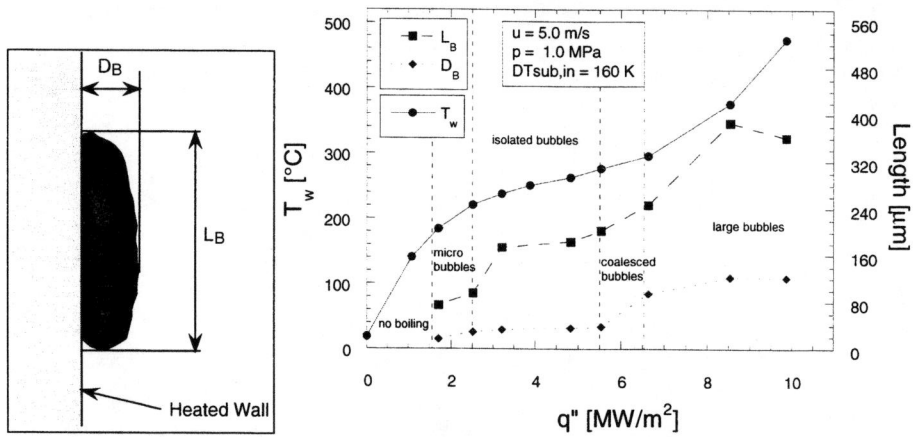

FIGURE 3. Bubble parameters. **FIGURE 4.** Measured wall temperature and bubble parameters.

Four different flow patterns leading to thermal crisis were singled out: *Microbubbles*, *Isolated Bubbles*, *Coalesced Bubbles* and *Large Bubbles*.

In the *microbubbles*, *isolated bubbles* and *coalesced bubbles* patterns the wall temperature exhibits a low slope with a weak influence of heat flux, typical of the subcooled flow boiling. During the *large bubbles* pattern, as the heat flux increases a bright red hot spot, typical of white-hot metals, appears on the heated wall and, after a period of time depending on test conditions, the burnout of the heater occurs. During the few millisecond preceding the appearance of the hot spot, the heater surface is interested by a *reduced boiling activity*: there are very few or, in some cases any, vapour bubbles. This *reduced boiling activity* persists during all the period of existence of the hot spot. Vapour bubble formation is almost absent on the surface with the highest temperature (the portion that emits the red light), whereas a very intense boiling occurs on the area just near the boundary between the hot spot and the remaining surface with lower temperature.

Bubble Sizes Behaviour As Function Of Thermal hydraulic Parameters

In this section we will report considerations on the trends of the bubble sizes as a function of some thermal-hydraulic parameters. As one expected bubble sizes increase as the subcooling decreases, for given heat flux, and increase with heat flux, other conditions being equal. Another parameter playing an important role in bubble sizes is the water velocity in the test channel. Figure 5 shows bubble sizes, L_B and D_B, versus the heat flux, grouped by the fluid velocity, other conditions being equal. Trends of L_B and D_B are increasing function of the heat flux, while the dependance on water velocity is inverse: the higher the velocity the smaller L_B and D_B.

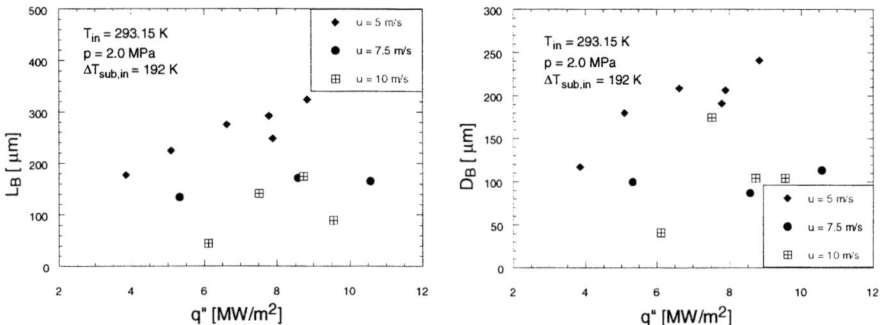

FIGURE 5. Influence of liquid velocity on bubble sizes.

CONCLUSIONS

High speed movies of flow pattern in subcooled flow boiling of water from onset of nucleate boiling up to physical burnout of the heater were recorded. Bubble dimensions were derived from images and reported as function of heat flux for different thermal hydraulic conditions. The video image analysis allowed a qualitative description of the near wall flow patterns for increasing values of heat flux up to CHF.

REFERENCES

1. Celata, G.P., Cumo, M., and Mariani, A., *Burnout in Highly Subcooled Water Flow Boiling in Small Diameter Tubes*, Int. J. Heat Mass Transfer, 1993, pp. 1269-1285.

2. Gallo Carrabba, D., *Studio sperimentale della crisi termica per il raffreddamento dei componenti sottoposti ad alto flusso di un reattore a fusione di tipo TOKAMAK*, Palermo: Degree Thesis Nuclear Engineering Department, 1998, pp. 91-136.

Nuclear Caloric Curve. A Systematic Study.

G.Immé[1,2,3], G.Raciti[1,2,3], G.Riccobene[1,2,3], F.P.Romano[1,2,3], A.Sajia[1,2], C.Sfienti[1,2], G.Verde[1,2,3], N.Giudice[1,2]

and the ALADiN Collaboration

(Milano,GSI-Darmstadt,NSCL-MSU,Heidelberg,Dresden,Warsaw)

[1]Dipartimento di Fisica Università di Catania - Corso Italia, 57, I-95129 Catania, Italy
[2]INFN, Sezione di Catania and Laboratori Nazionali del Sud - Catania, Italy
[3]CSFNSM, Catania, Italy

Abstract. Investigations performed by the ALADiN Collaboration about signals for a liquid-gas phase transition in nuclear matter are reported. Temperature-excitation energy correlation measurements on several systems at different incident energies are compared. Moreover the comparison between *isotope* and *excited states* temperatures, extracted from double ratios of isotope yields and population ratios of fragment unbound states, respectively, shows a discrepancy between the two methods that cannot be accounted for by the sequential feeding corrections. Instead, they seem to be related to the space-time evolution of the fragmentation process.

INTRODUCTION

The investigation on the nuclear liquid-gas phase transition reported very interesting results based on the nucleus-nucleus collision measurements at intermediate and relativistic energies. In particular, in the study of the Au+Au collisions at E_{inc}=100 - 1000 AMeV, by correlating the excitation energy with the isotope temperature (deduced from the double yield ratio between two isotope pairs) the so called "nuclear caloric curve" (fig.1) was obtained[1,2]. The similarity of this curve to first order liquid-gas transition in macroscopic systems, has initiated a widespread discussion which addresses both methodical aspects and questions of interpretation.

Therefore a systematic investigation of the calorimetric measurement dependence on the system size, incident energies and impact parameter is needed. In the present paper we discuss the results from the Xe, Au and U collisions on a variety of targets at incident energies between 50 and 1000 A MeV obtained at the GSI (Darmstadt) [3], Ca+Sc and Nb+Nb at 40 and 15 A MeV respectively at the LNS (Catania) and a few

CP513, *Nuclear and Condensed Matter Physics,* edited by A. Messina
© 2000 American Institute of Physics 1-56396-929-7/00/$17.00

recent data from Kr + Nb and Ar+ Sc collisions in the energy range between 35 and 120 A MeV at the NSCL (MSU) [4].

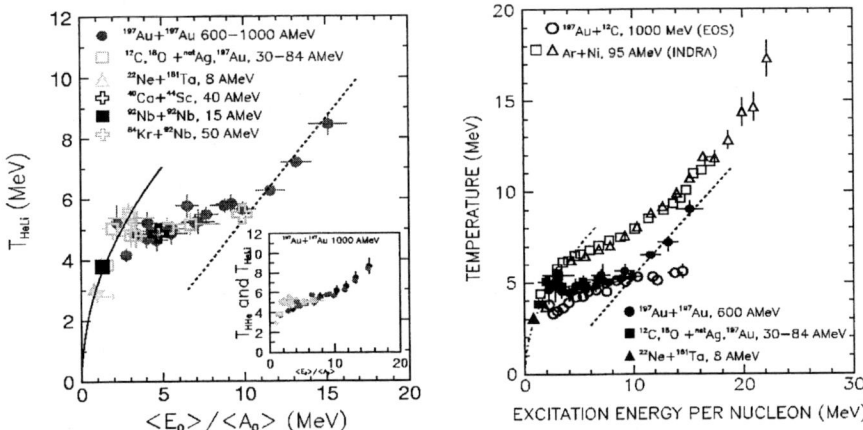

FIGURE 1. a)Caloric curve of nuclei determined by the dependence of the isotope temperature T_{HeLi} on the excitation energy per nucleon. The insert shows the comparison of $T_{He,Li}$ and T_{HHe} in 1 A GeV target fragmentation. b) Comparison of the ALADiN caloric curve (solid points) to the EOS collaboration [5] (open circles), for Au projectile spectators, and INDRA collaboration [6] (open triangles and squares), for Ar- quasi-projectile.

Nuclear caloric curve: experimental results.

Fig.1a shows the temperature - excitation energy correlation known as the *nuclear caloric curve*. Accordingly to ref. [7] temperatures are extracted from the double ratios of isotopic yields of isotopes that differ of one neutron ($^3He/^4He;^6Li/^7Li$) The excitation energy per nucleon was estimated by a detailed balance on the exit channels. In ref. [3] a universal behaviour of the caloric curve was found for Xe, Au, U projectile fragmentation suggesting a statistical process. The isotope temperatures T_{HeLi}, measured for the target spectator fragments in the Au+ Au reaction at 1000 AMeV, are in a very good agreement (fig. 1a) with the previous values obtained for the projectile spectators[2]. Moreover, the good agreement between the values of T_{HeLi} and T_{HHe}, (insert in fig. 1a), measured in the same reactions, shows that the increasing behaviour of the isotope temperature is not very much affected by the sequential feeding. Disentangling collective and thermal motion in the mid-rapidity source generated in central Au+Au collisions at incident energies of 50,100,150, 200 A MeV, the deduced caloric curve follows the *rise* of the one for the spectator fragmentation (fig. 2a), suggesting that in both cases the vaporization regime, even through different collision dynamics is reached. The results on the Ar projectile fragmentation at 95 A MeV [6], compared to our data (fig. 1b), show temperatures systematically higher by about 1-2 MeV. However, also these data show an upswing at an excitation energy of about 10 MeV per nucleon. The observation of the plateau of fig. 1 seems [8] to depend on the size of the system as suggested by the isotope temperatures, measured in Ca+Sc and Nb+Nb collisions and in Kr+Nb at 50 AMeV. In particular, they confirm the plateau-like behaviour at excitation energies between 5 and 10 A MeV for decaying systems ranging from almost 80 to 200 mass units, i.e. larger than the mass 35 estimated for the Ar projectile in the INDRA data, but similar to the estimate of the Au projectile size.

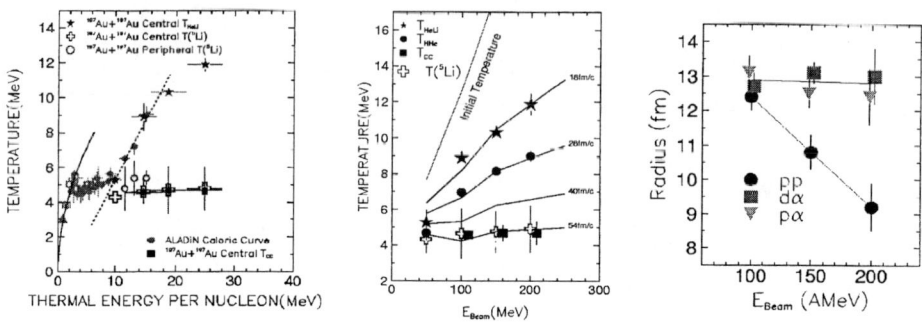

FIGURE 2. a)Caloric Curve for spectator fragmentation and central collisions. Both *isotope* and *emission* temperature are reported. b) Adiabatic expansion of an ideal gas. Points refer to the experimental values of fig.2a. c) Apparent radii extracted from both proton-proton and p-α, d-α correlation functions.

Isotope and emission temperatures and cooling of hot nuclei

Experimental correlation functions for p-α and d-α coincidences, both corresponding to decay channels of ^{5}Li fragments, show peaks due to the ground and excited (16.7 MeV) states. From the acceptance corrected yields the nuclear temperature was deduced [9] via the population ratio of these states. In fig. 2a isotope T_{HeLi} and emission temperatures from ^{5}Li decay in Au+Au collisions are compared. For peripheral collisions we notice a slightly increasing discrepancy between the two thermometers. For central collisions the differences between the two thermometers show up very pronounced.[10] Secondary decay correction cannot explain [11] this difference and other explanations have to be found. The dynamical evolution of the fragment formation [12] and emission [13,14] has been recently investigated. Although in ref. [12] it is suggested that the asymptotic fragments can be identified during the early stages of the process, a time delay between the emission of p, d, t and 3 He, on the one side and α and larger clusters on the other side, is predicted in Friedmann's *"Expanding Emitting Source"* model [13].Starting with a given volume with initial temperature deduced from the total available CM energy, subtracted of the radial flow, we follow (lines in fig. 2b) the temporal evolution of the cooling system, assuming an adiabatic expansion of an ideal gas with the radial flow velocity from the systematics of ref. [15].The comparison with the experimental temperatures, indicate that, while the $T_{He,Li}$ isotope temperatures reflect the ones at the first stages of the expanding system, the cluster ^{5}Li temperatures recall the system in the late stages. Whether the different emission time between particles and cluster [13] or the longest time (lower density) required by the excited fragments to get their properties are the reason of the different temperature measurements is still an open question. Moreover, the temporal evolution of the expanding system is also evident from the apparent radii extracted from proton-proton pairs (energy dependent values) and from p-α, d-α correlations (large and constant values)(fig. 2c). Deduced freeze out densities are ranging from $0.25\rho_0$ to $0.65\rho_0$, revealing the expansion of the source. Again the different radii values from p-p and p-α, d-α correlations could be related to the latest emission time of the α where lower densities are reached by the expanding system.

In summary the systematic study of calorimetric measurements on several systems of different size and excitation energy seems to explain the differences on the presently available caloric curves that can be attributed to the size of the system and to the underlying dynamical conditions. The time evolution of the multifragmentation process in terms of different degrees-of freedom freeze out at various stages seems intimately related to the collective expansion of the system and it is shown by the discrepancy between the isotope and emission thermometers. The different energy dependence of the two thermometers could enable us to sample the cooling curve of the expanding system.

REFERENCES

1. Pochodzalla J., *Prog. Part. Nucl. Phys.*, **39**, 443(1997).

2. Pochodzalla J. *et al.*, *Phys. Rev. Lett.*, **75**, 1040 (1995).

3. Schuettauff A. *et al*, *Nucl Phys.* A, **607**, 457 (1996).

4. Xi H. *et al*, *Phys. Rev.C*, **58**, 2636 (1998).

5. Hauger J. A. *et al*, *Phys. Rev. Lett.*, **77**, 235 (1996).

6. Ma. Y. *et* al, LPC pre-print LPCC 96-10 (1996).

7. Albergo S. et al., *Nuovo* Cimento A, **89**, 1 (1985).

8. Sammadar S. *K. et al.*, *Phys. Rev. Lett.*, **79**, 4962 (1997); De J. *N. et al,,* *Nucl. Phys.* A, **630**, 192 (1998).

9. Serfling V et *al*, *Phys. Rev. Lett.*, **80**, 3928 (1998).

10. Pochodzalla J. *et al.*, *Phys. Rev. C*, **35**, 1695 (1987); Schwarz C. *et al*, *Phys. Rev. C*, **49**, 3316 (1994).

11. Tsang M. B. *et al.*, *Proceedings of the 1st CRIS'96: Critical Phenomena and Collective Observables* (World Scientific) p.87, 1996.

12. Dorso C. O. and Aichelin J., *Phys. Lett. B*, **345**, 197 (1995).

13. Friedmann W. A., *Phys. Rev. C*, **42**, 667(1990).

14. Papp G. and Norenberg W, *XXII International Workshop on Gross Properties of Nuclei and Nuclear Excitations, Hirschegg, 87, 1994.*

15. Hsi W. C. *et al.*, *Phys. Rev. Lett.*, **73**, 3367 (1994); Lisa M. A. *et al.*, *Phys. Re v. Lett.*, **75**, 2662(1995).

Deep Water Cherenkov Light Scatter Meter

L. Pappalardo, C. Petta, G.V. Russo on behalf of NEMO Collaboration

Dipartimento di Fisica - Università di Catania
INFN - Sezione di Catania
Centro Siciliano di Fisica Nucleare e Struttura della Materia

Abstract. The relevant parameters for the site choice of an underwater neutrino's telescope are discussed. The *in situ* measurement of the scattering distribution of the Cherenkov light requires a suitable experimental setup. Its main features are described here.

NEMO UNDERWATER TELESCOPE FOR NEUTRINO'S DETECTION

The Choice Of The Optimal Site For The Deployment

Ultra-high energy neutrino's detection is possible detecting Cherenkov light coming from the rare secondary high energy charged leptons over a suitable km^3 scale underwater telescope[1]. High thickness of sea-water substrate acts both as detector and shield of cosmic charged radiation. The choice of its location is such an important task that much effort have been and is currently spent in the characterization campaign of several sites in the Mediterranean Sea, in order to identify the most suitable one.

Requirements

The following ones are the main requirements the optimum site has to fulfil: high depth (>3500 m); optimal underwater optical properties; low environmental background noise rates (K^{40} and bioluminescence); low sedimentation and bio-fouling, both corrupting track reconstruction resolution and detection efficiency; stability of hydro-dynamical conditions (small seasonal variations of water currents, absence of turbidity); close coastal infrastructures.

Test Planning In Mediterranean Sites

A test program has been planned and partially completed with a set of short/long term deep-sea measurements. Both the oceanographic parameters as sea currents, temperature, salinity, and the optical ones, as light attenuation and absorption vs. wavelength and depth are measured by using commercially available instrumentation. The measurement of the other parameters, as the optical noise, the sedimentation and

CP513, *Nuclear and Condensed Matter Physics*, edited by A. Messina
© 2000 American Institute of Physics 1-56396-929-7/00/$17.00

bio-fouling rates and the angular distribution of scattered light needs instead the use of customized experimental setup. The measurement program aims to characterize four Italian sites, whose co-ordinates are summarized in Table 1. Currently, the most promising site seems to be Capo Passero, where relatively high absorption and attenuation lengths (ca. respectively 70 m and 40 m at 440 nm, very good if compared with ca. 100 and 80 m for pure fresh water at the same wavelength) couple with low deep sea current, very small seasonal variations of current velocity and direction and very small sedimentation (7.62 10^{-3} mm^3/dm^3 per year at 2310 m depth) [2].

TABLE 1. Sites candidate to NEMO deployment.

Site	Coordinates	Depth (m)
Alicudi (Tyrrhenian Sea)	39° 05' N, 14° 20' E	3400
Capo Passero (Ionic Sea)	36° 30' N, 15° 50' E	3600
Ponza (Tyrrhenian Sea)	40° 40' N, 12° 45' E	3400
Ustica (Tyrrhenian Sea)	39° 05' N, 13° 20' E	3400

Deep-sea Optical Properties

Light traversing a water thickness x undergoes an attenuation given by:

$$I(x) = I_0 \cdot \exp\left(-\frac{x}{L} \right) \tag{1}$$

where L is the water attenuation length; its value depends on wavelength and characterizes the site. Its maximum value is given by pure water attenuation length L_W[3]. In ocean water, the attenuation length is reduced by the presence of organic and inorganic particulate (L_p), chlorophyll and phytoplancton (L_c) and dissolved yellowish organic matter (L_y)[4]. At 3000 m depth the contribution of chlorophyll and phytoplancton seems to be negligible.

Scattering is due to the medium inhomogeneities. We consider coherent and independent scattering, in which wavelength does not change and the particles are sufficiently far from each other (3 times the particle radius). The net effect is that intensities scattered by each particle must be added without regard to the phase. Moreover, if multiple scattering is allowed, we are facing with the radiative transfer problem. For unpolarized incident radiation and isotropic particles (scalar polarizability) we have the well-known Rayleigh scattered light intensity:

$$I(\vartheta) = \frac{2\pi \cdot [n(\lambda)-1]^2}{\nu \cdot \lambda^4} \left[1 + \cos^2(\vartheta)\right] \tag{2}$$

where n is the refractive index of water and ν the molecular concentration. The Rayleigh distribution fits quite well the experimental data for light scattering in pure or salted water (molecular scattering). The particulate presence in water makes more difficult and not ever analytically solvable the problem of the light absorption and scattering. A first-order description can be given by the Mie theory, concerning the problem of a diffusing sphere of arbitrary radius and refractive index. Even in this case, when an exact analytical solution can be established, its effective evaluation can require the use of complex and heavy computation.

THE SCATTER METER

Because of the hostile environment the apparatus must be very robust, compact and low power consuming. Moreover, if the measurement the scattering distribution both at small and large angles (Mie and Rayleigh components) is required, the detecting system must have an high dynamic range ($10^4 \div 10^5$) and a very good signal to noise ratio. These constraints strongly dictate the experimental set-up scheme

The apparatus is contained inside a suitably reinforced stainless steel container torus-like shaped ($\phi \sim 60$cm, h ~ 25 cm). In the surface of the central hole suitable optical windows allow the passage of incident beam into the water and the collection of the scattered light. The wall is made by a borosilicate glass tube (internal diameter 115 mm, thickness 13 mm) lined by a stainless steel tube.

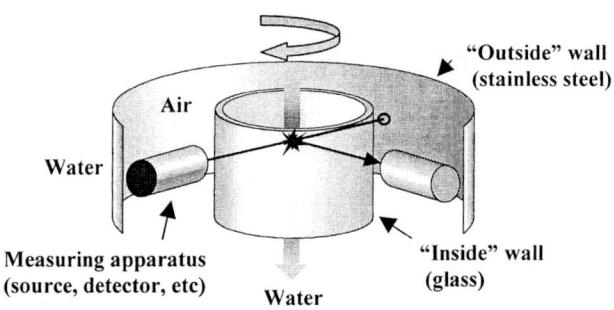

FIGURE 1. Schematic structure of the apparatus.

The source is a blue light emitting diode ($\lambda_{LED}= 4700$ Å) followed by an appropriate optical system. The diffused light is collected by a collimator (a converging lens and a pinhole) and it is sent onto the detector which can rotate around the system's symmetry axis, in order to measure the scattering distribution in almost the whole angular range.

The detector used is a miniature type (7mm bialkali photocathode, gain $3*10^5$) photomultiplier followed by a switched integrator which acts as a low noise amplifier and a synchronous detector controlled by the same low frequency signal generator which feeds the LED. A small photodiode on the transmitted light path acts as reference detector. Both the signal treatment circuitry and the rotation mechanism, with its related angular position measurement device, are under the control of a microcontroller which, through a FSK telemetry system, communicates with a personal computer on the support ship.

In order to lighten the water in the interaction region with a parallel light beam and to collect only the light diffused at an angle ϑ the "lens" effect due to the glass tube and the water contained therein must be taken into account. The input beam must have a suitable divergence in order that it is transformed into a "fan" with a very small divergence in the plane orthogonal to the symmetry axis of the system. Likewise, the pinhole of the collecting system must be placed before the focal plane of the lens in order to collect only the rays emitted in a small range of angles around ϑ.

By using a ray tracing software the behavior of the system has been estimated as a function of the various optical parameters. In particular the calculated theoretical angular resolution amounts to be $\Delta\vartheta = 0.4°$.

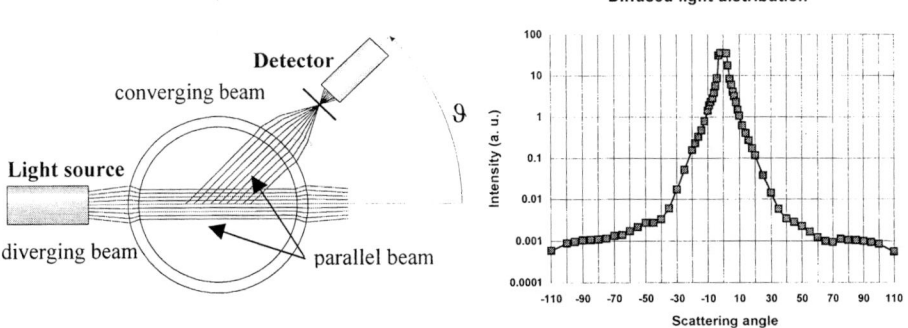

FIGURE 2. Scheme of the optics.   FIGURE 3. Diffused light intensity distribution.

Laboratory Measurements

Some preliminary measurements have been done in laboratory to test experimentally the assumptions on the dynamical range and the signal to noise ratio onto which the detecting scheme has been tailored, and the resolution of the optics.

The glass tube has been corked on the bottom and filled with mineral water. With a source output power of $\sim 10^{-5}$ W (equivalent to $\sim 3 * 10^{13}$ photons/sec in the interaction region), some distribution functions, as that of the figure 3, have been measured. The values of the photomultiplier anode currents (in the range $10^{-10} \div 10^{-5}$) confirm the estimated ones.

Moreover the distributions can be fitted quite well by using the phenomenological Henyey - Greenstein function

$$I(\vartheta) \propto \frac{1-g^2}{(1+g^2 -2g\cos(\vartheta))^{3/2}}. \tag{3}$$

with a value of 0.95 for the mean cosine g.

REFERENCES

1. NEMO Collaboration, *Letter of Intents*, Sept. 1998.

2. Riccobene, G., on behalf of NEMO Coll., *Proc. NNN99 Workshop*, Stoney Brook, NY, (Sept. 1999).

3. Smith, C., and Baker, K. S., *Appl. Optics* **20**, 177-184 (1981).

4. Price, B., *Appl. Optics* **36**, 1965-1975 (1997).

Study Of The $^7Li(p,\alpha)^4He$ Reaction At Astrophysical Energies Through The Trojan Horse Method

M.G. Pellegriti[1,2], M. Aliotta[1,2], S.Cherubini[3], M. Lattuada[2,4], D. Miljanic[5], R.G. Pizzone[1,2], S. Romano[2], N.Soic[5], C. Spitaleri[1,2], M. Zadro[5], R.A. Zappalà[6]

[1]Dip. di Metodologie Fisiche e Chimiche per l'Ingegneria, Università di Catania, Catania, Italy
[2]Laboratori Nazionali del Sud-INFN, Catania, Italy
[3]Institut de Physique Nucleaire, Université Catholique de Louvain, Louvain-la-Neuve, Belgium
[4]Dipartimento di Fisica, Università di Catania, Catania, Italy
[5]Institut Rudjer Boskovic, Zagreb, Croatia
[6]Istituto di Astronomia, Università di Catania, Catania, Italy

Abstract. The Trojan Horse Method has been applied to obtain information about $^7Li(p,\alpha)^4He$ reaction at astrophysical energies. The $^7Li(d,\alpha\ n)^4He$ reaction has been used and the two body reaction cross section for the $^7Li(p,\alpha)^4He$ has been extracted together with its astrophysical factor S(E).

INTRODUCTION

In the astrophysical direct reactions the Coulomb interaction decreases the measured low energy cross-section to the μbarn or pbarn, rising serious experimental problems. For this reason indirect methods have been recently used to measure around the Gamow's peak energy.

In this contest, the Trojan-Horse Method has shown to be a useful tool to extract information on nuclear cross sections in the low energy domain.

We present here the results obtained on the $^7Li(p,\alpha)^4He$ reaction at astrophysical energies.

This reaction is linked to the so called "Lithium depletion problem". The present abundance of lithium in stars is the result of the primordial nucleosynthesis during the Big Bang and the subsequent stellar evolution [1]. In order to reach a better understanding of the mechanisms leading to the lithium depletion, it is important to know the ^7Li destruction rate occurring through the $^7Li(p,\alpha)^4He$ process.

THE TROJAN-HORSE METHOD

The Trojan Horse method is based on a quasi-free break-up reaction mechanism [2].
We consider a particle X in the entrance channel impinging on a target nucleus C, that has a wave function whose behaviour underlines an a-s cluster configuration; under the Impulse Approximation hipothesis, the incident particle is considered to interact only with the cluster a of the target nucleus, while the other cluster is a spectator to the process $X+a \rightarrow Y+b$ (**Fig. 1**).

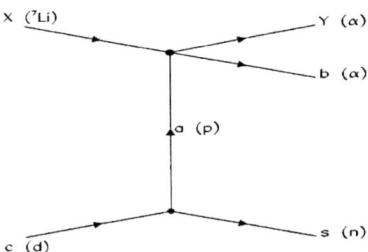

FIGURE 1. Quasi-free break-up scheme for the ^{7}Li+d reaction

In the Plane Wave Impulse Approximation, the three body cross section can be written as:

$$\frac{d^3\sigma}{dE_\gamma \, d\Omega_\gamma \, d\Omega_b} \propto (KF) \, |\Phi(\vec{p}_s)|^2 \left(\frac{d\sigma}{d\Omega}\right)$$

where KF is a kinematical factor, $\Phi(\vec{p}_s)$ is the momentum distribution of the spectator s inside the nucleus C and $\left(\dfrac{d\sigma}{d\Omega}\right)$ is the X-a two body cross section.

Thus, from the knowledge of KF and of the momentum distribution, it is possible to extract the two-body cross section from a measurement of the three-body one.
The process can be described as follows: at a beam energy that overcomes the Coulomb barrier in the incident channel, the projectile lets the particle a be brought into the nuclear interaction region, where the two-body reaction might be induced. Its cross section can be written as:

$$\left(\frac{d\sigma^N}{d\Omega}\right) \propto \frac{d^3\sigma}{dE_\gamma \, d\Omega_\gamma \, d\Omega_b} \left((KF) \, |\Phi(\vec{p}_s)|^2\right)^{-1}$$

In the present case, the ^{7}Li(d,α n)^{4}He three body reaction has been measured for the extraction of the ^{7}Li(p,α)^{4}He cross section.

EXPERIMENTAL SET-UP AND DATA ANALYSIS

The experiment was performed using the SMP Tandem Van de Graaff accelerator of the Laboratori Nazionali del Sud, Catania. ^{7}Li beams, at 19, 19.5, 20 and 21 MeV incident energies, were used to bombard a deuterated polyetilene target. In order to fully determine the kinematical properties of the spectator s (the neutron), the energies of the two α particles must be measured at specified angles (quasi-free angles).

Because of the high Q-value for the reaction $^2H(^7Li,2\alpha)n$ with respect to that of other possible reactions in the target, identification by means of ΔE-E telescopes was not necessary. The outgoing α particles were therefore detected in coincidence by three pairs of position sensitive detectors (PSD) placed on the two sides of the beam axis. The obtained nuclear cross section is shown in Fig. 2 a).

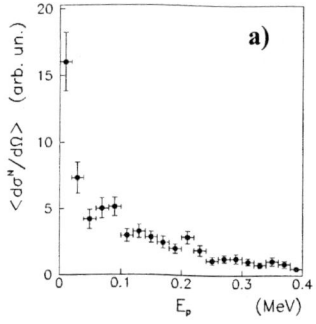 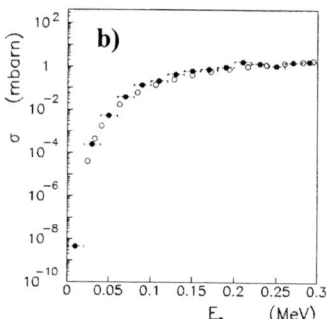

FIGURE 2 a) Nuclear cross section extracted according to the THM versus the proton energy, E_p. Dots represent the averaged results coming from measurements at 19, 19.5, 20 and 21 MeV beam energy; **b)** Cross section versus E_p obtained from a direct measurement [2] (open circles) compared with the THM results multiplied by the Coulomb barrier transmission coefficient (full dots).

Fig.2 b) shows the measured $d\sigma^N/d\Omega$ multiplied by the Coulomb barrier transmission coefficient, in order to have a straightforward comparison of the two-body cross section to the directly measured one:

$$\frac{d\sigma}{d\Omega} = \sum_l G_l \frac{d\sigma_l^N}{d\Omega}$$

ASTROPHYSICAL FACTOR

The astrophysical factor, given by the following expression:

$$S(E) = exp(2\pi\eta)E\sigma(E)$$

where η is the Sommerfeld parameter, is shown in Fig. 3 for the direct and the indirect data; it can be noticed the good agreement of these two sets within the experimental error and the presence of a very low energy (Ep= 10 keV) point for the indirect data.

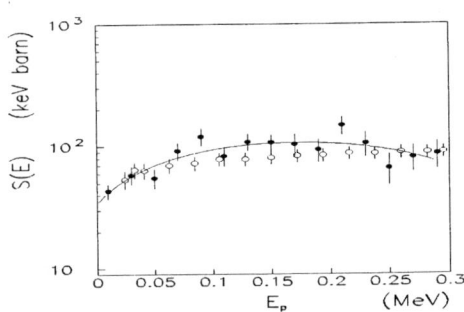

FIGURE 3 Astrophysical S(E)-factor for the ^{7}Li(p,α)^{4}He reaction. Data of the present measurement (full dots) are compared with those obtained in a direct measurement (empty circles) [2]

The obtained astrophysical factor at the Gamow energy (4.5 keV for this reaction), S(0)=36±7 keV*barn, can be compared with the one extrapolated from the direct data, S(0)=52±8 keV*barn.

CONCLUSIONS

The present work has shown the possibility of measuring the astrophysical S(E)-factor at energies relevant for astrophysical application, by means of the Trojan-Horse method.
We like to stress that our result does not need any electron screening correction, since the extracted cross section concern the interaction of two bare nuclei; thus the method could in principle be used to estimate the influence of the atomic electrons on the directly measured cross section at low energy.

REFERENCES

1. R.V. Wagoner et al., *Astrophys.J.*, **148** (1967) 3

2. S. Cherubini et al., *Ap. J.*, **457** (1996) 855

3. C. Rolfs et al., *Nucl. Phys.*, **A455** (1986) 179

NEMO Project: Studies on the Feasibility of a Km³- Scale Underwater Neutrino Telescope in the Mediterranean Area

Nunzio Randazzo[a] for the NEMO Collaboration[b]

[a]*Physics Department, University of Catania and INFN of Catania, Catania, Italy*

[b]*Physics Department, University of Bari and INFN of Bari, Bari, Italy*
Physics Department, University of Bologna and INFN of Bologna, Bologna, Italy
Physics Department, University of Cagliari and INFN of Cagliari, Cagliari, Italy
Physics Department, University of Catania and INFN of Catania, Catania, Italy
Physics Department, University of Messina and INFN of Catania, Messina, Italy
Physics Department, University "La Sapienza" of Roma and INFN of Roma, Roma, Italy
Laboratori Nazionali del Sud (INFN), Catania, Italy
Laboratori Nazionali di Frascati (INFN), Frascati, Italy
Fondazione "Ugo Bordoni", Roma, Italy
"Osservatorio Geofisico Sperimentale", Trieste, Italy
"Istituto per lo studio dell'Oceanografia Fisica" (IOF-CNR), La Spezia, Italy
"Centro Interdip. di Ricerche sulle Coste e sull'Ambiente Marino", University. Of Cagliari, Italy

Abstract. In the next decade the construction of a kilometre-scale Astrophysical Neutrino Observatory in the Mediterranean Sea could become a reality. Small-scale (10^4 m²) detectors are already under development (NESTOR, ANTARES). NEMO Collaboration proposes both to study the characteristics of some deep sea Mediterranean sites, close to the Italian coast, where the kilometre-scale Cerenkov Detector could be located and to develop technologies relevant for its design and optimization.

1. INTRODUCTION

The observation of the Universe has been carried out so far mainly looking at the electromagnetic emissions from the stars: very low frequency radio waves, optical frequencies and gamma rays allow us to investigate inside and outside our galaxy. Ultra GeV gamma rays have been observed from galactic objects like Crab Nebula and from the active Galactic Nuclei Markarian 421 and Markarian 501. Very high-energy particles, with energy in excess of 10^{20} eV, have been detected, but the emitting sources are unknown and the way they are produced is one of the open problems in Astrophysics and Astroparticle-physics.

Several models suggest, together with the emission of gamma's from the sources, the emission of neutrinos. If these extra-galactic 'accelerators' produce high-energy

CP513, *Nuclear and Condensed Matter Physics,* edited by A. Messina
© 2000 American Institute of Physics 1-56396-929-7/00/$17.00

photons through the production and the decay of neutral pions it is reasonable that the production and the decay of charged pions are further sources of neutrinos..

Neutrinos are massless, weakly interacting particles. They are not affected by strong, electromagnetic and gravitational interactions. As a consequence neutrinos do not interact and are not deflected during their journey towards the Earth and allow the identification of their sources through the "source tracking back": this is the neutrino astronomy basic assumption.

Moreover high energy gamma ray fluxes are reduced by its interaction with the low energy background radiation of the Universe. Neutrino fluxes are not expected to be reduced offering the possibility of studying regions of the space invisible to the gamma ray astronomy. Neutrino astronomy will complement and extend the traditional gamma rays astronomy.

One of the most common experimental techniques in High Energy neutrino Astrophysics is the Cherenkov detection of charged particles (i.e. muons) produced in neutrino charged-current interactions with matter. These secondary muons are roughly as energetic as the parent neutrino and travel in the same direction. When the interaction takes place in a transparent medium (radiator), the out-coming muon travels, inside the medium, faster than light and produces a light cone around its track, which is the signature of its presence.

Unfortunately the low interaction probability of neutrinos makes hunting for these particles a difficult challenge. Even if several astrophysical models preview that neutrino fluxes reaching the Earth should be more intense than gamma ray fluxes (at the same energy), a large amount of target matter is needed to detect them. An estimate based on the known values of charged-current cross sections (i.e. interaction probability) shows that a detector suitable for neutrino astronomy should reach a detection area 1 Km^2 large. Such a huge detector is usually called "Km^3-scale Telescope".

The typical configuration for a neutrino telescope is a lattice of optical modules (PMTs, i.e. photon detectors, and read-out electronics) connected each other in regularly spaced strings or towers. The strings are immersed into the target medium, which is also the Cherenkov radiator, and shielded from the cosmic background. Two approaches are, at present, the most interesting ones for the construction of the Km^3 Telescope. In the first approach, followed by AMANDA Collaboration in South Pole, strings are located inside holes drilled in ice pack at ~2 Km depth. In the second approach strings are deployed in water: deep Sea (DUMAND, ANTARES, NESTOR, NEMO) or lake (BAIKAL).

2. NEMO Collaboration R&D Activity

NEMO collaboration activity has started since middle of 1998 with the following purpose:

❑ Search for the best site for detector the deployment

❑ Search for the best detector configuration

❑ R&D for the electronics design

2.1 Site Characterization

The choice of the suitable place for the Telescope needs the knowledge of many parameters. The main characteristics required are: deepness, proximity to the coast, good optical transmission, low sea currents intensity, low biological activity, low optical background. At the beginning of its activity, the NEMO Collaboration has selected four sites close to the Italian coasts, that could be appropriate for the Km^3 astrophysical neutrino detector. Approximately the co-ordinates of these sites are:

❑ 35° 50' N, 16° 10' E in the Ionian Sea, South-Est of "Capo Passero"

❑ 39° 05' N, 13° 20' E in the Tyrrhenian Sea, North-Est of Ustica island

❑ 39° 05' N, 14° 20' E in the Tyrrhenian Sea, North of Alicudi island

❑ 40° 40' N, 12° 45' E in the central Tyrrhenian Sea, South of Ponza island

Several measurements have been carried out in these sites since June 1998, offering a useful map of seasonal changing, in particular in the Western Ionian area.

2.1.1 Optical Measurements

The optical properties of deep sea play a fundamental role in choosing the site. The detector effective area is not only directly determined by the extension of the instrumented volume but is strongly affected by the light attenuation in water. In order to measure attenuation and absorption, a compact size device is commercially available: the AC-9 trasmissometer manufactured by Wetlabs. The AC-9 measure attenuation and absorption independently, using two different light paths in 9 wavelengths (412, 440, 488, 510, 532, 555, 650, 676, 715 nm).

2.1.2 Sea Current Measurement

The site has to be "quiet", i.e. the water current intensity has to be small and sufficiently constant. In this case special requirements on the mechanical structure are not needed and detector deployment and positioning is easier. In low current intensity conditions the optical noise due to bioluminescence, often excited by variation of the water currents, is also reduced. The current velocities measured by the two instruments are in very good agreement. The deep Sea water current measured from August '98 to February '99 is quite constant in intensity (less than 3 cm/sec) and direction.

2.1.3 R&D For The Electronics Design

The electronics design must take into account also the difficulties in power transmission: about 10000 phototubes together with their read-out electronics should be deployed and biased. The NEMO Collaboration is developing a custom low-power microelectronics able to reach this goal.

Signals from the optical module cannot be transferred to the shore unaltered. It is necessary to transfer a compressed and codified representation of signals from the optical module to a concentrator module. The latter manages a group of optical modules, performs a further compression and provides the temporal alignment of the signals. A secondary trigger is foreseen in the same module to increase the S/N ratio ON-LINE. In the same way other hierarchically higher concentrator modules can compress data flow further, making it manageable.

REFERENCE

Collaborazione NEMO, "LOI" — July 1998

Collaborazione NEMO, "Addendum alla LOI" — Sept. 1998

G. Riccobene for NEMO Collaboration, "Deep site characterisation for the km^3 Telescope"

NNN99 Workshop – Stony Brook – NY – Sept. 1999

Low Noise Integrated Preamplifier For Application In Intermediate Energy Physics Experiments.

N.Randazzo[(1),(2)], G.V.Russo[(1),(2)], D.LoPresti[(1)], C.Petta[(1),(2)], S.Reito[(1)],
L.Todaro[(3)]; G.Fallica[(3)],G.Valvo[(3)]; M.Lattuada[(2),(4)], S.Romano[(4)],
A.Tumino, [(2),(4)] V. Bonvicini, [(5)] A. Vacchi[(5)]

1.Istituto Nazionale di Fisica Nucleare, Sezione di Catania, Italy
2.Dipartimento di Fisica, Università di Catania, Italy
3.ST Microelectronics - Catania, Italy
4.Istituto Nazionale di Fisica Nucleare, Laboratorio Nazionale del Sud - Catania, Italy
5.Istituto Nazionale di Fisica Nucleare, Sezione di Trieste, Italy

Abstract. Only a few charge preamplifiers suitable for use with Large Area Silicon Detectors have been realised as integrated circuits. In this paper we presents some of them that includes in the same chip both a charge preamplifier and a shaper, dramatically cutting power dissipation, size and cost. The resolution achieved is very good. It is intended in particular for use in Intermediate Energy Physics Experiments.

1. INTRODUCTION

High Energy Physics Experiments (HEPE) have given a great boost to the analogue VLSI for front-end devices applied to Solid State Detectors, the main reason being the large number of channels required for such experiments. On the contrary, Intermediate Energy Physics Experiments (IEPE) have been performed, up to now, with a reduced number of channels. In these cases physicists have used a limited number of expensive modular devices. However in recent years the rise of new demands has required a number of channels comparable to those used for HEPEs. Some first steps have led to the realisation of some reduced hybrid or SMD circuits versions. More recently some monolithic front-end devices suitable for IEPE have been presented [1],[2],[3].

The growing number of channels, more than 100, in the multidetector systems, sets essentially three problems:

Power Dissipation. For noise considerations it is better to put the detector and the front-end amplifier as close as possible. However the heat transferred from the amplifier to the detector can give problems of drift and can make its resolution worse. This problem is intensified when the detectors are in a vacuum;

CP513, *Nuclear and Condensed Matter Physics,* edited by A. Messina
© 2000 American Institute of Physics 1-56396-929-7/00/$17.00

Overall Size. The segmentation of the multidetector systems is limited by the preamplifier sizes;

Cost. The price of a preamplifier depends on the technology involved. A VLSI preamplifier costs much less than a hybrid one or a preamplifier unit. Besides, since the power required is much less, the power supply costs less too!

2. Design considerations of the read-out system for solid state detectors.

Generally speaking, a readout system for a particle detector consists of a preamplifier that makes the integral of the charge released by the particle in the detector, giving a signal whose amplitude is proportional to its energy losses. A suitable filter (shaper) is provided to reduce noise. Fig.1 shows a schematic drawing of this system.

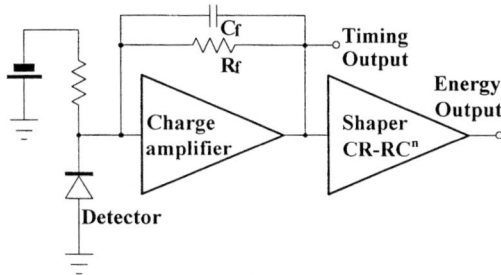

FIGURE 1. The read-out system for solid state detector

3. Noise Considerations

Noise analysis depends both on design and process parameters. We made the choice to use a 1.2 μm CMOS, double-poly, double-metal process. We have investigated three kinds of noise contributions : flicker, ENC_f; thermal, ENC_t; and shot, ENC_s; bound, respectively, to the gate capacitance of the input transistor, the drain-source current and the leakage current of the detector. Some parameters depend on assigned constraints such as the detector capacitance, the time of differentiation-integration and the number of integrators. Fig. 2 show an example of simulation performed for optimization of the input device.

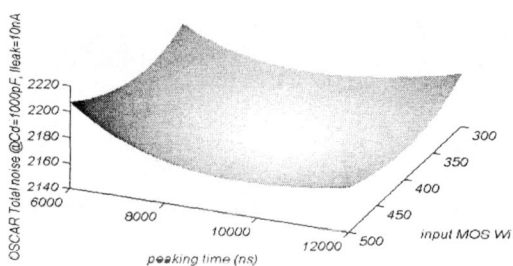

FIGURE 2. Input transistor optimisation: noise vs. peaking time and first transistor gate width

4. Realized Chips

In this chapter we briefly show some realizations:.

4.1 CASH_06 Chip

FIGURE 3. Microphotograph of CASH_06 chip

TABLE 1. CASH_06 Chip main characteristics [2]

Technology	CMOS 1.2 μm
Power Supply	5 V
Power Dissipation	6 mW/channel
Input Range	70 MeV
Shaping Time	2.4 μs
Sensitivity	30 mV/MeV
Noise	2000 e⁻ @ 200 pF
Size	12 mm²
Number of channels	6

FIGURE 4. Microphotograph of OSCAR chip

TABLE 2. OSCAR Chip main characteristics [3]

Technology	CMOS 1.2 μm
Power Supply	5 V
Power Dissipation	10 mW/channel
Input Range	100 MeV
Shaping Time	800 ns
Sensitivity	20 mV/MeV
Noise	1900 e⁻ @ 200 pF
Size	10 mm²
Number of channels	6

5. Conclusions

The most important characteristics of these realisations are the good performance compared with the low cost, dissipation and size which are appreciable in IEPEs, especially when the number of channels is very high.

REFERENCE

1. Randazzo ,N. et al., IEEE Trans. In Nuclear Science, Vol. 44, 1997, pp. 31-35.

2. Randazzo ,N. et al., NIM, A420, 1999, pp. 279-287.

3. Randazzo ,N. et al., IEEE Trans. In Nuclear Science, In press

Heavy Flavour Electroweak Measurements

Alessia Tricomi

ALEPH Collaboration
University of Catania, CSFNSM, INFN Bari and Catania, Italy

Abstract. The electroweak properties of the heavy-quark production have been measured both at LEP and SLD. Progress in the flavour separation technique has significantly improved the measurement accuracies. We review recent measurements of the partial decay widths, R_b and R_c, and of the Forward-Backward asymmetries at LEP and of the Left-Right Forward-Backward asymmetries at SLD, and their comparison with the Standard Model expectations.

INTRODUCTION

Both LEP and SLD have produced a great number of electroweak measurements: e.g., the Z^0 lineshape parameters, the hadronic and leptonic cross-sections and their forward-backward and polarization asymmetries. These measurements constitute the building blocks of electroweak physics at the Z and a stringent test of the Standard Model of electroweak interactions by overcostraining its parameter. Among the electroweak quantities measured both at LEP and SLD, there are six that involve the production of bottom and charm quarks: the partial decay width, R_b and R_c, the forward-backward asymmetries, A_{FB}^b and A_{FB}^c, and the left-right forward-backward asymmetries, A_{LRFB}^b and A_{LRFB}^c (only at SLD), which can all be expressed in terms of the chiral couplings:

$$R_q = \frac{\Gamma\left(Z^0 \rightarrow q\bar{q}\right)}{\Gamma\left(Z^0 \rightarrow hadrons\right)} \propto \left(c_V^q\right)^2 + \left(c_V^q\right)^2 \propto \left(c_L^q\right)^2 + \left(c_R^q\right)^2 , q = b,c \qquad (1)$$

$$A_{FB}^f = \frac{\sigma_F - \sigma_B}{\sigma_F + \sigma_B} = \frac{3}{4} A_e A_f \qquad (2)$$

$$A_{LRFB}^f = \frac{\left[\sigma_F^L - \sigma_B^L\right] - \left[\sigma_F^R - \sigma_{BB}^R\right]}{\left[\sigma_F^L + \sigma_B^L\right] + \left[\sigma_F^R + \sigma_{BB}^R\right]} = P_e A_f \frac{2|\cos\theta|}{1 + \cos^2\theta} \qquad (3)$$

where σ is the differential cross-section of the process $e^+ e^- \rightarrow Z^0 \rightarrow f\bar{f}$, θ is the polar angle, P_e is the initial state longitudinal polarization ($P_e \neq 0$ only at SLD), and

$A_e(A_f)$ is the initial (final) state coupling asymmetry defined as:

CP513, *Nuclear and Condensed Matter Physics,* edited by A. Messina
© 2000 American Institute of Physics 1-56396-929-7/00/$17.00

$$A_f = \frac{2c_V^f c_A^f}{\left(c_V^f\right)^2 + \left(c_A^f\right)^2} = \frac{\left(c_L^f\right)^2 - \left(c_R^f\right)^2}{\left(c_L^f\right)^2 + \left(c_R^f\right)^2} \tag{4}$$

Asymmetries and partial widths are therefore sensitive to different aspects of the coupling.

In the following sections we report on the new measurements made at LEP and SLD.

MEASUREMENTS OF Rb AND Rc

Since the beginning of LEP and SLC, measurements of the decay partial widths, R_b and R_c, are performed with ever increasing precision. For an experimental point of view, the measurements of partial widths essentially need efficient and self-calibrating flavour-tags. They fall in two categories. In the first, called single-tag measurement, a method to select b or c events is devised, and the number of tagged event is counted. This number must then be corrected for backgrounds from other flavours and for the tagging efficiencies to calculate the true fraction of hadronic Z decays of that flavour. The dominant systematic errors come from the knowledge of the tagging efficiency In the second technique, called double-tag measurement, the event id divided in two hemispheres and the partial widths can be determined solving the two equations:

$$\frac{N_t}{2N_{had}} = \varepsilon_b R_b + \varepsilon_c R_c + \varepsilon_{uds}\left(1 - R_b - R_c\right),$$

$$\frac{N_{tt}}{N_{had}} = C_b \varepsilon_b^2 R_b + C_c \varepsilon_c^2 R_c + C_{uds} \varepsilon_{uds}^2 \left(1 - R_b - R_c\right) \tag{5}$$

in which N_t is the number of tagged hemispheres, N_{tt} the number of events with both hemispheres tagged, N_{had} the total number of hadronic Z decays, ε_x and C_x are respectively the tagging efficiencies and the hemisphere correlations for the different flavours.

In the double-tag measurements the tagging efficiencies are determined directly from the data and this reduces the systematic errors.

Many techniques have been used to measure R_b and R_c from the LEP Collaborations and from SLD. They exploit different properties of b and c events, like the presence of large transverse momentum leptons, long lifetime of B hadrons (high impact parameter and well displaced secondary vertices), consistency of the event shape with the presence of heavy boosted objects (event shape tag) and the presence and decay of D mesons. In Tab.1 a schematic summary of the different techniques used both for R_b and R_c measurements are displayed (the * denotes the best analyses), while in Fig.1 a summary of the results up to this summer is shown compared to the SM expectations. A detailed description of each analyses can be found in ref. [1, 2].

TABLE 1. Techniques used for R_b and R_c measurements

Measurements	Experiments	Tagging Type	Tagging properties
R_b	ALEPH	Multi	Lifetime+Mass
	DELPHI*, OPAL	Multi	Lifetime+Mass+Event Shape
	L3	Double	Lifetime
	SLD	Double	Lifetime+Mass
R_c	ALEPH	Single	Lepton
	ALEPH, DELPHI, OPAL	Single	Charm Counting
	ALEPH, DELPHI, OPAL	Double	Excl. D, Excl./Incl. D+slow pion
	SLD*	Double	$M_{VTX}+P_{VTX}$

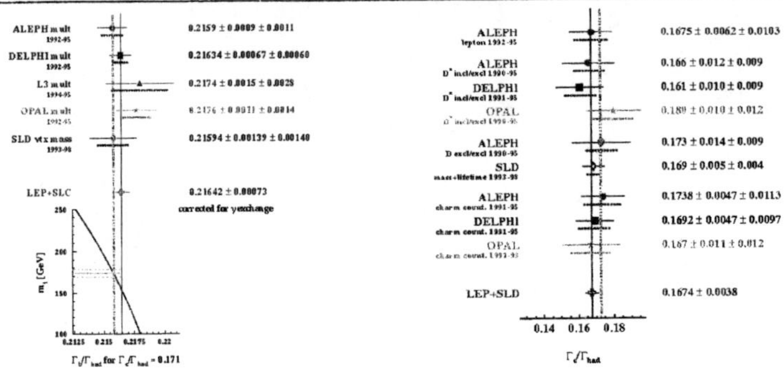

FIGURE 1. R_b (left) and R_c (right) measurements.

ASYMMETRY MEASUREMENTS

The measurements use event tagging procedures similar to the R_b and R_c analyses, but also need the estimate of the quark charge and direction. A summary of the different techniques used for the asymmetry measurements is shown in Fig. 2.

	Analysis	Tag	Quark Sign	Exp.
A_b	Jet/Vtx Charge	Lifetime	Jet/Vtx Charge	ADLOS
	Kaon Tag	High Mvtx	Kaon	S
	Lepton Tag + NN	e,μ	e,μ	DLS AO
A_c	Lepton Tag + NN	e,μ	e,μ	DLS AO
	Inclusive Charm	Low Mvtx High Pvtx	Kaon or Vertex Charge	S
	Inclusive Slow Pion	Veto	Slow Pion	S
	Exclusive Charm	D	D^{**}, D^{*}, D^{0}	ADOS

FIGURE 2. The different technique used for the asymmetry measurements.

Several new measurements both of the b and c asymmetries have been reported to the summer conferences. ALEPH has presented a preliminary measurement of both the b and c Forward-Backward asymmetries using leptons. Three neural networks are used to optimize the flavour separation and the $b\rightarrow l/b\rightarrow c\rightarrow l$ discrimination, improving on both statistic and systematic errors [3]. SLD has added a *Vertex-Charge* A_b measurement, a *Kaon A_b* measurement and a *Vertex-Charge A_c* measurement [4].

In Fig.3 the comparison between the forward-backward asymmetries, A_{FB}^{b} and A_{FB}^{c}, measured at LEP, the coupling asymmetries, A_b and A_c measured at SLD, and SM expectations is shown.

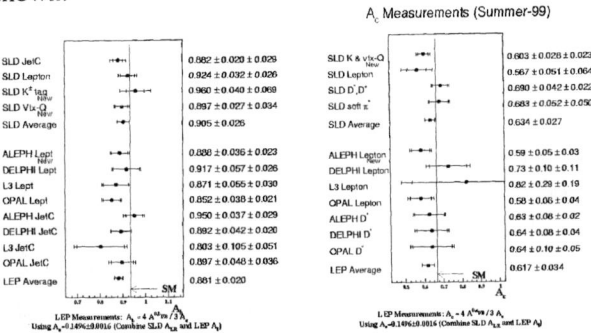

FIGURE 3. Bottom and charm asymmetry measurements compared to the SM expectations.

The quark asymmetries still present the largest pull in the SM fit. No new data are available to solve this "puzzle", however several new analyses are still improving, in particular new multivariate analyses seem promising in helping to extract the most from data.

ACKNOWLEDGMENTS

I am very grateful to the LEP and SLD Collaborations and the LEP Electroweak Working Group. I thank also the CSFNSM for the financial support.

REFERENCES

1. ALEPH Collaboration, *Phyics Letters* **B 401**, 150 (1997), ibidem 163; DELPHI Collaboration, *to appear in Eur. Phys. Journal*, CERN-EP/98-180; L3 Collaboration, L3-2420; OPAL Collaboration, *Eur. Phys. Journal* **C8**, 217 (1999); SLD Collaboration, *Phys. Rev. Lett.* **80**, 660 (1998).

2. ALEPH Collaboration, *Eur. Phys. Journal* **C4**, 557 (1998);DELPHI Collaboration, *to appear in Eur. Phys. Journal*, CERN-EP/99-066; OPAL Collaboration, *Eur. Phys. Journal* **C1**, 439 (1998); SLD Collaboration, SLAC-PUB-7880.

3. ALEPH Collaboration, *Contributed paper to EPS-HEP 99 Tampere*, ALEPH 99-076.

4. SLD Collaboration, *Contributed paper to EPS-HEP 99 Tampere*, SLAC-PUB-8201, SLAC-PUB-8200, SLAC-PUB-8199.

Search for Higgs bosons with the ALEPH detector

Alessia Tricomi

ALEPH Collaboration

* University of Catania, CSFNSM, INFN Bari and Catania, Italy

Abstract. The data collected at a centre-of-mass energy up to 196 GeV and up to 188.6 GeV by the ALEPH experiment at LEP, are analysed to search for the neutral Higgs bosons in the Standard Model and the MSSM and for pair-produced charged Higgs bosons. No evidence for a signal is found. When combined with the lower energy ALEPH data, this observation results in 95% confidence level lower mass limits.

INTRODUCTION

At LEP2, the Standard Model Higgs boson production is dominated by the Higgs-strahlung process, $e^+e^- \to HZ$, with smaller contributions from the WW- and ZZ-fusion processes to the $H\nu\bar{\nu}$ and He^+e^- final states. The MSSM neutral Higgs bosons are produced via two complementary reactions: the Higgs-strahlung process $e^+e^- \to hZ$ and the associated pair production $e^+e^- \to hA$. The cross section of the Higgs-strahlung process is proportional to $\sin^2(\beta - \alpha)$, where $\tan\beta$ is the ratio of the vacuum expectation values of the two Higgs doublets and α is the mixing angle in the CP-even Higgs sector. The cross section of the pair production is proportional to $\cos^2(\beta - \alpha)$. In the mass range relevant to LEP2 searches, the Higgs boson decays mostly into $b\bar{b}$ and to a lesser extent into $\tau^+\tau^-$. The searches described in this note cover most of the topologies arising from the HZ process, with $H \to b\bar{b}$ or $\tau^+\tau^-$, and $Z \to e^+e^-$, $\mu^+\mu^-$, $\tau^+\tau^-$, $\nu\bar{\nu}$ or $q\bar{q}$ and also from the hA process, with h and A decaying to $b\bar{b}$ or $\tau^+\tau^-$. In the following, the h notation stands both for the Standard Model Higgs boson and the lighter CP-even neutral Higgs boson of the MSSM.

Searches for neutral Higgs bosons have already been performed by ALEPH up to a centre-of-mass energy of 184 GeV. No evidence for signal was found and a lower limit of 87.9 GeV/c^2 was set at 95% confidence level on the Standard Model Higgs boson mass [1,2]. In the MSSM, masses of h and A lower than 72.2 GeV/c^2 and 76.1 GeV/c^2 respectively were excluded at the 95% confidence level [3,4].

A total integrated luminosity of 176.2 pb^{-1} was recorded by ALEPH in 1998 at a centre-of-mass energy of 188.6 GeV and an integrated luminosity of 27.3 and 26.2

CP513, *Nuclear and Condensed Matter Physics*, edited by A. Messina
© 2000 American Institute of Physics 1-56396-929-7/00/$17.00

pb^{-1} at centre-of-mass energy of 191.6 and 195.6 GeV respectively was recorded during the first part of 1999 data taking period. This higher centre-of-mass energy and integrated luminosity increase substantially the experimental sensitivity for the detection of the Higgs boson.

On the other hand, due to the absence of direct experimental information about the Higgs sector, the investigation of the implications of more complicated Higgs sector must be taken into account.

The most important phenomenological consequence of an extended Higgs structure is the appearance of additional physical scalar states [5]. For example, with the addition of one more doublet of complex scalar fields, five physical states remain after spontaneous breaking of the $SU(2)_L \times U(1)_Y$ symmetry to give mass to $W^\pm$ and Z gauge bosons: three neutral and a pair of charged bosons. Among the possible choices, multi-doublet models are theoretically interesting because they automatically lead, at tree level, to $m_W = m_Z \cos\theta_W$ and to the absence of flavour changing neutral currents, two major constraints which must be satisfied by any extension of the Standard Model to agree with the experimental observations.

Pair production of charged Higgs bosons occurs mainly via s-channel exchange of a photon or a Z boson; in two-doublet models, the couplings are completely specified in terms of the electric charge and θ_W, making the production cross section depend only on one additional parameter, the charged Higgs boson mass m_h. As expected in most implementations of multi-doublet models [5], it is assumed that H^+ decays, with negligible lifetime, predominantly into $c\bar{s}$ or $\tau^+\nu_\tau$ (and the respective charge conjugates for H^-). Additional decay channels, such as those involving neutral Higgs bosons, are not considered here. Since the relative weight of the two main channels depends on the details of the model, no assumption is made about the decay branching fractions, and three different selections are developed to address the possible final states $c\bar{s}s\bar{c}$, $c\bar{s}\tau^-\bar{\nu}_\tau / \bar{c}s\tau^+\nu_\tau$ (hereafter referred to as $c\bar{s}\tau^-\bar{\nu}_\tau$) and $\tau^+\nu_\tau\tau^-\bar{\nu}_\tau$. Under the same hypothesis, the negative results of the searches performed using all the data collected at centre-of-mass energies ranging from 130 to 184 GeV allowed ALEPH to exclude charged Higgs boson masses less than 59 GeV/c^2 at 95% C.L., independently of the final state [6].

SEARCH FOR NEUTRAL HIGGS BOSONS

The analysis strategy is described in detail in [1,4,7].

Event selections were previously developed for the various topologies arising from the hZ and hA processes. These selections address the $h\ell^+\ell^-$ channel (where ℓ denotes either an electron or a muon), the $h\nu\bar{\nu}$ channel, the $hq\bar{q}$ channel (excluding h decays to τ pairs), the $h\tau^+\tau^-$ channel and the $\tau^+\tau^-q\bar{q}$ channel which complements the $hq\bar{q}$ channel when h decays to a $\tau^+\tau^-$ pair. Dedicated selections are devoted to the four b quarks and $b\bar{b}\tau^+\tau^-$ channels arising from the pair production process.

The analyses have been refined for the Higgs search up to $\sqrt{s} = 186$ GeV and in some cases new selections are adopted.

The various hZ selections are tuned to give optimal performance for a Higgs boson of mass 95 GeV/c^2, which is near the expected experimental sensitivity. The corresponding signal used to optimise the hA selections is for m_A =85 GeV/c^2 and $\tan \beta = 10$.

For the new data sample collected this summer at $\sqrt{s} = 191.6$ GeV (27.3 pb^{-1}) and at $\sqrt{s} = 195.6$ GeV (26.2 pb^{-1}), the expected numbers of background events, the expected signal for a Higgs boson with m_h=95 GeV/c^2, and the number of selected candidates are listed in Table 1 for the hZ analyses and in Table 2 for the hA analyses of ref. [7]. No departure from the Standard Model expectations is observed.

The negative results of all the searches are combined to produce exclusion confidence levels in the MSSM plane $[m_h, \sin^2(\beta - \alpha)]$. The hZ searches dominate at large $\sin^2(\beta - \alpha)$, leading to the SM result for $\sin^2(\beta - \alpha)= 1$, while at low $\sin^2(\beta - \alpha)$ the exclusion is driven by the hA searches.

The limit obtained for the Standard Model Higgs boson is $m_h \geq 94.9$ GeV/c^2, with the expected limit at 97.4 GeV/c^2. This is derived as described in [7], with the proper treatment of the overlap between the hZ and hA four jet and tau selections. For the h boson of the MSSM the limit is 83.8 GeV/c^2, for all values above of $\tan \beta$ above 1, with 85.0 GeV/c^2 expected. The corresponding excluded region translated in the $[m_h, \tan \beta]$ plane of the MSSM is also shown in Fig. 1.

SEARCH FOR CHARGED HIGGS BOSONS

To search for pair produced charged Higgs boson and to ensure a good discovery potential independent of the branching fraction $B(H^+ \to \tau^+ \nu_\tau)$ for the decay of the Higgs boson into $\tau \nu$, three selections are defined for the topologies $\tau^+ \nu_\tau \tau^- \bar{\nu}_\tau$, $\bar{c} s \tau^+ \nu_\tau$ and $c \bar{s} s \bar{c}$. As explained in detail in [6,8,9], the most relevant selection criteria for the three selections are chosen in order to achieve, on average and in case no signal is present, the best 95% C.L. limit on the Higgs boson production cross section. To do so, each selection is optimised individually with the most optimistic $B(H^+ \to \tau^+ \nu_\tau)$ in each case (0%, 100% and 50% for the $c \bar{s} s \bar{c}$, $\tau^+ \nu_\tau \tau^- \bar{\nu}_\tau$ and $\bar{c} s \tau^+ \nu_\tau$ channels, respectively, for which the contribution of the other two analyses is minimal). The new future of the analyses performed on the data collected at centre-of-mass energy of 188.6 GeV/c^2 are described in detail in [9].

The results of the three selections are combined with those obtained using 130-183 GeV data to set a 95% C.L. lower limit on the charged Higgs boson mass, as described in [9], and are displayed in Fig. 2, where the curves corresponding to expected and observed confidence levels of 95% exclusion are drawn.

Charged Higgs bosons with masses less than 65.5 GeV/c^2 are excluded at 95% C.L. independently of $B(H^+ \to \tau^+ \nu_\tau)$, corresponding to an expected exclusion of 69.5 GeV/c^2.

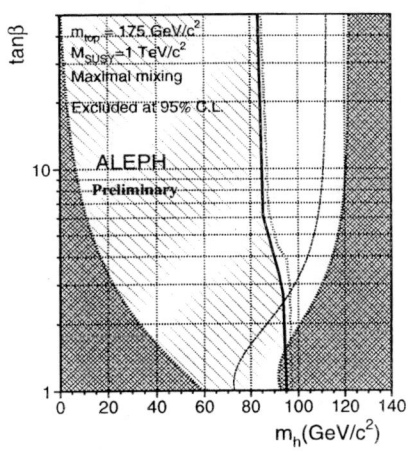
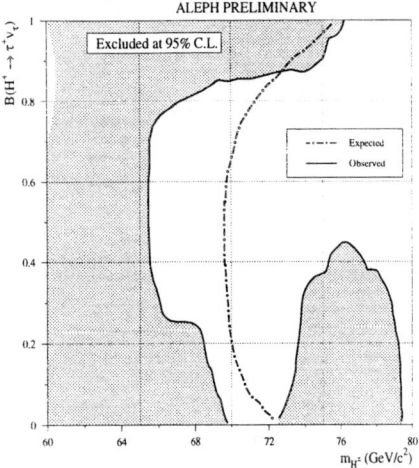

FIGURE 1. Expected (dashed) and observed (solid) 95% confidence level curves in the $[m_h, \tan\beta]$ plane of the MSSM. The dark regions are not allowed theoretically. The combined experimental excluded region is essentially identical in the case of no stop mixing, for wich the theoretically forbidden region is also indicated (dashed-dotted curve).

FIGURE 2. Limit at 95% C.L. on the mass of charged Higgs bosons as function of $B(H^+ \to \tau^+ \nu_\tau)$. Shown are the expected (dash-dotted) and observed (full) exclusion curves for the combination of the three analyses, and the full 130-189 GeV data set. The shaded area is excluded at 95% C.L.

ACKNOWLEGMENTS

I am very grateful to the ALEPH Collaboration and in particular to the Higgs Task Forks Group. I thanks the conference organizer for their hospitality. I thank also the CSFNSM for the finantial support.

REFERENCES

1. ALEPH Collaboration, Phys. Lett. **B412** (1997) 155.
2. ALEPH Collaboration, Phys. Lett. **B447** (1999) 336.
3. ALEPH Collaboration, Phys. Lett. **B412** (1997) 173.
4. ALEPH Collaboration, Phys. Lett. **B440** (1998) 419.
5. J. F. Gunion et al., *"The Higgs Hunter's Guide"*, Frontiers in Physics, Lecture Note Series, Addison Wesley, 1990.
6. ALEPH Collaboration, *Phys. Lett.* B **450**, 467 (1999).
7. ALEPH Collaboration, ALEPH-PHY 99-075, ALEPH-CONF 99-047.
8. ALEPH Collaboration, *Phys. Lett.* B **418**, 419 (1998).
9. ALEPH Collaboration, ALEPH-PHY 99-070, ALEPH-CONF 99-044.

A Neural Network For Off-Line Z Classification And Energy Calibration

S. Tudisco [a,b] and C.M. Iacono Manno [a]

[a] INFN-laboratorio Nazionale del Sud, Via S.Sofia 44,95123 Catania, Italy
[b] Dipartimento di Fisica dell'Università di Catania, Corso Italia 57, 95129 Italy

Abstract. In this work a neural network has been used to reconstruct the residual energy after the first stage and classify the atomic number of the particles detected in a Silicon-CsI ΔE-E telescope. The adopted net is described and the whole procedure has been compared with the standard calibration methods for the E stage.

INTRODUCTION

The great number of detection modules used in the most sophisticated experiments in heavy ion physics at low and intermediate energy, requires large area and low cost detectors. Considerable thickness is also necessary to stop the light charged particles emitted with energies up to several hundred MeV. The relatively simple handling and the light output performance of CsI(Tl) crystals coupled with photodiodes make them very suitable for sophisticated detector assemblies where they are often used as the second stage of a ΔE-E telescope. Nevertheless, their non-linear response requires a detailed calibration for each of the nuclear species to be detected, over the whole energy range. To perform this calibrations is also necessary the classification each nuclear species, and this is usually done by a "manual" procedure (for instance by graphic contours). Therefore, this conventional kind of analysis requires a long time, particularly in the 4π-detector systems [1]. In this paper a new technique is used to identify the atomic number of the incident particles in the detector and to calibrate the light response of CsI(Tl) [2]. A neural network is used to compute the correlation between the energy loss in the first stage of the detector (ΔE) and light response of CsI (L) and then obtain in output the atomic number (Z) and the residual energy (E) of particle identified in the telescope.

NET RESULTS

A feedforward architecture was chosen because it is able to associate as input and output two different vectors (ethero – associative net). The Kolmogorov theorem [3] states that a net with only one hidden layer is able to approximate any function in L_2 with whatever accuracy, other hidden layers are often added to reduce the number of nodes with little loss in the accuracy. After some trials, we adopted a configuration

CP513, *Nuclear and Condensed Matter Physics,* edited by A. Messina
© 2000 American Institute of Physics 1-56396-929-7/00/$17.00

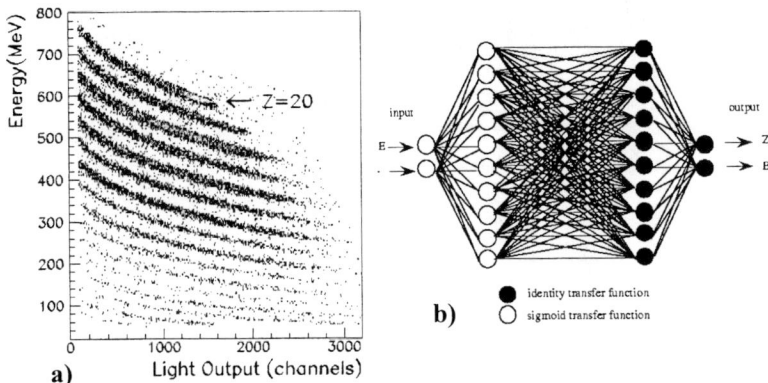

a) Light Output (channels)

b) identity transfer function / sigmoid transfer function

FIGURE 1. a) ΔE versus Light matrix from silicon-CsI(TLC) detector telescope. b) Net Architecture. Filled nodes correspond to sigmoid activation function; empty nodes correspond to identity activation function.

with 10 nodes in each of the two hidden layers (see figure 1b). It is a trade-off between complexity and accuracy. To perform our analysis we used data from the $^{40}Ca + ^{40}Ca$ reaction at 25 MeV*A collected with the TRASMA apparatus [4]. The single telescope module consists of silicon strip detectors, 300 μm thick, as first stage, followed by a 6 cm thick CsI(Tl) crystal (with a photodiode readout) as second stage. All the telescope modules are placed at forward angles. To properly calibrate the ΔE stage different reference points were used: the elastic scattering point of the $^{40}Ca + ^{40}Ca$ at 25 MeV*A and the maximum energy released in the ΔE stage for each Z (punch-through). These last points were identified using the ΔE-Time identification scatter plot. Training was performed with 340 patterns; each pattern consists of four parameters; the energy of ΔE stage and the light output of CsI are input values; the atomic number Z and the residual energy E are the output values. The latter has been evaluated using an energy loss program, the others are taken from the ΔE-Light identification scatter plot (figure 1a). About 20 points were considered for each Z. Another set of 250 patterns has been used for verification during the learning step.

Net performance related to the atomic number recognition is shown in figures 2a and 2b, where all data sets coming from the ΔE-Light scatter plot have been treated. In order to better evaluate this separation we show in figure 2b the charge obtained from the projection of the Z versus energy scatter plot. A constant value marks the distance between two adjacent peaks and a sharp separation is obtained for the whole Z range.

To check the validity of the residual energy output obtained with our neural network approach a good calibration of CsI detector performed in a conventional way is necessary. The standard way to perform this, essentially consists of two steps:
- selection, for each atomic number, of some points with a well known energy;
- performing a fit using the most appropriate parameterization.
The first step often requires several specific runs of measure, also using different kinds of reactions. Alternatively, where high-energy resolution is not required, the ΔE stage

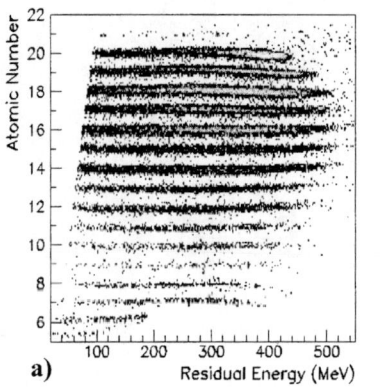 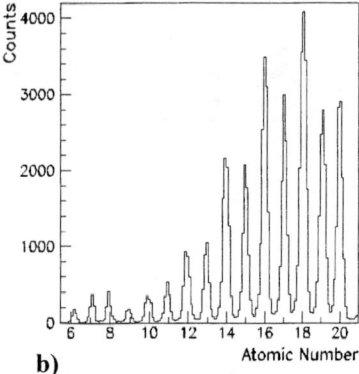

FIGURE 2. a) Atomic number Z versus residual energy E evaluated by the net. b) Z axis projection of the Z versus E matrix.

calibration can be used. For this work only the ΔE stage calibration was available. Concerning the second step, the determination of parameters covering all the Z values in the whole energy range is difficult. This is strictly related to the detector's geometrical shape and construction, but it also depends on the accuracy of the known energy reference points. Often some different parameterizations are needed to cover the whole E and Z ranges of the detected particles.

In this work, to compare the results from the neural network approach with the standard calibration procedure, several fits were performed using some of the relationships reported in literature [5,6,7,8]. The fits have been performed for each Z and the best chi-square has been found using the following relation [8]:

$$E = a(Z) + b(Z) \cdot L + \frac{c(Z)}{d(Z) + L} \tag{1}$$

where a, b, c and d are the free parameters of the fit, E is the residual energy and L the light output of the CsI(Tl). The energy reference points used for the fit are the same used for the learning step. Figure 3 shows the particle spectra obtained from the conventional calibration procedure (filled histogram) and the particle spectra obtained with our new neural network approach (normal histogram). A good agreement between the two approaches has been found once again. However, the neural approach is more general compared to the standard fit procedure. In fact, for each charge, the fit procedure gives different values for the fit parameters.

CONCLUSIONS

In this work a new approach based on a neural network to Z classification and energy calibration has been investigated. The sharp separation obtained in the whole range between two following Z values, makes this technique suitable for automatic

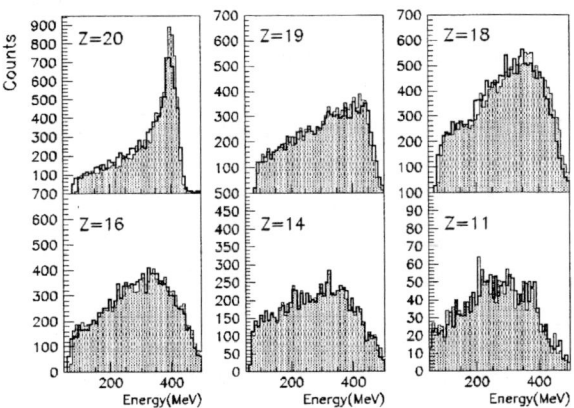

FIGURE 3. Particle energy spectra for different Z values. Comparison between net (filled histogram) and energy calibration (normal histogram) results.

selection procedures with no graphic contour. This can reduce the time required for the off-line analysis, particularly in the 4π detector systems. If many detectors have the same characteristics, the learning step can be performed only once for one of them and the resulting net can be used for the whole set of detectors. The quite good resolution of particle energy spectra suggests a widespread use in off-line applications. On the other hand, this approach can be also used as a second level-on line trigger where a high precision is not required. For instance, the construction of the total energy and/or the total charge is useful to select the multi-fragmentation events. The energy resolution achieved by the net can be improved using dedicated calibration runs giving more accurate energy reference points for the learning step. This could lead alternatively to a simpler net configuration or to a more accurate performance compared to the current one.

REFERENCES

1. S. Aiello et al., *Nucl. Phys.* B583 (1995) 461c.

2. C.M. Iacono Manno, and S. Tudisco; in press on *Nucl. Instr. and Meth. A*

3. R. Hecht-Nielsen, Neurocomputing , Adaptive Pattern Recognition and Neural Networks, Addison-Wesley Publishing Company-1989

4. A. Musumarra et al., *Nucl. Instr. and Meth.* A370 (1996) 558.

5. C.J.W Twenhofel et al., *Nucl. Instr. and Meth.* B51 (1990) 58.

6. D. Horn et al., *Nucl. Instr. and Meth.* A320 (1992) 273.

7. P.F. Mastinu et al., *Nucl. Instr. and Meth.* A338 (1994) 419.

8. B. Ocker Ph.D Thesis, pp-Korrelationen in der Reaktionen ^{197}Au + ^{197}Au bei einschubenergien von 100,150 und 200 AMeV. DISS 98-22 November 1998 GSI.

The Role Of Clusters
In 4N Self-Conjugated Nuclei

A.Tumino [1,2], O. Goryunov [4], M. Lattuada [1,2], V. Ostashko [4], S. Romano [2],
C. Spitaleri [2,3], D. Vinciguerra [1,2]

[1]Dipartimento di Fisica, Università di Catania, Catania, Italy
[2]Istituto Nazionale di Fisica Nucleare-Laboratori Nazionali del Sud, Catania, Italy
[3]Dipartimento di Metodologie Fisiche e Chimiche per l'Ingegneria, Catania, Italy
[4]Institute for Nuclear Research, Academy of Sciences of Ukraine, Kiev, Ukraine

Abstract. We have performed an experimental study on the $^{12}C + {}^{12}C$ and $^{16}O + {}^{12}C$ interactions with the aim of exploring the structure of clusters in 4N self-conjugated nuclei. The populated exit channels, like $^{12}C + \alpha + {}^{8}Be_{gs}$ in the $^{12}C + {}^{12}C$ interaction and $^{12}C + 4\alpha$ in the $^{16}O + {}^{12}C$ one, feed on through the formation of excited states of ^{12}C, ^{16}O and ^{24}Mg which are described in terms of clusters of their decay products.

INTRODUCTION

The description of nuclei in terms of α clusters has a long history, dating back to the birth of nuclear physics itself, even before the discovery of neutron. Since then, the idea of clustering has been developed in many models on the attempt to solve difficulties imposed by the many body problem applied to the study of 4N self-conjugated nuclei; today there is a wealth of experimental information on these nuclei which suggests the need of clustering concept to account for some nuclear properties the Shell Model fails in reproducing. Recently, some unexpected results at high excitation energy, where the high level density should forbid the formation of any correlated structure, as the cluster one, have aroused a renewed interest. A wide resonance at $E_{c.m.} = 32.5$ MeV and $\Gamma = 5$ MeV, was observed in the excitation function for the $^{12}C + {}^{12}C \rightarrow {}^{12}C(0_2{}^+) + {}^{12}C(0_2{}^+)$ reaction [ref.1]; some models developed to account for the peculiar characteristics of this resonance [ref. 2] assumed the existence of a very stretched configuration in ^{24}Mg, that was described as a linear chain of six α-particles. Moreover the evidence of decay through formation of the mentioned level of ^{12}C, whose association to a system of three loosely bound α-particles is known, was a strong argument in favour of this description. The observation of a similar resonant behaviour, at the same energy, in other decay channels, involving poorly deformed configurations as the $^{16}O_{gs} + {}^{8}Be_{gs}$ one we studied [ref. 3-4], claimed the need to assume at this excitation energy a mixing of different configurations with different degrees of deformation. But the hypothesis of the α-chain cannot be ruled out; it could

CP513, *Nuclear and Condensed Matter Physics*, edited by A. Messina
© 2000 American Institute of Physics 1-56396-929-7/00/$17.00

break into intermediate chains as the one of two α's, which corresponds to the ground state of ^{8}Be, and a 4α-chain. The Cluster Model predicts the existence of such a configuration in ^{16}O above the threshold of decay into four α's (14.4 MeV); this configuration is expected to be associated with a rotational band with a band head around 17 MeV. There is some experimental evidence of few members of this band drawn from the study of the α + ^{12}C → ^{8}Be$_{gs}$ + ^{8}Be$_{gs}$ reaction [ref. 5].

Study Of Three Body Channels In The ^{12}C + ^{12}C Interaction

We performed an experimental study of ^{12}C + ^{12}C interaction at a beam energy corresponding to the top of the wide resonance in ^{24}Mg with the aim of looking at other possible decays of the resonance towards rotational states of ^{16}O [ref. 6]. The experiment was performed at the LNS in Catania using the SMP Tandem Van de Graaff accelerator to produce a beam of ^{12}C at 65 MeV. The experimental set-up consisted of dual detectors placed on both sides of the beam direction and covering the angles between 12° and 55°. Coincidences between a dual detector and at least one half of another were recorded in order to involve at least a ^{8}Be stored in all events. ^{8}Be was identified by the off-line reconstruction of the relative energy of the two particles detected in the two parts of each dual detector, under the assumption of equal masses; a sharp peak centred at about 90 keV shows up, corresponding to the ground state of ^{8}Be. For coincidence events involving two ^{8}Be in their ground state, the Q-spectrum is drawn in figure 1; in spite of the poor statistics a very resolved peak is present at around −15 MeV. It corresponds to the formation of the ^{8}Be$_{gs}$ + ^{8}Be$_{gs}$ + ^{8}Be$_{gs}$ channel. Unfortunately it was not possible to study the correlation between particles in the final state, and thus to identify the ^{16}O states involved in the process, because of the insufficient statistics. As concerns coincidence events between a ^{8}Be and an unidentified particle detected in one part of a dual detector, all the kinematics were reconstructed by assuming a given mass for this particle. Figure 2 shows the two Q-spectra obtained under the assumption of mass 4 and mass 12 respectively for the unidentified particle. Of course they reflect the excitation spectrum of ^{12}C. In both spectra there is evidence of production of ^{12}C in its ground state and in the 2_1^+ state at 4.43 MeV of excitation energy, corresponding to the two levels under the particle decay threshold; moreover the spectrum of figure 2a shows that the undetected ^{12}C is produced also in the 0_2^+ state at 7.65 MeV and in the 3_1^- one at 9.63 MeV, both above the mentioned threshold. The 3_1^- state is described in terms of three loosely α-particles placed at the corners of an equilateral triangle; as mentioned above, a similar description is given for the 0_2^+ one. These two levels feed the ^{8}Be + 4α channel, but the poor statistics prevents us from observing any decay from ^{16}O states. For decay channels involving stable particle states of ^{12}C it was possible to study the correlation between final particles; the ^{12}C - α relative energy was plotted as a function of the ^{8}Be − α relative energy for events falling under each one of the two Q peaks corresponding to ^{12}C$_{gs}$ and ^{12}C(2_1^+). In this representation a vertical alignment of events corresponds to production and decay of ^{12}C states while events falling on horizontal straight lines reveal the formation and decay of ^{16}O states.

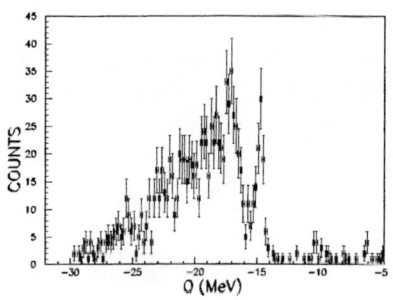

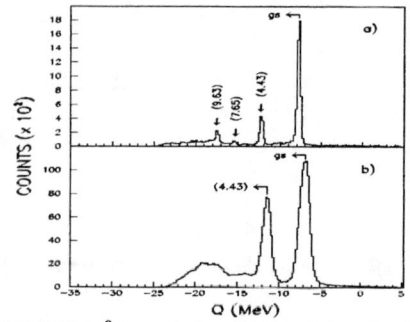

FIGURE 1. $^8Be_{gs}$-$^8Be_{gs}$ coincidences

FIGURE 2. $^8Be_{gs}$- unidentified particle coincidences

Examples of such representation are shown in figure 3; besides two vertical lines due to ^{12}C break-up from the 0_2^+ and the 3_1^- states, horizontal ridges show up, revealing the existence of a two steps process through formation of $^{16}O^*$ + $^8Be_{gs}$ system subsequently decaying into ^{12}C + α + $^8Be_{gs}$ channel. Several ^{16}O states appear to be excited in the reaction: excitation energies of about 10.6, 11.3, 14.3, 15 and 16.5 MeV show up and decay into $^{12}C(0_2^+)$ from a state at 18.2 MeV even occurs. Some of these states lie above the 4α threshold (14.4 MeV) but there is no clear correspondence with the members of the 4α chain band. These members have been recognised to decay into four α's through formation of unbound ^8Be or $^{12}C^*$ [ref. 7], having a much higher deformation than that of ^{12}C particle stable states.

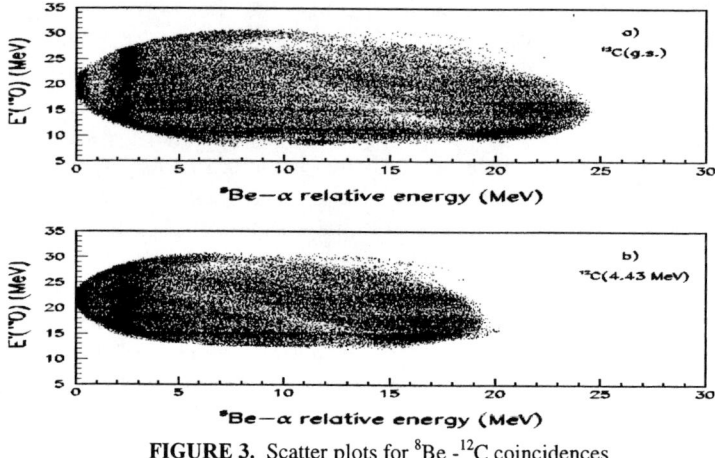

FIGURE 3. Scatter plots for ^8Be -^{12}C coincidences

The ^{16}O + ^{12}C Interaction To Study The ^{16}O Rotational Bands

We performed a second experiment at LNS in Catania to study the ^{16}O + ^{12}C interaction at 109 MeV with the aim to cover the same range of ^{16}O excitation energy and extend its upper limit. Our detection system consisted of a triple telescope on one side with respect to the beam direction, composed of an ionisation chamber and two position sensitive detectors (PSD), and a system of four arrays on the other side, each one containing twelve strips. We looked at the coincidences between an identified particle detected in the telescope and at least a particle in one strip. Some preliminary results bear witness the population of the 4α decay channel; figure 4 shows the Q spectrum obtained from the selection of events corresponding to the identification of a ^{12}C in the telescope in coincidence with three unidentified particles detected in the strips. The Q value is calculated under the assumption of mass 4 for all the particles detected. Two sharp peaks are drawn: the first peak centred at 14.4 MeV corresponds to the ^{12}C$_{gs}$+ 4α channel while the second one regards the excitation of ^{12}C at the 2_1^+ level (4.43 MeV). From the study of correlation between the 4α's a sequential decay through ^{8}Be+^{8}Be or ^{12}C*+α channels seems to be present; moreover these channels are fed on through the decay of an ^{16}O nucleus, produced in the first step of the reaction. As it is shown in figure 5 a wide excitation energy range for ^{16}O is involved in the process; work is in progress to identify states belonging to the 4α chain band.

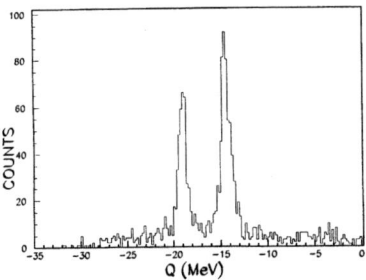

FIGURE 4. ^{12}C – 3α coincidences

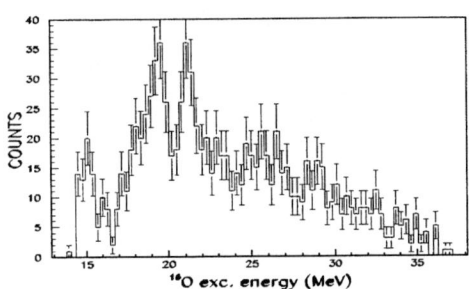

FIGURE 5. ^{12}C – α coincidences

CONCLUSIONS

Some decay channels populated in ^{12}C - ^{12}C and ^{16}O - ^{12}C interactions have been analyzed. These data bring further information on the decay modes of the wide resonance at 46.4 MeV in ^{24}Mg; the decay through unbound ^{16}O states adds to the known ones. Some of these states are described in terms of poorly deformed cluster systems, while some other above the threshold of decay in 4α's seem to belong to the 4α chain rotational band. The degree of deformation of cluster configurations involved plays an important role to understand the resonant properties in the different channels.

REFERENCES

1. A.H. Wuosmaa et al., *Phys. Rev. Lett.* **68**, 1295 (1992)
2. S. Marsh and W.D.M. Rae, *Phys. Lett. B* **180**, 185 (1986)
3. M. Aliotta et al. *Z. Phys. A* **353**, 119 (1996)
4. M. Lattuada et al., *Il Nuovo Cimento A* **110**, 1007 (1997)
5. P. Chevallier et al., *Phys. Rew.* **160**, 827 (1967)
6. E. Costanzo et al., *Eur. Phys. J. A* **5**, 69-75 (1999)
7. A.H. Wuosmaa, *Z. Phys. A* **349**, 249 (1994)

Neutron Production in Coincidence with Fragments from the $^{40}Ca + H$ Reactions at $E_{lab} = 357$ and 565 A MeV

C. Tuvè[1], S.Albergo[1], D. Boemi[1], Z.Caccia[1], C.-X.Chen[3], S.Costa[1], H.J.Crawford[4], M.Cronqvist[2], J.Engelage[4], L.Greiner[4], T.G.Guzik[3], A.Insolia[1], C.N.Knott[5], P.J.Lindstrom[2], J.W. Mitchell[8], R.Potenza[1], G.V.Russo[1], A.Soutoul[7], O.Testard[7], A.Tricomi[1], C.E.Tull[3], C.J.Waddington[5], W.R.Webber[6] and J.P.Wefel[3]

[1] University of Catania, INFN and CSFNSM- I - 95129 Catania, ITALY
[2] Lawrence Berkeley National Laboratory, Berkeley, CA, USA
[3] Louisiana State University, Baton Rouge, LA, USA
[4] Space Science Laboratory, University of California, Berkeley, CA, USA
[5] University of Minnesota, Minneapolis, MN, USA
[6] University of New Mexico, Las Cruces, NM, USA
[7] Service d'Astrophysique, C.E.N. Saclay, FRANCE
[8] NASA/Goddard Space Flight Center, Greenbelt, MD,USA

Abstract. In the frame of the Transport Collaboration neutrons in coincidence with charged fragments produced in the $^{40}Ca + H$ reaction at $E_{lab} = 357$ and 565 $AMeV$ have been measured at the Heavy Ion Spectrometer System (HISS) facility of the Lawrence Berkeley National Laboratory, using the multifunctional neutron spectrometer MUFFINS. The detector covered a narrow angular range about the beam in the forward direction ($0^{o} - 3.2^{o}$). In this contribution we report absolute neutron production cross sections in coincidence with charged fragments ($10 \leq Z \leq 20$). The neutron multiplicities have been estimated from the comparison between the neutron cross sections, in coincidence with the fragments, and the elemental cross sections. We have found evidence for a pre-equilibrium emission of prompt neutrons in superposition to a 'slower' deexcitation of the equilibrated remnant by emission of nucleons and fragments, as already seen in the inclusive rapidity distributions.

INTRODUCTION

We have studied the reaction $^{40}Ca + H$ at 357 and 565 A MeV. The experiment was performed at the Heavy Ion Spectrometer System (HISS) facility of Lawrence Berkeley National Laboratory. The complete experimental setup is described in detail in [1]. We used a liquid hydrogen target 0.254 ± 0.004 g/cm^2 thick. The

CP513, *Nuclear and Condensed Matter Physics*, edited by A. Messina
© 2000 American Institute of Physics 1-56396-929-7/00/$17.00

Neutron Spectrometer called MUFFINS was located on the incident beam line away from the charged particle detectors. MUFFINS is a modular detector made of several individual discs of NE102A plastic scintillator [2,3]. We have measured the energy spectra, angular and rapidity distributions and total inclusive cross sections for neutrons emitted in these collisions. Some interesting insights on the reaction mechanism have been obtained by looking at the rapidity distributions [4].

In this paper we will present neutron production in coincidence with charged fragments. The neutrons were detected in coincidence with fragments of charge $Z \leq 20$ identified as the charge "islands" seen in the scatter plots of the raw analog-to-digital converter (ADC) response for the two charge measuring detectors located immediately after the target.

RESULTS

The neutron cross sections are reported in the upper panels of Fig. 1 as a function of Z for the two beam energies. In the lower panels of Fig. 1 we report the corresponding elemental production cross sections for the same reaction and energies [5,6]. Also reported in each panel of Fig. 1 are the corresponding theoretical

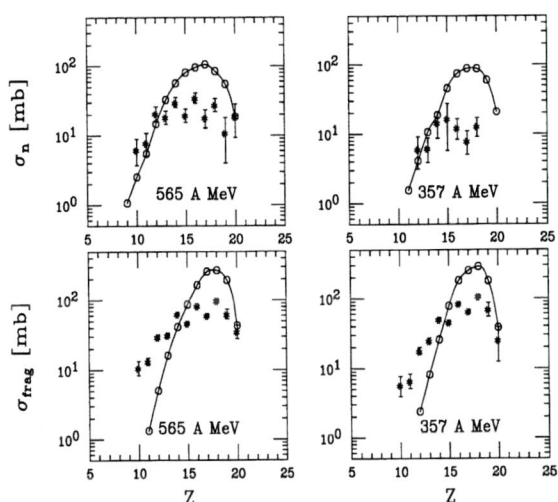

FIGURE 1. Upper left panel: Measured neutron cross sections σ_n (asterisks) for the $^{40}Ca + H$ reaction at $E_{lab} = 565 \ A \ MeV$, in coincidence with charged fragments, compared with the BNV + phase space coalescence model (circles joined by a solid line). Upper right panel: same for $E_{lab} = 357 \ A \ MeV$. Lower panels: Elemental production cross sections σ_{frag} (asterisks) for the same reaction and energies in comparison with calculations (circles joined by the solid line).

327

cross sections, calculated in the frame of the Boltzmann - Nordheim - Vlasov approach (BNV) [7]. It is worthwhile to note that the cross section for $Z = 20$ is a measure of the neutron stripping cross section in the considered reaction. At 565 $A\ MeV$ we have detected neutrons in coincidence with fragment charges in the range $20 \leq Z \leq 10$, while at 357 $A\ MeV$ the coincidence was taken with fragment charges in the range $18 \leq Z \leq 12$, due to lower statistics. For $Z \leq 9$ (at 565 $A\ MeV$) or $Z \leq 11$ (at 357 $A\ MeV$) it was impossible to identify the fragment charge and consequently the neutron cross section in coincidence with these fragments. The experimental neutron production appears to be too small with respect to the number of free neutrons allowed by the size of the remnant fragment detected in coincidence. We have inferred the mean neutron multiplicity M_n vs. the remnant charge through the ratios between neutron and elemental cross sections $M_n = \sigma_n / \sigma_{frag}$, at the two beam energies. In Fig. 2 we report M_n vs Z_f. σ_{frag} has been obtained summing up all isotopic cross sections of Table I of ref. [5], but excluding, for $Z=17$, 18, and 19, the case in which a fragment with $N = 20$ has been produced. The mean neutron multiplicity shows an increasing trend as Z decreases. Lower multiplicities are found at the lower energy. However, as it is

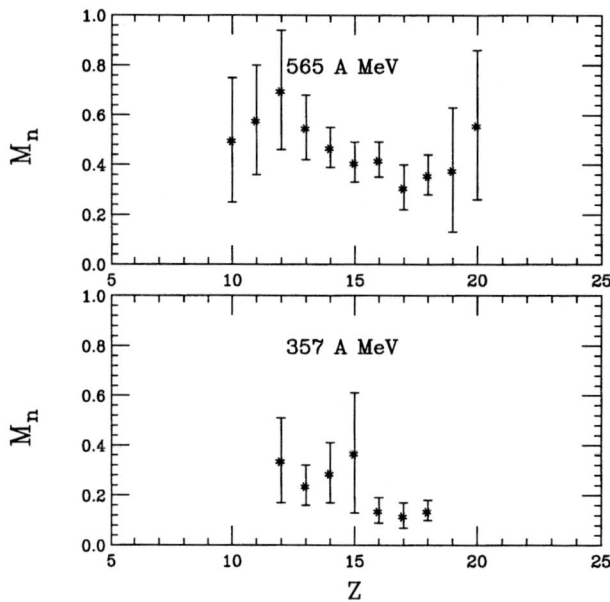

FIGURE 2. Mean neutron multiplicity versus Z for the $^{40}Ca+H$ reaction at $E_{lab} = 565\ A\ MeV$; (b):the same at 357 $A\ MeV$.

possible to infere from the data plotted in Fig. 2, the neutron multiplicity is always much smaller than the "missing neutrons", defined as the number of neutrons necessary to form a ^{40}Ca from the remnant.

We cannot explain this observed "neutron defect" as due to clustering of neutrons in light fragments, particularly for the ones detected in coincidence with high-Z fragments, that correspond to peripheral collisions in which there are very few available nucleons, with unfavorable phase space conditions, to allow the necessary rate of light fragment emission. On the other hand, due to reverse kinematics, we expect most of the neutrons emitted by the remnant, which travels with a rapidity close to the beam rapidity, to enter the geometrical acceptance of MUFFINS. A reasonable way to explain the observed neutron defect seems to be the existence of a double reaction mechanism [8]. So, as suggested also by other recent studies of reverse kinematics nuclear collisions [9], we interpret our data as the result of a two-step reaction: the "neutron defect" is caused by those neutrons emitted in a pre-equilibrium stage. Since the system is not yet thermalized and so the energy is not shared between a large number of degrees of freedom they can take a larger fraction of the transverse momentum and escape out of the MUFFINS angular coverage, effectively reducing the neutron multiplicity. Around $0°$ we detect on the contrary mainly neutrons emitted by the excited remnant, at low energy in the source frame, which show indeed the rapidly decreasing p_t distribution already discussed [4].

In conclusion, we have presented new neutron production data in coincidence with fragments emitted in the reaction $^{40}Ca + H$ and suggested an interpretation for the inferred mean multiplicity and energy dependence of the neutron cross sections. Our data show evidence of a two-step reaction mechanism already discussed in connection with very asymmetric nuclear collisions [9,10]. We interpret the observed neutron defect as due to a pre-equilibrium emission of energetic neutrons that escape from the angular coverage of our detector. The neutrons detected around $0°$ are, therefore, mainly emitted by the remnant in the second step of the reaction.

REFERENCES

1. Albergo, S., et al.,Transport Collaboration,*Radiation Measurements* **27**, 549 (1997)
2. Albergo, S., et al.,*Nucl. Instr. Meth.* **A311**, 280 (1992).
3. Albergo, S., et al.,*Nucl. Instr. Meth.* **A362**, 423 (1995).
4. Tuvè, C., et al., Transport Collaboration,*Phys. Rev.* **C 56**, 1057 (1997)
5. Chen, C.-X., et al., Transport Collaboration, *Phys. Rev.* **C 56**, 1536 (1997)
6. Knott, C.N.et al., Transport Collaboration, *Phys. Rev.* **C 53**, 347 (1996)
7. Bonasera, A., Gulminelli, F. and Molitoris, J.,*Phys. Rep.* **243**, 1 (1994)
8. Tuvè, C., et al., Transport Collaboration,*Phys. Rev.* **C 59**, 233 (1999)
9. Hauger, J.A., et al., EOS Collaboration, *Phys. Rev.* **C 57**, 764 (1998)
10. Mahi, M., et al., *Phys. Rev. Lett.* **60**, 1936 (1988)

MISCELLANEOUS

Analysis Of The Electroretinogram In Normals And Pathological Subjects[+]

M. Anastasi, G. Andaloro, R. Barraco, L. Bellomonte, M. Brai

Università di Palermo e Unità INFM, Palermo, Italy

Abstract. The Electroretinogram (ERG) is a composite signal reflecting the complex response of the retinal different parts. It is known that the rod and cone systems contribute to this response. The most used method to analyze the data is the study of the b-wave amplitude and peak time change against the intensity of the stimulus theoretically investigated by an approach based on a simplified picture of a single responding photoreceptorial population in scotopic or photopic condition. The aim of the present paper is to investigate the dependence of the amplitude and the implicit time of the a- and b-wave of the ERG on the input luminance level to obtain hints devoted to recognize the different contribution to the response of the two systems, rods and cones, in normals and pathological subjects.

INTRODUCTION

The scotopic ERG is a composite signal that reflects the response of different parts constituting the retina. The clinical ERG is the recording of electrical potentials evoked by a flash of light and picked up at the cornea. It consists of various components which arise in different layers of the retina and it has different features if the signal originates in the cones or in the rods. Particularly, the rods saturate at luminance level of about 3-4 cd m^{-2}, while there is no evidence of a contribution of the cones to the scotopic ERG below 10^{-3} cd m^{-2} .

The main ERG components linked to the photoreceptor activity are the a- and b-waves, described by their amplitudes and implicit times. Both the parameters show modifications depending upon the quality and level of the stimulus. For instance, under fully dark-adapted conditions, their values are determined by the responses of the rod related units, but the modifications at increasing intensity could reflect the activity of both rod and cone systems. The dynamics related to the change of amplitudes and implicit times of the individual retinal units are not completely understood. It is supposed that the a-wave is generated extracellularly along a radial path from the cell body of the photoreceptor and consists of two components arising from the cones and rods. The b-wave is believed to be generated by the bipolar cells in the inner nuclear layer in some way driven by the Müller cells[1]. Regardless of where

---

[+] Partially supported by Comitato Regionale Ricerche Nucleari e Struttura della Materia (CRRNSM)

it is generated, the b-wave is the major component of the ERG. It reflects the post-synaptic summed neuronal activities of the inner nuclear layer and thus is an important measure.

We, here, discuss the data of the amplitude and implicit time of a- and b-waves in normal subjects and in patients with pathologies selectively affecting either or both the rod and the cone systems: congenital stationary night blindness (CSNB) and rod monochromatism (RM) in which the scotopic and the photopic systems are severely damaged; ERGs of subjects affected by retinitis pigmentosa in which both the photoreceptorial populations suffer, at different extent, modification.

MATERIALS AND METHODS

We studied a group of 32 subjects so arranged: 20 normal subjects; 6 patients affected by retinitis pigmentosa (RP), 3 autosomal dominant RP (ADRP), 3 autosomal recessive RP (ARRP); 3 patients affected by rod monochromatism (RM); 3 patient with congenital stationary night blindness (CSNB).

The scotopic ERG was detected by means of Henkes corneal electrodes, amplified (1-100 Hz band pass) and stored for later retrieving. The repetition rate of the stimulus was 15 s and its duration was 1.5 ms. Each record was obtained by averaging three responses. The b-wave amplitude was determined from the trough of the a-wave to the peak of the b-wave. The a- and b-wave implicit times were determined by measuring the time between the beginning of the record triggered by the stimulus to the peaks of the a- and b-wave.

We obtained the stimulus intensity-response function by a best fit of the Naka-Rushton (NR) equation to the experimental data,

$$R(I) = R_{max} \frac{I^n}{(I^n + k^n)} \tag{1}$$

In equation (1) I is the light intensity of the stimulus, R(I) represents the response amplitude to an impulse of luminance I, R_{max} is the saturation value of the response, k measures the required luminance to obtain a response with one-half amplitude of R_{max}, n is a dimensionless slope variable.

According to current models any curve fitting the temporal aspect of the retinal responses to increasing luminance must decrease steadily from an initial value τ_0 to a minimal value τ_c. The dynamics can be explained by double saturation law functionally represented by the Gompertz function (GZ):

$$T(I) = \tau_c * e^{(\beta \exp(-1/I_T))} \tag{2}$$

where T is the b-wave implicit time and I is the stimulus intensity. Provided one homogeneous photoreceptorial system is activated, τ_c is the minimal delay obtained at the saturation level of the population of the coupled cells; $\tau_0 = \tau_c * e^\beta$ corresponds to the implicit time at the threshold luminance intensity. The function (2) has only a point of inflexion at the value of luminance $\log I = \log I_T$.

RESULTS AND DISCUSSION

Table 1 reports the mean values of the fitting parameters appearing in eqs. (1) and (2). In all cases the fits were quite good. The S.E. of the amplitude parameters were less than 10% for normals and of the order of 20-30% for the pathological groups.

The patients affected by RP (ADRP and ARRP) present a reduction of R_{max} without significant shift of logk. In RM, the maximal amplitude is lightly reduced and logk shows a clear displacement towards the low luminance levels; the CSNB NR maximal amplitude is dramatically reduced, the curve is displaced towards the high luminance levels.

TABLE 1. Parameters of amplitude and time fitting curves

Group	R_{max} (μV)	k (ulog)	τ_c (ms)	τ_0 (ms)
normals	488±17	-1.85±0.06	51.5±5	103±10
RM	194 – 446	-2.15±0.1	60±5	118±17
CSNB	83 – 260	-0.9±0.3	43±2	69±5
ADRP	86 – 456	-1.6±0.5	54±6	104±10
ARRP	208 - 454	-1.7 ±0.1	58±7	108±4

The scattergram, figure 1, shows an increase of the amplitude and a decrease of the delay of the a- and b-wave peaks at increasing luminance level. This dependence is not monotonous but there is a systematic jump at about -1.6 ulog relative intensity. This evidence suggests that two systems are responding, one incoming at low luminance and the other involved at higher light intensity.

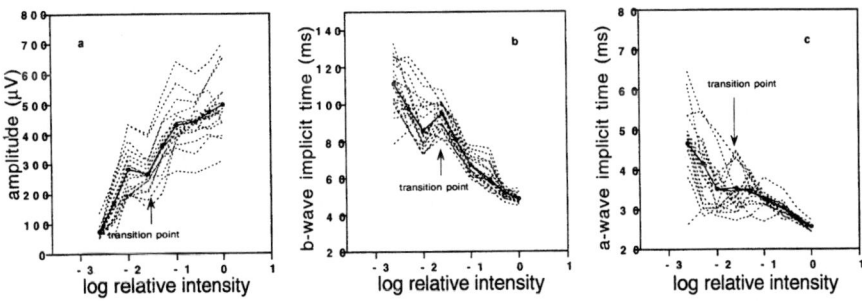

Figure 1. Scattergram of data of the normal amplitude (a) and the a- and b-wave implicit times (b, c). The dashed lines in the figure represent the interpolation curves of 9 amplitude and implicit time values of the 20 normal subjects, the continuous line is the curve interpolating the mean values.

In our experiments the standard condition of the intensity of the luminance was 1.3 cd s m^{-2}; at this value of luminance there is an overlapping region where both photoreceptor populations are activated. The ERG, hence, would consist of signals belonging to the activated rods at the lowest intensities and to both rods and cones at higher intensities. At high levels of flash intensity the retinal response shows an initial saturation followed by a new increase indicating the activation of another system. At low intensities, each photoreceptor acts as an isolated detector of light quanta and the time constant (RC) is due to the single resistivity and capacity of each uncoupled unit. At increasing luminance, the unit coupling activates and the RC decreases until all photoreceptorial units are in speaking terms and its value remains constant.

The compared study of the b-wave amplitude and of the a- and b-peak time vs stimulus intensity gives integrated information about the rod/cone interactions. The selected pathologies help to further confirm the above models. In fact, the RM[2] is a congenital disease in which the cones are functionally inactive even if they are anatomically well represented in the retina. The CSNB[3] seems to be due to a defect of transmission at the level of the rod/bipolar synapses. The RP[4] is a disease affecting both rods and cones, heterogeneous from the point of view of the heredity, of the rhodopsin gene abnormality, of the clinical picture and the natural course.

The GZ model is well supported by a non linear system in which the latency is linked to the K^+ ion currents and to the membrane capacity. In RM, the GZ parameters are associated to the rod model, hence $\tau_0=118$ ms corresponds to the RC of one rod. The minimal delay, obtained at the saturation level, $\tau_c=60$ ms, gives the equivalent RC of n rod-type response generators coupled. At more intense light, the ERG is no longer pure scotopic since the cone system contributes to the response. In CSNB, we estimate the parameters of the cone-type response generators ($\tau_0=69$ ms, $\tau_c=43$ ms). The data obtained in ADRP support the hypothesis that in autosomal dominant RP, the surviving individual generators are relatively normal, the damage arising from an even loss of rods and cones. In fact, the temporal parameters have values comparable to the normal ones and the NR parameters indicate only a decrease of the responsiveness, due to loss in the number of photoreceptors, while a non significant shift in logk is observed. Unlikely, the ARRP data of the amplitude versus light intensity show a homogeneous distribution of photoreceptor loss on the retinal surface. In conclusion, our data is in good agreement with the hypothesis about the alteration of the activation stages of the phototransduction in RP. We suggest the deconvolution of the analysis of the ERG data in two curves, low luminance, rod dominated, and high luminance, cone dependent region

REFERENCES

1. Falk, G. (1991). Retinal physiology. In Heckenlively, J.R. & Arden, G.B. (eds) Principles and practice of clinical electrophysiology of vision. (pp 69-84). St Louis Mosby-Year Book.

2. Heckenlively, J.R. (1991) Cone dystrophy and dysfunction. n Principles and practice of clinical electrophysiology of vision. Heckenlively, J.R. and Arden, G.B. eds. Mosby publisher. 537-543.

3. Carr, R.E. (1991). Congenital stationary night blindness. In Principles and practice of clinical electrophysiology of vision. Heckenlively, J.R. and Arden, G.B. eds. Mosby publisher. 713-720.

4. Berson, E.L. (1993). Retinitis Pigmentosa. Investigative Ophthalmology & Visual Science, 34, 1659-1676.

LEM - Electromagnetic Fields Measurement Laboratory

A. Annino, F. Falciglia, F. Musumeci, M. Oliveri, G. Privitera*, A. Triglia

*Dipartimento di Metodologie Fisiche e Chimiche per l'Ingegneria - * Borsista CRRNSM*
Università degli Studi di Catania - Viale A. Doria 6 - I 95125 Catania
Work supported by CRRNSM - E mail: atriglia@dmfci.ing.unict.it

Abstract. The widespread presence of electromagnetic waves and the relative problems regarding them have favoured the constitution of the LEM at the DMFCI in Catania University, where competence has been developing in this sector for about 10 years.
Full operativeness has been reached as far as the electromagnetic field measurements in anthropized environments are concerned. Other research will be undertaken as soon as further funds are available.
Some problems connected with the perfecting of measurements instruments and the results of emission measurements of cellular telephones are presented.

In 1999, at the Physical and Chemical Methodology Department of the Engineering Faculty of Catania University (DMFCI), it has been set up the Electromagnetic Field Measurement Laboratory (LEM), prevalently dealing with the measurement and control of the electromagnetic radiation presence on the territory.

The preliminary conditions that conduced to the LEM constitution are connected on the one hand, to the demand of people and the scientific community of more careful control of radiation levels, on the other, to the research interests and competence present inside the DMFCI. In fact, the researchers who founded the LEM do their research and teaching on electromagnetic subjects 1,2,3]. Especially research was conducted on the biological system and electromagnetic interaction, the use of electromagnetic fields in greenhouse cultivation and monitoring and cataloging of electromagnetic waves on Sicilian territory.

The analysis of specific needs connected to the territory in particular and national and international issues, concerning the ever increasing presence of electromagnetic fields and their effects on biological system, have brought the authors to determine the following purposes and aims for the LEM:

electromagnetic monitoring in anthropized environment;

full compliance measurement;

training of specialized technics and advisors on electromagnetic field measurement;

information center on the regulations regarding field measurement, health and risk due to the biological system electromagnetic field exposure;

promotion and realization of research on biological and medical effects of electromagnetic fields.

Due to the vastness and multiplicity of the purpose, it has been given priority to the territory measurement of electromagnetic fields, full operativeness has been gained in

CP513, *Nuclear and Condensed Matter Physics,* edited by A. Messina
© 2000 American Institute of Physics 1-56396-929-7/00/$17.00

this sector with the acquisition of instruments and operative methodologies, as specified in the present regulations (D.M. 381/98, ICNIRP, CENELEC, etc...). The completion of the other purposes is connected with the possibility of obtaining adequate specific funds and human resources.

Below some of the preliminary results perfecting methods of measurement and instruments will be reported; moreover will be presented the variability, in the axial direction, of the electrical field emitted from a cellular phone as function of the relative distance between the source (the cellular phone) and the probe.

Measurements were conducted on standard ETACS and GSM mobile telephones. The telephone and the meter were positioned at a 3.5 m distance from the ground level to reduce the reflection from the floor as much as possible. The transmission time was about 60 seconds.

The measurement was performed with the EMR-300 supplied by Wandel & Goltermann. This instrument gives the integral of all electric fields present at the measurement point in the range of frequency between 100 kHz to 3GHz. For this reason it has been already checked that the possible interference effect due to the presence of other sources were below the noise level. The meter was connected to a computer by an optical fibre for data acquisition.

As the probe was calibrated for continuous electrical fields, in the case of GSM telephones which use pulsed fields, the data has been corrected by a factor k = 1.59 provided by the manifacturer. In this case, the measured values, as presented by the instrument, are modified according to the k factor correction. The error, occurring during the acquisition of the electric field values, supplied by the manufacturer, was ∆E= ±1.40 V/m.

Below will be shown the electrical field values emitted by a GSM, as a function of the calling time (fig. 1) and of the distance between the source (GSM) and the probe (fig. 2).

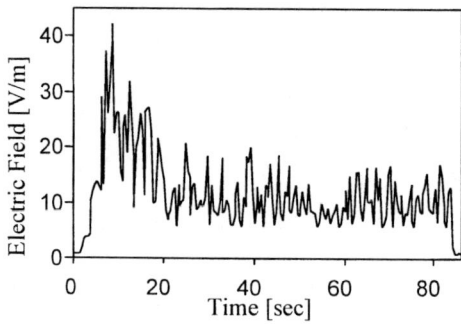

FIGURE 1. Electric field values varying as a function of the calling time at a distance of 5 cm between the source and the probe.

As shown in fig. 1, in the first interval time (from 10 to 20 seconds), the intensity of the electric field emitted by the telephone reaches the highest values, $\approx$ 45 V/m; then it becomes lower, ≈ 25 V/m. This variation is due to the fact that the telephone fits its emission power as function of the distance from the nearest base station.

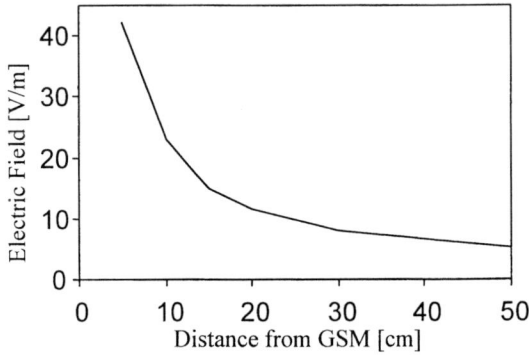

FIGURE 2. Maximum electric field emitted from a GSM telephone as function of distance.

In fig.2 the values of the maximum electric field reached during the first 10 seconds as a function of distance are displayed. We report measurements at the distances of 5, 10, 15, 20, 30 and 50 cm. The distance between the signal source and the probe was not inferior to 5 cm because the probe geometry does not allow lower distances. We can see that the maximum electric field decreases from ≈ 45 V/m, at a 5 cm distance, to ≈ 5 V/m, at a 50 cm distance.

According to Italian D.M. n° 381/98 the electric field limit must be inferior to 20 V/m. The values must be calculated as an average on a period of six minutes and over an equivalent area to the transversal section of the human body.

The ICNIRP Standard establishes that the reference electric field level for pulsed fields results ≈1300 V/m [4].

CENELEC Standard establishes that the electric field level for pulsed fields ≈ 1200 V/m [5].

Therefore, standard GSM mobile telephone measurements result substantially lower than the electric field reference levels fixed by ICNIRP and CENELEC.

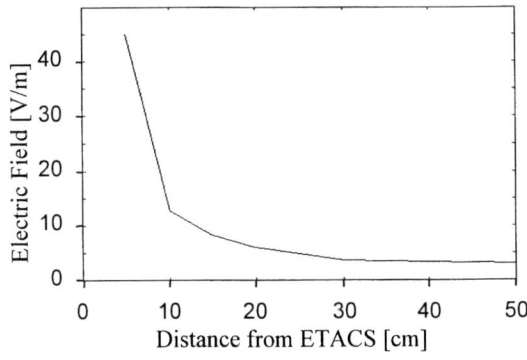

FIGURE 3. Electric field emitted from an ETACS telephone as function of distance.

In the fig.3 the electric field intensity generated by a standard ETACS mobile telephone is shown. In this case it is not necessary the correction of the measured values because the ETACS uses a modulation phase technique for transmission. The range value varies from ≈ 45 V/m at a 5 cm distance to ≈ 3 V/m at a 50 cm distance.

In this case the ICNIRP and CENELEC Standards establish an electric field reference level equal to ≈ 41 V/m [4,5].

Therefore close to the telephone the field values are slightly higher than ICNIRP and CENELEC reference levels and twice those indicated in the Italian legislation, 20 V/m.

The results show the existence of remarkable level of the electrical field close to the cellular telephones and emphasize one aspect of the issue as regards the risks deriving from the biological system exposition to electromagnetic fields. So research should continue along these lines.

REFERENCES

1. Van Wijk, R., Musumeci, F., Scordino, A., and Triglia, A., *Journal of Photochemistry and Photobiology B* **49**, 142-149 (1999)
2. Musumeci, F., Scordino, A., Triglia, A., Blandino, G., and Milazzo, I., *Europhysics Letters* **47**, 736-742(1999)
3. Ciccazzo, A., "Confronto tra le normative per la protezione dalle NIR e misure dei livelli di campi elettromagnetici esistenti" Tesi di Laurea - Facoltà di Ingegneria - Università di Catania - A.A 1998/99
4. ICNIRP, "Guidelines for limiting exposure to time varying electric, magnetic and electromagnetic fields (up to 300 GHz)", Health Physics Society, 1998
5. CEI ENV50166-2, "Human exposure to electromagnetic fields - High frequency (10 kHz to 300 GHz), edited by CEI, Milano, Italy, 1995

Radon as Tracer of Diffusive Motions in Atmosphere

A. Biagi, N. Giudice, G. Immé, <u>D. Morelli</u>, S. Urso

Dipartimento di Fisica dell'Università di Catania
CSFNSM-Catania
INFN- Sezione di Catania

Abstract. We report on continuous measurements of hourly concentration of Radon in outdoor air, carried out in Catania in the 1997-1998 period The equipment allows determination of Radon concentration using α–particle spectrometry. The result show dependence on the meteorological conditions and especially on atmospheric stability. A typical behavior of Radon trend with a maximum near the sunrise and a minimum in the late afternoon is found for each month of the year. A good correlation is found between the variation of concentration of Radon and air pollution like CO. The Radon variation can be a valid index of dispersive power of atmosphere. A theoretical study is developed for confirming experimental observations.

INTRODUCTION

The dispersive power of atmosphere is highly conditioned by meteorological and thermodynamic conditions of the lower atmospheric layers: horizontal transport is due to the anemological field, vertical diffusion instead is governed by turbulent phenomena that can be both thermic, due to free convection: rising of air particles owing to the heating of the soil; and mechanical, forced convection: due to the friction between ground and moving air masses, as well as to the presence of obstacles along the path.

The determination of dispersive power is then a complex problem owing to the self-complexity of atmospheric phenomena and to the difficulty connected with the determination of turbulent conditions both trough experimental measures and computational modeling. Limiting investigation field to the dispersion due to thermoconvective turbulent processes is possible to study these processes in connection to the temporal variation of gas Radon and its daughters[1,2]. We report on continuous measurements of radon variation, carried out daily in the urban area of Catania. They have shown the radon concentration as a suitable index of air vertical diffusion, that we can use to validate atmospheric dispersion models.

CP513, *Nuclear and Condensed Matter Physics*, edited by A. Messina
© 2000 American Institute of Physics 1-56396-929-7/00/$17.00

EXPERIMENTAL PROCEDURE FOR MONITORING RADON

The ^{222}Rn, noble gas and then chemically inert, is free to diffuse or to be transported without loose its physical characteristics. After the production in soil or rock, the Radon can leave terrestrial crust for molecular diffusion or convection and enter into atmosphere, where, for decaying, gives rise to its daughters that, chemically active, cling to aerosol particles in atmosphere, and therefore meteorological processes govern their behavior and distribution.

^{222}Rn concentration can be determined by its daughters α–particle spectrometry. For such purpose a device for Radon monitoring was carried out at the Physics Department of Catania. Outdoor sampling was done at height of 3 meters above the ground level. The system consists of an automatic air sampling device and a chamber in which an aspiration filter and a silicon detector are located. The system is optimized to assure, trough an electronic regulator of air flux, a constant flux of 3 l/min. The particulate where Radon's daughters are attached is collected on a Millipore filter (with pore diameter of 0.8 micron and aspiration surface of 314 mm^2) that tolerates long aspiration without clogging. The silicon detector is located in front of the filter at a distance of 12 mm, to minimize the energy lost in the air interposed between them and to maximize the geometric efficiency. The signals, trough standard electronics, are carried out via a multi-channel analyzer card to a personal compute. Hourly storage of the α-spectra is automatically made.

Results and Discussion

With the hypothesis of constant air flux during the sampling time, the Radon daughters concentration is determined assuming their radioactive equilibrium in air, instead Radon concentration is calculated assuming an unbalance among it and its daughters in atmosphere. The trend of Radon hourly concentration measured in several periods shows a great variability both daily and monthly. On daily scale the monthly mean day shows in all periods, a characteristic trend in which there is a maximum concentration in the early morning and a minimum in the late afternoon.

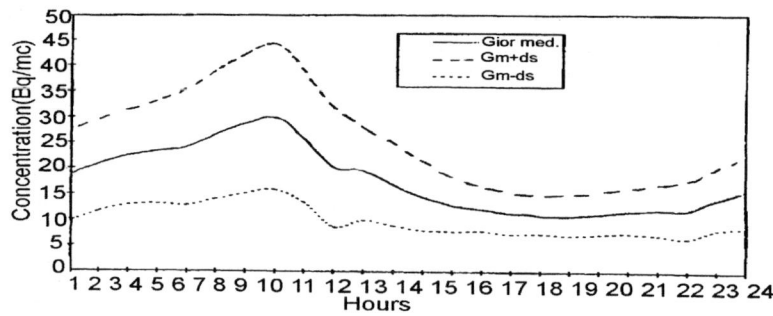

FIGURE 1. ^{222}Rn concentration and its standard deviation

346

The trend, known as "Radon wave" is shown in fig.1.

In order to compare the radon concentration trend with other pollutants[3], we have choice to compare it with the daily trend of CO concentration, which source is at ground level (traffic) and moreover it is characterized by long time of persistence in atmosphere, and reaction velocity not very high. From the ratio (Fig.2-a) between daily time mean trend of CO concentration and that of traffic, a trend is point out that qualitatively well reproduces the characteristics of Radon daily time concentration trend. We have also considered the nightly concentration of CO meaning between 22:00 and 05.00 of morning and the variation in radon concentration between the minimum of afternoon and the maximum of the following morning, obtaining a good correlation (R = 0.87) (Fig.2-b).

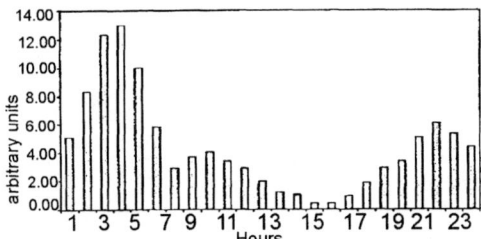 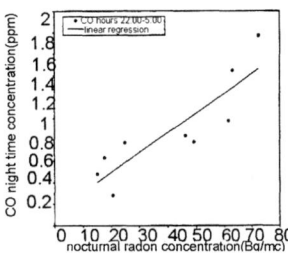

FIGURE 2. a) Ratio between CO concentration and urban traffic fraction, b) Correlation between CO nocturnal concentration and nocturnal radon concentration

DISPERSION MODELS

Computational models have been formulated [4] for studying atmospheric diffusion. All the models considered start from the solution of the diffusion equation:

$$\frac{\partial c}{\partial t} = \nabla \cdot (\mathbf{k} \cdot \nabla c)$$ (1)

where c is the concentration and $\mathbf{k}$ is the diffusion coefficient. If a pollutant is released continuously from a point source at ground level at a rate Q, with the hypothesis of uniform atmosphere, steady wind of constant velocity u and with crosswind and vertical gaussian distributions of concentration, the solution is

$$c(x, y, z) = \frac{Q}{\pi \sigma_y \sigma_z} \exp\left[-\frac{1}{2}\left(\frac{y^2}{\sigma_y^2} + \frac{z}{\sigma_z^2} \right) \right]$$ (2)

where horizontal and vertical standard deviations depend on atmospheric stability. Using the Pasquill atmospheric classification, they can be determined trough the Briggs formulation[5].

347

Considering a cloudless day a daily typical cycle of atmospheric stability starts with an adiabatic, locally superadiabatic, lapse rate during the daytime due to the solar heating, which is associated with well-developed turbulence. One or two hours before sunset, the air near the soil begins to cool, then the lapse rate decreases and a surface inversion develops. This inversion reaches the maximum near sunrise. After sunrise the ground is warmed by the sun, the inversion is gradually destroyed and superadiabatic lapse rates again appear. With these consideration is possible to simulate the daily concentration trend of a pollutant emitted at ground level. In fig.3 is shown a typical trend, that is very similar to the Radon concentration one.

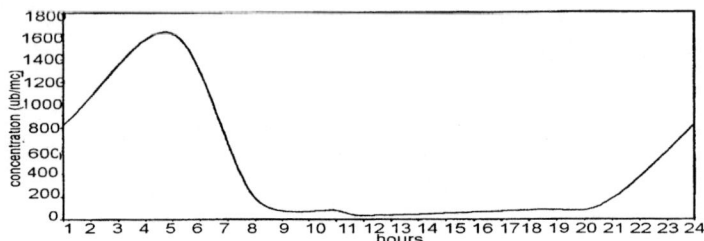

FIGURE 3. Simulated daily concentration trend.

CONCLUSIONS

This study had as aim the investigation of suitability of radon as a tracer of atmospheric vertical diffusion. Continuous measurements proved it to be a good method to estimate atmospheric dispersive power. It was also shown that the radon concentration could be used to confirm the validity of theoretical models on the atmospheric diffusion.

REFERENCES

1. Kataoka T., Tsukamoto O., Yunoki E., Michihiro K., Sugiyama H., Shimizu M., Mori T., Sahashi K., and Fujii S., *Variation of ^{222}Rn Concentrations in Outdoor Air Due to Variation of Atmospheric Boundary Layer, Radiation* Protection Dosimetry Vol. 45, 1992, pp. 403-406

2. Porstendorfer J., Buttrweck G., Kesten J., and Reineking A., *The Use of Natural Radioactive Noble Gases Radon and Thoron as Tracer for the Study of Turbulent Exchange in the Atmospheric Boundary Layer – Case Study in and Above a Wheat Field,* Atm. Env., 1994, pp. 1963-1969

3. Marcazzan G. M., Mantegazza F., and Astesani R., *Radon as Indicator of Atmospheric Dispersion: Correlation with Carbon Oxide Level at Milan*, Harwood Acad. Publ. GmbH, 1995, pp. 151-158.

4. Zannetti P., *Air Pollution Modeling,* Computational Mechanics Publications, 1990

5. Tirabassi T., *Analytical Air Pollution Advection and Diffusion Models,* Water Soil and Air Pollution,, 1989,Vol 47, pp 19-24

Pulse Phase Resolved Spectroscopy of the Cyclotron Line in Centaurus X-3: Magnetic Field Structure of the Neutron Star.

L. Burderi[1], T. Di Salvo[2], N. R. Robba[2], A. La Barbera[2]

[1] *Osservatorio Astronomico di Roma, via Frascati 33 - 00040 Monteporzio Catone, Roma, Italy*
[2] *Dipartimento di Scienze Fisiche ed Astronomiche, Università di Palermo, via Archirafi 36 - 90123 Palermo, Italy*

Abstract. We report spectral analysis of the X-ray pulsar Centaurus X-3 out of eclipse observed by BeppoSAX. The spectrum (0.12-100 keV) is described by an absorbed power-law with a high energy rollover plus an iron emission line, and a soft excess. An absorption feature at ~ 30 keV is interpreted as a cyclotron line. Phase resolved spectroscopy reveals asymmetric variations of the cyclotron resonance energy that is decreasing along the main peak in the pulse profile, starting from the ascent. This can be explained with an offset of the dipolar magnetic field with respect to the neutron star centre.

The X-ray source Cen X-3 is a High Mass X-ray Binary with a spin period of ~ 4.8 sec and an orbital period of ~ 2.1 days [2]. The 1–40 keV X-ray spectrum has been fitted by a power law with a high energy cutoff, iron line and soft X-ray absorption [5]. The analysis of *Ginga* data [3] showed that the high energy cutoff is better modeled by a Lorentzian due to cyclotron resonant scattering, rather than by the usual e-folding energy cut–off. However this fit does not provide firm evidence of the cyclotron resonant scattering because the energy of the line (~ 30 keV) is very close to the upper end of the *Ginga* LAC detectors. The presence of a cyclotron line at ~ 30 keV has been recently confirmed [4].

Observations and Analysis

BeppoSAX observed Cen X–3 on 1997 February 27 and 28. The source was out of eclipse in the last 30 ks of the observation. We used these data for the analysis. We performed standard temporal analysys obtaining the intrinsic pulse period $P_0 = 4.8146 \pm 0.0005$. Figure 1 shows the pulse profiles in different energy bands. The high energy part of the spectrum (above ~ 10 keV) is not well fitted

CP513, *Nuclear and Condensed Matter Physics*, edited by A. Messina
© 2000 American Institute of Physics 1-56396-929-7/00/$17.00

by an absorption feature alone (as in the analysis of *Ginga* data [3] . The broader band of the BeppoSAX instruments shows that both an absorption feature and an exponential cutoff are needed to model the spectrum. We fitted this spectrum using a power law with a high energy cutoff plus a blackbody to model a soft excess, modified at low energies for absorption by cold matter and multiplied by an absorption cyclotron line of Gaussian shape, plus a gaussian iron emission line, see Table 1. We extracted phase resolved spectra in the energy interval 1.8 − 100 keV. The four phase intervals are shown in Figure 1 , panel c). The data below 1.8 keV were excluded because of their low statistics, and the spectra were fitted with the same model used for the whole data set without the blackbody component that is not significant above 1.8 keV. The detection of the cyclotron line was not statistically significant in the phase interval 0.6 − 1.1 and we fixed the energy and the width of the line to the values obtained for the pulse averaged spectrum to derive an upper limit on the equivalent width. On the other hand we clearly detected the line in the remaining phase intervals. The centroid of the line shows a strong asymmetric dependence on the pulse phase having its maximum in the ascent and decreasing along the pulse profile. In Figure 2 we plot the ratio between the data and the continuum (that does not include the cyclotron line) as derived from the best fit. We interpreted the absorption feature at energies ~ 30.6 keV as cyclotron scattering by a magnetic field of $B = 3.47 \times 10^{12}$ Gauss, after correction for gravitational redshift. The phase resolved spectra show a strong dependence of the line energy on the pulse phase. In principle small variations (up to few keV) of the cyclotron energy with the angle between the magnetic axis and the line of sight are expected [1] that are symmetric with respect to the peak of the pulse profile. On the other hand we found variations in the cyclotron energy up to 8 keV, from ~ 36 to ~ 28 keV, that are asymmetric along the pulse profile. In particular the line energy decreases along the peak starting from the ascent. This behaviour can be explained assuming that the centre of the magnetic dipole is offset with respect to the neutron star centre. In this case the strength of the surface magnetic field decreases along the magnetic cap originating the observed asymmetric line variations. In Figure 3 we plotted the surface magnetic field strength, as function of the pulse phase, for a magnetic dipole inclined ($20°$) with respect to the spin axis and displaced ($0.1\ R_{NS}$) with respect to the neutron star centre. These values are in good agreement with the measured cyclotron energies.

REFERENCES

1. Mészáros, P., Nagel, W., 1985, ApJ, 298, 147.
2. Nagase, F. 1989, PASJ, 41, 1.
3. Nagase, F., *et al.*, 1992, ApJ, 396, 147.
4. Santangelo, A., *et al.*, A&A, 340, L55.
5. White, N. E., Swank, J. H., Holt, S. S. 1983, ApJ., 270, 711.

TABLE 1. Results of the fit of the pulse phase averaged spectrum in the energy range 0.12-100 keV. Uncertainties are at the 90% confidence level for a single parameter. Power-law normalizations are in unit of 10^{-2} ph keV^{-1} cm^{-2} s^{-1} at 1 keV.

Parameter	Value
$N_H (\times 10^{22}$ cm$^{-2})$	1.95 ± 0.03
kT$_{BB}$ (keV)	0.110 ± 0.012
BB Normalization	$0.26^{+0.29}_{-0.13}$
Photon Index	1.208 ± 0.007
E_{cut} (keV)	13.79 ± 0.13
E_{fold} (keV)	8.39 ± 0.19
W$_{smooth}$	3.1 ± 0.3
Normalization	0.708 ± 0.008
E_{cyc} (keV)	30.6 ± 0.6
σ_{cyc} (keV)	5.9 ± 0.7
EW$_{cyc}$ (keV)	11.4 ± 2.0
E_{Fe} (keV)	6.666 ± 0.025
σ (keV)	0.302 ± 0.035
I_{Fe} (ph cm^{-2} s^{-1})	$(7.7 \pm 0.5) \times 10^{-3}$
Eq. Width (eV)	107
χ^2/d.o.f.	651/646

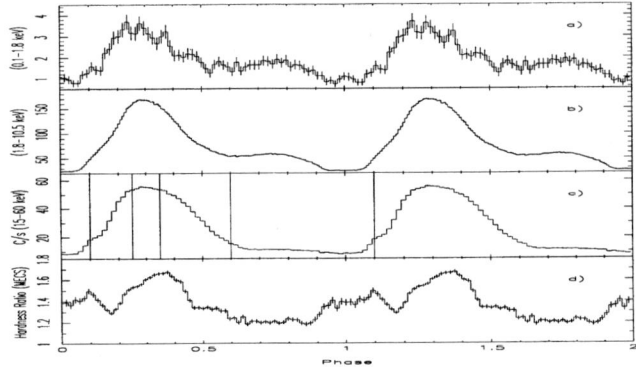

FIGURE 1. Panel a): Cen X-3 light curve folded at the spin period reported in the text (corrected for the orbital motion) in energy range 0.1–1.8 keV. Panel b): Same as a) in the energy range 1.8–10.5 keV. Panel c): Same as a) in the energy range 15–60 keV. The four phase intervals used to extract the pulse phase resolved spectra are also shown. Panel d): Hardness ratio 4-11 keV/1-4 keV.

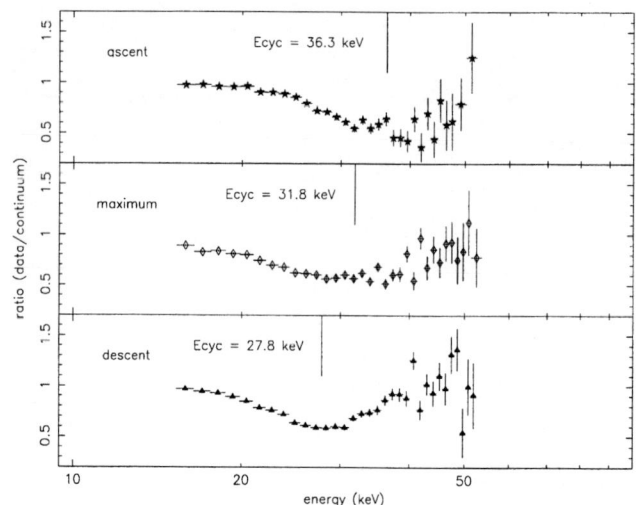

FIGURE 2. Ratio of the total flux over the flux of the continuum (the best fit model without the cyclotron line) as function of energy in the phase intervals ascent (upper panel), maximum (middle panel) and descent (lower panel) of the pulse profile (c.f. Figure 1). For each phase interval, the energy of the cyclotron line is also shown.

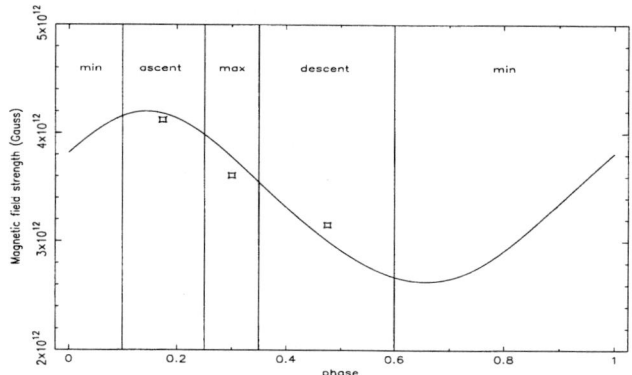

FIGURE 3. The solid line represents the magnetic field strength as a function of the pulse phase for a magnetic dipole offset by 0.1 R_{NS} with respect to the center of the neutron star. The data points are the measured magnetic field strength as derived from pulse phase resolved spectral analysis. Typical errors on the estimate of the magnetic field strength are $\pm0.15 \times 10^{12}$ Gauss. The intervals refers to the pulse shape shown in Figure 2, panel c).

The gamma-ray pulsar *Geminga*: radio quiet or radio loud?

L. Burderi[1], F. Fauci[2]

[1] *Osservatorio Astronomico di Roma, via Frascati 33 - 00040 Monteporzio Catone, Roma, Italy*
[2] *Dipartimento di Scienze Fisiche ed Astronomiche, Università di Palermo, via Archirafi 36 - 90123 Palermo, Italy*

Abstract. For long time *Geminga* has been considered a radio-quiet isolated neutron star. Recently, a weak pulsed radio emission from this source at ~ 100 MHz has been reported. We observed *Geminga* with the Arecibo radio telescope at frequencies between 318 and 1400 MHz. No pulsed emission was detected. The high sensitivity of the Arecibo telescope allowed us to put an upper limit to the spectral index in the radio band: $\alpha \leq -3.6$.

Introduction

After the detection by the ROSAT satellite of X-ray pulsation with a period of ~ 237 ms [1], and the subsequent detection of pulsed γ-ray emission at the same period by EGRET observations [2], and archival COS-B data [3], *Geminga* has been recognized as an Isolated rotating Neutron Star (INS). No radio emission has been detected up to 1997 [4], [5], [6], [7]. The spin period and its derivative have been determined with timing analysis on EGRET data over a period of ~ 1.5 yr [8], and combined COS-B/EGRET data over a period of ~ 24 years [9]. A recent classification of the Isolated Neutron Stars (ISN) is based on the presence of pulsed radio emission [10]. The class of Radio-Loud ISN has > 700 isolated radio pulsar detected up to now. The class of Radio-Quiet ISN has seven candidates of which only RXJ 0002+6246 and *Geminga* show pulsations in the X-ray band. The X-ray emission is of thermal origin at $\sim 5.6 \times 10^5$ °, emitted from part of the surface of a cooling neutron star [11]. In the optical range, the presence of a spectral feature has been interpreted as an ion (H or He) cyclotron frequency caused by a NS surface magnetic field of $\sim 3 - 5 \times 10^{11}$ G (for pure Hydrogen ions) in line with the value deduced from the dynamical parameters of the pulsar [12]. The presence of such a strong field, would have made the lack of pulsed radio emission problematic, for spin period and magnetic field strength that are typical of radio pulsars, although some quenching mechanisms for the radio pulses have been proposed [13]. The

CP513, *Nuclear and Condensed Matter Physics*, edited by A. Messina

Radio-Quiet nature of *Geminga* has been challenged by the reported detection of a weak pulsed radio emission at 102.5 MHz [14], [15], [16].

Observations and Analysis

The observations were done at the Arecibo Observatory in the period 1982 - 1984 at 318, 430, and 1400 Mhz. The sampling times were 1, 2, or 8 msec. In most observations the back end of the receiver consisted of two filter banks (one for each antenna polarization) each with 32 filter channels (Band Width 250 kHz). The data were barycentrized adopting the position estimate at the epoch of the observation RA(2000): 6h 33m 54.16s DEC(2000): +17° 46' 11.6". This position differs from the recent, accurate to $\sim$ 40 mas estimate [19] by ΔRA:0.007s and ΔDEC:1.31". This difference produce a phase shift in the folded pulse profile $\sim 10^{-3}$ times less than the used phase bin. The signals from different filter channels same-frequency pairs were added togheter using a dedispersion algorithm to correct for the arrival time delay determined by the different frequencies of each filter channel. The dispersion space was searched in a range 0.0 – 10.0 pc·cm^{-3} (step 2 pc·cm^{-3}), compatible with the expected distance of $\sim$ 150 pc [17]. The data were folded modulo a series of trial periods around the period expected from the timing analysis on combined COS-B/EGRET data taking into account the period derivative, which seems to be stable over $\sim$ 23 years [9]. The ephemeris used were $T_0 = 2,446,600$ JD, $P = 0.2370957461262(7)$ s, $\dot{P} = 1.0974012(6) \times 10^{-14}$ s s^{-1}, $\ddot{P} = -8.3(2) \times 10^{-27}$ s s^{-2}, where T_0 is the epoch, P, $\dot{P}$, and $\ddot{P}$ are the period and its first and second derivative respectively, and the numbers in parenthesis are the 95% error on the last digit. For each trial period a χ^2 was computed. As an example, in Figure 1 the χ^2 *vs.* trial periods curve is shown for the data set at 425 MHz (for which the sensitivity is the highest). Because the period step adopted in the folding, $\Delta P = 6.25 \times 10^{-7}$, is bigger than the error in the period expected at the epoch of observations (taking into account the errors in the estimate of the period derivatives), the χ^2 curve was expected to have its maximum in one channel, namely that corresponding to the predicted period at the epoch of the observation. As we can see in Figure 1, no statistically significant periodicities were detected in any of the data sets. The upper limits for each frequency together with the 60 mJy detection at 102.5 MHz are reported in Figure 2. Because of the intensity variations reported [15], we assigned an error bar from 5 to 500 mJy to the 60 mJy detection at 102.5 MHz. In the same figure we show the less steep power-laws compatible with our upper limits and a 102.5 MHz detection at a level of 500 and 5 mJy respectively. According to our analysis the upper limit for the spectral index is in the range $-7.3 \leq \alpha \leq -3.6$.

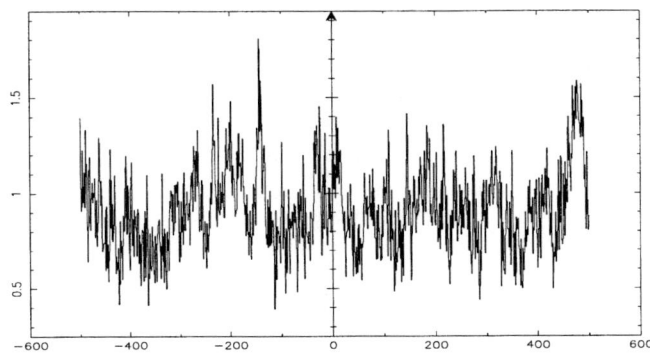

FIGURE 1. χ^2 *vs.* trial frequencies curve for the data set at 425 MHz. The frequency were the pulsation of *Geminga* was expected is the central one (indicated with 0). In the abscissa is indicated the quantity $(\nu_{trial} - \nu_{expected})/\nu_{step}$, where ν_{trial} is the frequency at wich the data were folded, $\nu_{expected}$ is the frequency at which the *Geminga* pulsations were expected, and $\nu_{step} = 1.7 \times 10^{-5}$ is the adopted frequency step. In the ordinate is reported the normalized χ^2.

Discussion and Conclusion

No pulsed emission was detected from *Geminga* in the period 1982-1984 at various radio frequencies between 318 and 1400 MHz with a high sensitivity level. Thus we are left with one of the following possibilities:

i) The detection of pulsed radio emission at 100 MHz ([14], [15], [16]) will be confirmed. This implies that *Geminga* is genuinely Radio-Loud although its radio spectrum is exceptionally steep: spectral index $-7.3 \leq \alpha \leq -3.6$ while for more than 500 radio pulsar detected up to now all the spectral indexes are ≥ -3 [20]. Such a steep spectrum have to be explained by any pulse emission model.

ii) The detection of pulsed radio emission at 100 MHz will not be confirmed. *Geminga* is a Radio-Quiet INS. Radio quenching by copious $e^\pm$ production in the case of an highly inclined rotator seemed a viable explanation for the lack of radio emission [13].

iii) The radio quenching by $e^\pm$ pairs in *Geminga* is variable, consistent with claims that features in the pulse profiles (dips) associated with pair production are variable [11]. In this scenario the quenching plasma occasionally clears away, causing the radio emission episodes observed [15].

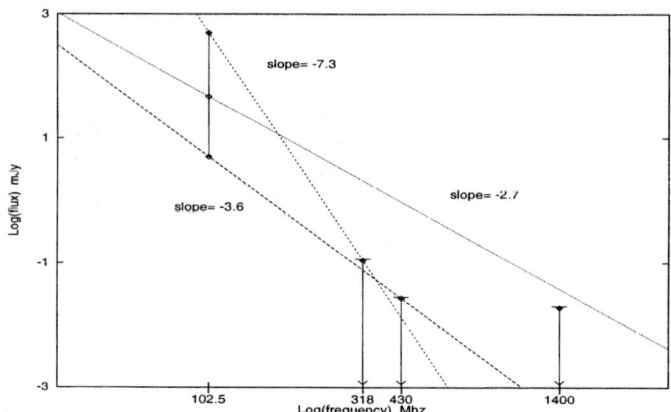

FIGURE 2. Detections (Malofeev and Malov, 1997) and upper limits (this work) in different frequency bands for the *Geminga* pulsar. The dotted lines are the less steep spectra compatible with the upper limits and a 102.5 MHz detection at 500 and 5 mJy respectively. The dashed line is a typical pulsar spectrum (spectral index 2.7).

REFERENCES

1. Halpern, J. P., Holt, S. S., 1992, *Nature*, 357, 222.
2. Bertsch, D. L., *et al.*, 1992, *Nature*, 357, 306.
3. Bignami, G. F., Caraveo, P. A., 1992, *Nature*, 257, 287.
4. Bignami, G. F., Gavazzi, G., Harten, R. H., 1977, *A&A*, 54, 951.
5. Manchester, R. N., Taylor, J. H., 1981, *Astron. J.*, 86, 1953.
6. Fauci, F., Boriakoff, V., Buccheri, R., 1984, *Nuovo Cimento*, 7C(6), 597.
7. Spoelstra, T. A. Th., Hermsen, W., 1984, *A&A*, 135,135.
8. Mayer–Hasselwander, H. A., *et al.*, 1994, *ApJ*, 421, 276.
9. Mattox, J. R., Halpern, J. P., Caraveo, P. A., 1998, *ApJ*, 493, 891.
10. Caraveo, P. A., Bignami, G. F., Trumper, J. E., 1996, *Astron. Astrophys. Rev.*, 7, 209.
11. Halpern, J. P., Wang, F. J. -H., 1997, *ApJ*, 477, 905.
12. Mignani, Caraveo, and Bignami, 1998, ?, ?, ?.
13. Halpern, J. P., Ruderman, M., 1993, *ApJ*, 415, 286.
14. Kuzmin, A. D., Losovsky, B. Ya., 1997, *IAU Circ.* No. 6559.
15. Malofeev, V. M., Malov, O., I., 1997, *Nature*, 389, 697.
16. Shitov, Yu. P., Pugachev, V. D., 1997, Preprint n. 33, Lebedev Physical Inst.
17. Caraveo, P. A., Bignami, G. F., Mignani, R., Taff, L. G., 1996, *ApJ*, 461, L1.
18. Bignami, G. F., Caraveo, P. A., 1996, *ARA&A*, 34, 331.
19. Caraveo, P. A., *et al.*, 1998, *A&A*, 329, L1.
20. Taylor, J. H., Manchester, R.N., Lyne, A. G., 1993, *Ap.J.Suppl.*, 88, 529

Air Quality Control in Confined Spaces by Means of Natural Ventilation

G. Cannistraro[1], G. Galli[1], C. Giaconia[2], S. Magazù[1] and A. Piccolo[1]

[1]Dipartimento di Fisica,Università degli Studi di Messina
Contrada Papardo, Salita Sperone, 31 – 98166 Messina, Italy
[2]Dipartimento di Energetica ed Applicazioni di Fisica, Università degli Studi di Palermo
Viale delle Scienze – 90128 Palermo, Italy

Abstract. In this work a simplified methodology for predicting the effectiveness of natural ventilation in assuring minimum ventilation rates for internal comfort is reported.

INTRODUCTION

Natural ventilation, so far applied to residential buildings, is quickly extending also to commercial buildings; this is mainly due to the energy saving expected from a reduced use of commercial devices which control forced ventilation. The reduction or total elimination of forced ventilation devices certainly entails an improvement of the indoor comfort due to the absence of mechanical noise and to the reduction of the bacterial pollution source associated with filters. The natural ventilation (depending on external atmospheric conditions) and the forced one (activated by means of mechanical devices) can both be employed in the same building equipped with an intelligent control system able to compute the minimum ventilation rates assuring a given indoor air quality and to establish if the required air flow rates may be obtained by suitable remote control of the openings before forced ventilation be activated.

In the present work a simple procedure for testing the efficacy of natural ventilation to meet the comfort air change requirements of an occupiable space of given characteristics is presented.

AIR QUALITY AND VENTILATION REQUIREMENTS

To maintain inside an occupied space a good air quality level, a certain ventilation rate of external air has to be supplied for diluting to acceptable concentrations the pollutants released by internal sources. Current standards[1] prescribe ventilation requirements on the basis of the following arguments:
– the occupants of a space are not the exclusive sources of pollution;
– materials in buildings contribute considerably to the total pollution load;
– the quality of outdoor air crucially determines the quality of the indoor air;
- each confined space is characterized by a ventilation effectiveness.

CP513, *Nuclear and Condensed Matter Physics*, edited by A. Messina
© 2000 American Institute of Physics 1-56396-929-7/00/$17.00

– the health risk associated with exposure to specific air pollutants must be negligible;
– the perceived air quality must be comfortable.

TABLE 1. Typical Pollution Loads.

Source	Pollution Load
Sedentary Person (1 met)	1 olf
Person in action (6 met)	11 olf
Smoker	6 olf
Offices	0.3 olf/m^2 floor

The Fanger[2] methodology for determining the minimum ventilation rates for internal comfort is based on the "olf" and "decipol" units which quantify the air pollution in the same way it is perceived by humans indoors and outdoors.

TABLE 2. Daily Average Values for the Perceived Air Quality

Environment	Perceived Air Quality
At Mountain and Sea	0 decipol
In Town with Excellent Air Quality	< 0.1 decipol
In Town with Discrete Air Quality	0.2 decipol
In Town with Poor Air Quality	> 0.5 decipol

The relation between the perceived air quality expressed in decipol (Tab. 2) and the percentage of dissatisfied unadapted persons (PD) is:

$$C_i = 112[\ln(PD) - 5.98]^{-4} \tag{1}$$

By normative, comfort air quality corresponds to PD ≤ 20%. This formula allows then to select the desired air quality level in a ventilated space to be used as an input in the relation giving the corresponding minimum air flow rate:

$$Q_c = 10 \frac{G}{C_i - C_o} \frac{1}{\varepsilon_v} \tag{2}$$

where Q_c is the ventilation rate required for comfort (l/s), G is the sensory pollution load (olf), C_i is the perceived desired indoor air quality (decipol) and C_o is the perceived outdoor air quality intake (decipol).

NATURAL AIR FLOW RATES ANALYSIS

As a model for our analysis a block shaped building has been considered (fig. 1). The geometrical and environmental characteristics selected (nor reported) simulate a multi-space office building situated in inner city areas with high terrain roughness and a consistent density of surrounding buildings. In particular, the sketch meets the normative requirements[1] for naturally ventilated spaces. For each room a comparison is made between the required ventilation for comfort, Q_c, and the flow rates Q_w

supplied by natural ventilation for assigned atmospheric conditions (angle of incidence, θ, and velocity, v, of external wind).

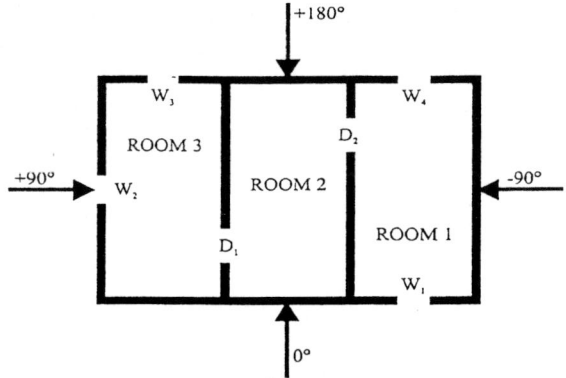

FIGURE 1. Sketch of the Examined Floor Building with the Layout of Windows and Doors.

The Q_w (θ, v) curves have been obtained utilizing the BREEZE simulation program (ver. 5.1) which requires as an input the pressure coefficients on the external envelope surface of the building; these last, in turn, have been calculated by means of the M. Grosso's alghoritms[3].

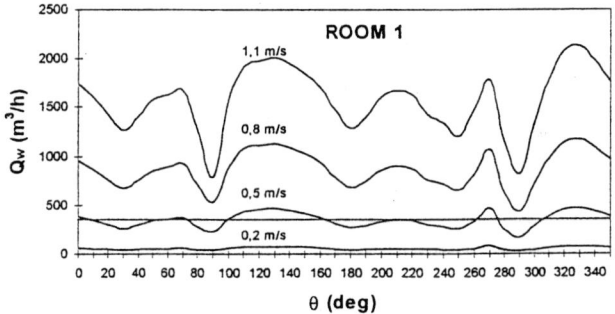

FIGURE 2. Ventilation Rates as a Function of Incidence Angle at Selected Wind Velocities.

The ventilation rate required for comfort is determined by using equation (1) and (2) with a maximum value of 20% of dissatisfied people. For values reported in Tab. 1 and 2 it results for each room Q_c=356 m³/h. In Fig.1 this value is compared with the flow rate Q_w for all incidence angles at some selected velocities. The ventilation behavior of room 2 and 3 is qualitatively analogous exhibiting a considerably dependence on the direction of the incidence wind. At some "preferred" incidence angles, in fact, the air change rates for comfort are achieved at relatively low wind velocities; for other angles, the comfort level is never attained even at the maximum wind velocities considered. In these cases, natural ventilation results ineffective to supply the design air flow rates and the mechanical ventilation must be started up

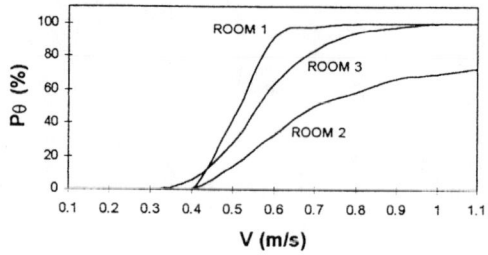

FIGURE 3. Percentage of wind Incidence Directions Producing Comfort Conditions as a Function of Wind Velocity

In Fig.3 the percentage of wind incidence angles producing comfort conditions are reported as a function of the wind velocity for each room. It results that the "response" of room 1 is the best while for room 2 the effectiveness of natural ventilation is largely reduced, lacking in this space direct openings to the outdoor. It is noteworthy to observe that, even if rooms 1 and 3 have the same number of windows and doors and the same value of open area to the outdoor, they behave in a different way against blowing wind. Evidently, the geometrical disposition of the windows constitute an important factor that must be carefully taken into account in the design phase. We note, at last, that the wind velocity v≈0.4 m/s represents a cut-off value for each room. For wind velocities equal or lower to this limit, natural ventilation fails in maintaining the required air flow rates.

CONCLUSIONS

The ventilation characteristics of a model building located in a specific surrounding environment have been studied. These characteristics have been used to predict the atmospheric conditions (wind incidence angle and wind velocity) under which natural ventilation may properly be employed in lieu of the mechanical one to assure comfort conditions for the occupants of the examined space. This approach is based on the well known comfort equation for indoor air quality and constitutes a simplified and powerful tool for optimizing the ventilation equipment of buildings.

REFERENCES

1. BSR/ASHRAE Standard 62-1989R. "Ventilation for acceptable Indoor Air Quality", Atlanta 1996.

2. Fanfger, P. O., *Energy and Buildings* **12**, 1-6 (1988).

3. Grosso, M., *Energy and Buildings* **18**, 101-131 (1992).

Design of a Testing Chamber for the Study of Material Emission Rates

G.Cannistraro[1], G.Galli[1], C.Giaconia[2], S.Magazù[3], A.Piccolo[1]

[1]Facoltà di Ingegneria dell'Università di Messina, 98166 S. Agata Messina (Italy)
[2]Dipartimento di Energetica ed Applicazioni di Fisica Università di Palermo,90100 Palermo (Italy)
[3]Dipartimento di Fisica dell'Università di Messina and INFM, 98166 Messina (Italy)

Abstract. In the present work an experimental device for the detection of pollutants released in indoor environment is presented. The system, constituted by a flux-chamber equipped with sensors for the control of relevant thermophysical parameters (temperature, relative humidity, air velocity), allows to trace the release curves of pollutants emitted from samples of widely employed furnishings.

INTRODUCTION

Indoor air quality problems associated with building furnishings present a major challenge due to ever-changing construction and design renovations and reconfigurations. The modern building furnishings may emit levels of contaminants which are irritating to occupants, especially where ventilation is compromised. Given that the concentration of these pollutants is, generally, greater indoors than outdoors and that people spend most of their time indoors as well, a knowledge of the source emissions and contaminant levels from furnishings is of prime interest to those responsible for providing good indoor air quality. Irritating emissions of VOCs (Volatile Organic Compounds) come from the resins, glues and other chemicals that hold the pressed-wood products (particle-board, fiberboard, hardwood plywood, chipboard and hardboard) together. A number of emitted VOCs have potential health effects as formaldehyde, acetonitrile, acetone, 2-propanol, methyl acetate, butanal, 2-butanone, acetic acid, methylpentane, pentene, 2-furaldehyde, hexanal, toluene, heptane, octane, benzaldehyde and dimethylhexadiene. Over 900 different VOCs have been identified in the office environment; 250 of these compounds regularly occur at concentrations greater than 1 ppb and are 2 to 20 times greater indoors than outdoors. A number of these VOCs (benzene, xylene, toluene, trichloroethylene, methylene chloride) can cause irritation, induce central nervous system depression and increase the risk of cancer; however acceptable levels for total VOC emissions have not been determined. It should be noticed that changes in temperature, humidity, occupant activities, and ventilation will bring about changes in the ambient pollutant concentration. For example high heat and relative humidity will increase the emission rate of formaldehyde/VOCs from pressed-wood. Note also that natural wood

CP513, *Nuclear and Condensed Matter Physics*, edited by A. Messina
© 2000 American Institute of Physics 1-56396-929-7/00/$17.00

products, used as an alternative to pressed-wood, can also emit VOCs which act as irritants and sensitizers. Formaldehyde is the dominant VOC released by office furnishings. At low concentrations (0.05 to 1.5 ppm), formaldehyde can be irritating to the mucous membranes of the eyes and upper respiratory tract, and the skin. Sensitive individuals have experienced nausea and vomiting, headaches, mental confusion, dizziness, depression, poor coordination, allergic reactions and asthmatic attacks. Long-term exposure to low concentrations have been associated with impairment of memory, equilibrium and dexterity. Higher concentrations of formaldehyde (5 to 30 ppm) can cause serious cardiovascular and pulmonary effects. The concentration range typically found in office environments is from 0.01 to 0.5 ppm. It makes up the greatest proportion of the emissions from fabrics/fabric treatments.

IAQ practitioners use the ASHRAE (American Society for Heating, Refrigerating and Air Conditioning Engineers) 62-1989 Standard. These exposure limits are based on providingventilation rates which will reduce the contaminant levels to 1/10th the ACGIH (American Conference Of Governmental Industrial Hygienists) TLVs (Threshold Limit Values). OSHA and NIOSH standards/criteria are not applicable to office worker exposures and the ASHRAE guidelines are somewhat arbitrary and will not cover the sensitive individual.

CHAMBER TEST DESIGN

A glass-steel test chamber for the detection of material emission rates under different conditions of temperature, relative humidity and air velocity has been designed (see fig.1). It is a robust chamber made of acid proven stainless steel with a large thermal capacity with a negligible adsorption capability. The chamber has a volume of ~0,6 m^3 and presents proper openings for the supply of clean air and for the quick insertion and remove of different test samples. The clean air supply is provided by cylinder air of a pure medical breathing grade and the temperature control is achieved by a heater electronically monitored. The chamber is developed for measuring chemical emissions from any surface in order to provide a simple, but well documented emission testing facility capable of testing construction products in a climate where the important climatic parameters like temperature, ventilation rate and air speed may be varied independently around typical indoor values. The chamber is designed to meet the requirements of the important methods for quantifying air pollution. In this investigation human subjects acted as air pollution judges, and chemical characterization of the air pollution will be performed on filter samples. Carpet, linoleum, wall paint and sealant can be tested and results will be reported in a next work. The chamber allows to produce information of volatile organic compound emissions in function of time. The emission data can be used in characterization of emission sources, in toxicological evaluation of material emissions, and as base data for modeling the decay of the emissions. Preliminary results indicate that the decay is usually rapid for liquid surface treating agents like paints, varnishes and floor oils. The TVOC emission rates have usually in 28 days decreased at least 90 per cents. The emission decay for solid materials is slower, but the maximum emission levels are also lower than for fast evaporated liquid agents. The thickness of the material shows also an effect on the emissions.

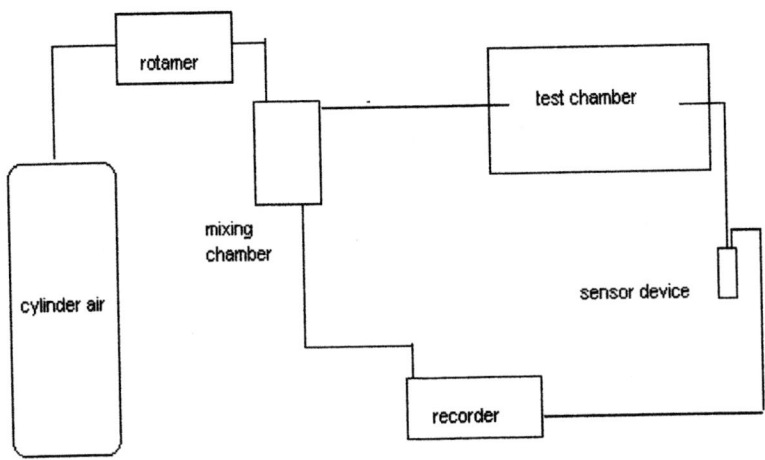

FIGURE 1. Schematic drawing of the experimental test set-up.

CONCLUSIONS

The characterization of emissions from building materials in real rooms is very difficult since many parameters influence the VOC concentrations, including furnishings, fluctuating temperatures, relative humidity, air changes, indoor activities and sink effects. With the purpose to monitor the release curves of pollutants emitted by furniture we designed a test chamber.

What emerges from the preliminary tests performed is that source control and ventilation are the best way to reduce contaminant concentrations indoors. In particular Source Control allows to eliminate sources of emissions, to substitute high emitters with low emitters, to isolate through containment and/or encapsulation, to condition after installation (bake-out, sealants), to assure the proper housekeeping and maintenance. Ventilation Control can increase the effectiveness through design and layout changes, and through intervenes with air filtering and purification (carbon, HEPA). These techniques necessarily require as a pre-requisite the knowledge of the pollutant release curves from furniture samples. In this frame it is relevant to build a proper device and to establish a methodology which allow accurate experimental measurements. In the present paper we propose to design an easy to handle and to maintain chamber. The design of the test chamber provides a constant and efficient air velocity over the entire test area. A system for labelling building materials according to their impact on the indoor air quality is also introduced. The system takes into account the emission decay of volatile organic compounds (VOCs) form new building materials according to their impact on comfort and health. The principle is to determine the time value, $t(Cm)$, required to reach the relevant indoor air value, Cm, of a given VOC in a standard room. $t(Cm)$ is a measure of the duration in witch a new building material may cause increased exposure and enhance the probability of increased indoor air quality problems, unless special precautions are made. The emission profiles of indoor pollutants are measured and the time values of potentially irritating voices are determined. In the next future, the results will be analysed by using mathematical modelling of the emission profiles.

REFERENCES

1. Cannistraro, G., Giaconia, G., Magazù, S., Pietrafesa, M., and Rizzo,G., *Clean Air* **2**, 365, (1994).

2. Cannistraro, G., Giaconia, G., Interdonato, S., Magazù, Piccolo, A., S., Pietrafesa, M., and Rizzo, G., *Atti Accademia Peloritana dei Pericolanti, Classe I di Scienze Fis. Mat. e Nat.* **LXXII**, 369, (1994).

A Simple Stochastic Model for a Complex Ecosystem

M. A. Cirone [1], F. de Pasquale [2] and B. Spagnolo[1,3]

[1] *Istituto Nazionale per la Fisica della Materia, Unità di Palermo*
D.E.A.F., viale delle Scienze, Palermo
[2] *Dipartimento di Fisica, Università di Roma "La Sapienza", P.le A. Moro 2, Roma*
[3] *Dipartimento di Energetica ed Applicazioni di Fisica, Università di Palermo,*
viale delle Scienze, Palermo

Abstract. We investigate the role of the noise on the transient dynamics of a simple ecosystem composed of many interacting species. We assume that the species interact randomly. The influence of the environment is modelled by a multiplicative noise. We study the dynamics of the ecosystem for two values of the growth parameter. We obtain the transient and the asymptotic behaviour of the average population of the species and the distributions of the population and of the local field. Numerical results are compared with analytical asymptotic behaviours of the time average of the i-th species.

Several works have been recently devoted to the study of population dynamics of an ecosystem composed of a large number of species [1–4]. In this paper, we investigate the dynamics of an ecosystem of N = 100 species, interacting randomly each other. We consider an N-species Lotka-Volterra model with a Malthus exponential growth parameter and a Verhulst modelization for the saturation effects [5]. For simplicity, all species are assumed equivalent, i. e., the characteristic parameters of the ecosystem are independent of the species. The random variability of the environment is also taken into account in our model by introducing a multiplicative noise in the equations describing the populations dynamics. The Ito stochastic differential equation which describes the evolution of the ecosystem is

$$ dn_i(t) = \left[\left(\gamma + \frac{\epsilon}{2} \right) - n_i(t) + \sum_{j \neq i} J_{ij} n_j(t) \right] n_i(t) dt + \sqrt{\epsilon} n_i(t) dw_i, \quad i = 1, ..., N \quad (1) $$

where $n_i \geq 0$ is the number of elements of the i-th species, γ is the Malthus' factor, J_{ij} is a non symmetric random interaction matrix and ϵ is the noise strength. In Eq.(1) $dw_i(t) = w_i(t + dt) - w_i(t)$ are the increments of the Wiener process with the usual statistical properties [6]

$$ < dw_i(t) >= 0; \quad < dw_i(t) dw_j(t') >= \delta_{ij} \delta(t - t') dt. \quad (2) $$

CP513, *Nuclear and Condensed Matter Physics*, edited by A. Messina
© 2000 American Institute of Physics 1-56396-929-7/00/$17.00

The random interaction among the species is obtained by constructing an ensemble of ecosystems with different interaction matrices. In our model, the values of J_{ij} are generated randomly with normal distribution, whose average and variance are $\langle J_{ij} \rangle = 0$ and $\sigma^2_{J_{ij}} = J/N$ respectively. The values $n_i(t = 0)$ are fixed for all the ecosystems and they are also generated with normal distribution with average value $\langle n_i(0) \rangle = 1$ and variance $\sigma^2_{n_i} = 0.01$.

We focus on the time average of the ith population species $\bar{N}_i(t)$

$$\bar{N}_i(t) = \frac{1}{t} \int_0^t dt' n_i(t').$$ (3)

In the long time regime the asymptotic expression of $\bar{N}_i(t)$ is [4]

$$\bar{N}_i(t) \simeq \frac{1}{t} \ln \left[n_i(o) e^{\sqrt{\epsilon} w_{max_i}(t) + J \eta_{max_i}(t)} \int_0^t dt' e^{\gamma t'} \right].$$ (4)

In Eq.(4) the influence of other species on the differential growth rate of the time integral of the ith population is described by the local field $h_i(t)$ [4,7]

$$h_i(t) = \sum_{j \neq i} J_{ij} N_j(t) = J \eta_i(t)$$ (5)

and $w_{max_i}(t) = \sup_{0 < t' < t} w(t')$, $\eta_{max_i}(t) = \sup_{0 < t' < t} \eta(t')$. The time average of the ensemble average of the ith population $\bar{N}_i(t)$ for zero value of the growth parameter γ and for $\gamma > 0$ are respectively

$$\left\langle \bar{N}_i(t) \right\rangle \simeq \frac{1}{t} \left[N_w \sqrt{\epsilon t} + \ln t + \langle \ln [n_i(o)] \rangle \right], \quad \gamma = 0,$$ (6)

and

$$\left\langle \bar{N}_i(t) \right\rangle \simeq \frac{1}{t} \left[N_w \sqrt{\epsilon t} + (\gamma + N_\eta) t + \left\langle \ln \left[\frac{n_i(o)}{\gamma} \right] \right\rangle \right], \quad \gamma > 0,$$ (7)

where N_w and N_η are variables with a semi-Gaussian distribution [2]. It is interesting to note that the statistical properties of the time average of the i-th population process are determined asymptotically from the statistical properties of the processes $w_{max_i}(t)$ and $\eta_{max_i}(t)$.

In Fig. 1 (a,b) we report the transient behaviour of the average population of the first species $\langle n_1(t) \rangle$. In the absence of noise, the populations extinguish if $\gamma = 0$ (no Malthus' growth) or reach an equilibrium value for $\gamma > 0$. It is evident that the presence of noise gives rise to fluctuations of $\langle n_1(t) \rangle$. In Fig. 1 (c,d) the transient behaviour of the time integral of the population of the first species $t\bar{N}_1(t)$ for the same values of the growth parameter, namely $\gamma = 0, 1$ are shown.

When the ecosystem relaxes towards an equilibrium population we have a typical long time tail behaviour ($t^{-1/2}$ dependence) [2]. Our numerical results are consistent with the analytical asymptotic expressions of $\bar{N}_1(t)$ (Eq.s(6,7)).

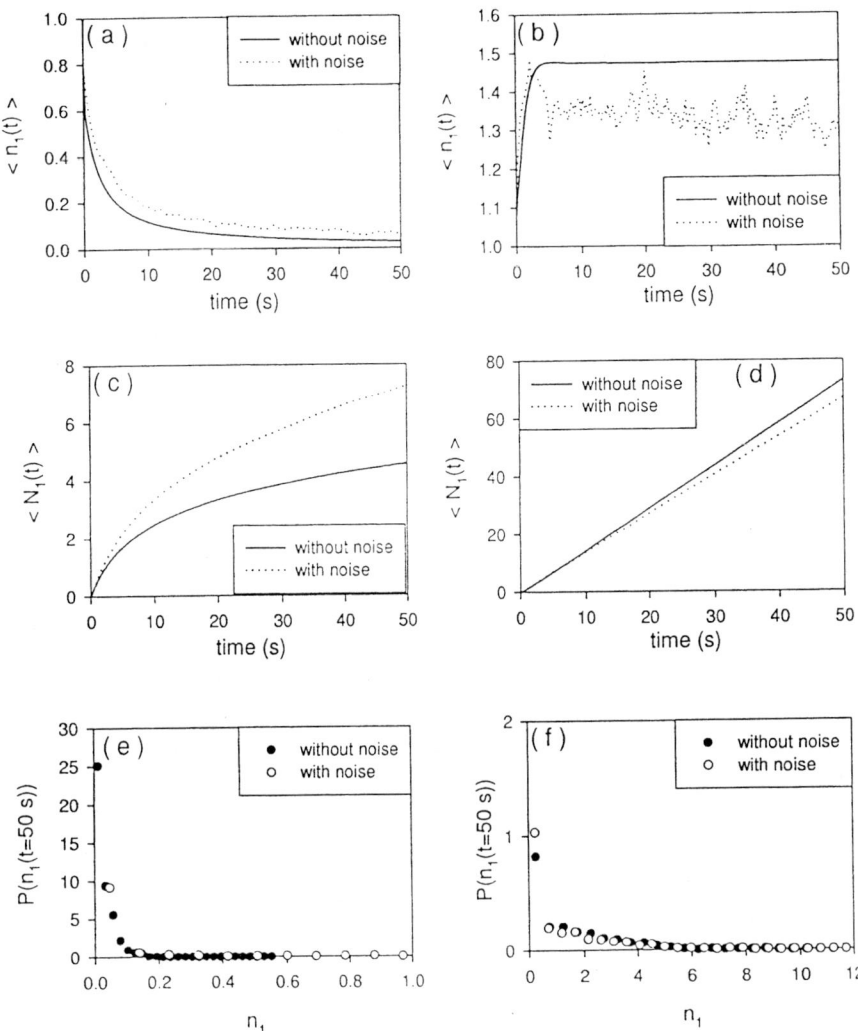

FIGURE 1. The transient behaviour of the average population $\langle n_1(t) \rangle$ (a,b) and its time integral $\langle N_1(t) \rangle$ (c,d), for $\gamma = 0$ (left) and $\gamma = 1$ (right), $J = 1$, with ($\epsilon = 1$) and without ($\epsilon = 0$) noise. The probability distribution of $n_1(t)$ for the same parameters is shown in (e,f).

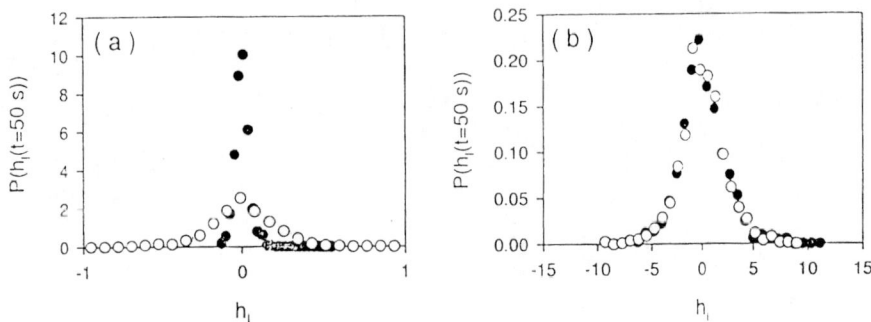

FIGURE 2. Probability distribution of the local field $h_l(t)$ (a,b) for the same parameters of Fig.1.

Transition from stability to instability (divergence of populations) takes place at a critical interaction intensity $J_c \simeq 1.2$.

The presence of external noise has a significant effect on the population dynamics: (i) noise acts against the natural extinction of species; (ii) the probability distributions of the species and of the local field are Gaussian-shaped only for weak interactions between the species, in the absence of noise and for $\gamma = 0$; in all the other cases we have analyzed, the distributions are not Gaussian; (iii) the probability distribution of the local field in the presence of noise is generally wider than the distribution without noise; only for $\gamma = 1$ and $J = 1$ the two distributions are similar.

ACKNOWLEDGMENTS

This work has been supported by INFM and by MURST.

REFERENCES

1. H. Rieger, *J. Phys. A* **22**, 3447 (1989).
2. S. Ciuchi, F. de Pasquale and B. Spagnolo, *Phys. Rev. E* **54**, 706 (1996).
3. M. Marsili, in *Scale Invariance and Beyond*, edited by B. Dubrulle, F. Gramer and D. Sornette, EDP Science, Springer-Verlag, Berlin, p. 173 (1997).
4. F. de Pasquale and B. Spagnolo, in *Chaos and Noise in Biology and Medicine*, edited by M. Barbi and S. Chillemi, World Scientific, p. 305 (1998).
5. N. S. Goel, S. C. Maitra and E. W. Montroll, *Rev. Mod. Phys.* **43**, 231 (1971).
6. C. W. Gardiner, *Handbook of Stochastic Methods*, Springer-Verlag, Berlin (1985).
7. M. Mezard, G. Parisi and M. A. Virasoro, *Spin Glasses Theory and Beyond*, World Scientific Lect. Notes in Physics 9, Singapore (1987).

TL And TSC Solid State Detectors In Proton Therapy

G.A.P. Cirrone[3,6], M.G. Sabini[3,6], M. Bruzzi[7], M. Bucciolini[1,4] , G. Cuttone[3] , A. Guasti[2,4], S. Lo Nigro[6], S. Mazzocchi[1,4], S. Pirollo[7], L. Raffaele[5,3], S.Sciortino[7]

1 Dipartimento di Fisiopatologia Clinica, Università degli Studi di Firenze
2 U.O.Fisica Sanitaria, Ospedale di Careggi, Firenze
3 INFN-Laboratori Nazionali del Sud, Catania
4 INFN-Sezione di Firenze
5 Istituto di Radiologia,Università di Catania
6 Centro Siciliano di Fisica Nucleare e Struttura della Materia - Catania
7 Dipartimento di Energetica, Università di Firenze

Abstract. The necessity to develop methods and techniques for a better determination of absorbed dose in the radiotherapy field stimulates new clinical applications of solid state detectors. In this work we have studied the possibility to use of TLD-100 and synthetic CVD diamond detectors as dosimeters for high-energy proton beams.

INTRODUCTION

Radiotherapy with external beams of charged particles of protons or heavier ions addresses supremely the two main objectives of radiotherapy: the exact match between the radiation field and the tumor volume, and the reduction of dose to adjacent tissue. The main goal of radiation therapy is to develop methods and techniques in order to improve the accuracy in delivering a sufficiently high radiation dose to the target volume, while maintaining the dose to normal surrounding tissues as low as possible. Another related objective of radiotherapy is to macth the spatial distribution of the delivered dose with the tumour volume as accurately as possible. Now that new beam delivery techniques are available, there is a demand of specially tailored solid state detectors and for their characterization with high LET radiation. Particularly ThermoLuminescent emission (TL) and Thermally Stimulated Current (TSC) are physical processes that should be widely used in this field. Our R&D program has been then based on the study of the TL and TSC response of two different materials: LiF:Mg,Ti and Synthetic Diamond film grown by Chemical Vapour Deposition (CVD). The first material has been widely studied for dosimetric applications with conventional radiations but few data are available with high energy protons. CVD diamond film needs a wider dosimetric characterization due to their novelty. In the following some results obtained will be reported.

CP513, *Nuclear and Condensed Matter Physics*, edited by A. Messina
© 2000 American Institute of Physics 1-56396-929-7/00/$17.00

MATERIAL AND METHODS

A group of 300 virgin Lif:Mg,Ti (TLD-100) dosimeters (3x3x0.88 mm) purchased from the Harshaw Company and one CVD metalled diamond film (5x5x0.6 mm) produced by a manufacturer at the best of the state of the art have been employed for the experiments. The annealing of the TLD-100 material before the irradiations, and sensitivity factor S_i for each dosimeters, were done using the standard procedure reported in literature [1]. All the samples have been irradiated in the ^{60}Co gamma beam of the Radiotherapy Unit of the Florence University. The irradiation set-up has been chosen according to the AAPM calibration protocol [2]. Dosimeters have been always analysed with a HARSHAW 5500 reader, at least 48 hours after the irradiation. The reading cycle was the same for all the samples, consisting in a linear warming up with ramp of 15 °C/s starting from 50 °C up to 300 °C. The glow curve was acquired during a time of 20 s, considering as TL signal the integral of the fourth and fifth peak. The samples considering have been divided in three batches. The first one (A) is considered as the reference. The batch B samples have been submitted to a controlled thermal stress procedure, while the batch C samples, have been submitted to a radiation damage treatment. The thermal stress procedure to whom the dosimeters of the batch B have been submitted can be so described: 1 hour at 400°C; cooling to room temperature with a rate of about 1°C/min; 2 hours at 100°C and cooling to room temperature [3]. The radiation damage was obtained by submitting the dosimeters of the batch C to a dose of 500 Gy from 21 MeV protons. The irradiation was performed with the Tandem accelerator of the Istituto Nazionale di Fisica Nucleare-Laboratori Nazionali del Sud (INFN-LNS, Catania). The diamond sample has been irradiated with 27 MeV proton beams from the TANDEM accelerator of INFN-LNS and with 20 MeV electron beams from a LINAC accelerator available at Radiotherapy Unit of Florence University with dose varying from 2 Gy to 10 Gy. The sample has been placed in a PMMA phantom at depth of 1mm for the proton irradiation and at a water equivalent depth of 5cm for electron irradiation. The dose absorbed by diamond has been studied with TSC analysis measuring the current emitted by the sample during a constant heating rate of 0.1 °C/s, from room temperature to 400 °C [4]. The bias voltage applied to diamond is 100 V.

RESULTS AND DISCUSSION

TLD-100 dose response curves have been measured both in ^{60}Co and in proton beams. The selected dose interval is 0.05-15 Gy, where the maximum value is a dose typically furnished in a single fraction for the treatment of uveal melanoma in proton beams. Five TLD-100 for each batch have been placed in the phantom. [60]Co dose measurements have been performed using a Farmer type ionisation chamber. The TLD-100 characterisation in proton beam has been carried out at OPTIS beam line of the Paul Scherrer Institute of Villigen (CH). The measurements have been performed in a fully modulated 62 MeV proton beam at a depth of 15 mm in PMMA, with an effective residual energy of about 40 MeV. Five TLD-100 for each batch have been placed in the PMMA phantom at the isocenter and aligned along the beam axis using the lasers routinely employed in patient positioning. The dose rate during the irradiation was 1 Gy/s. The absorbed dose to water in the proton beam, has been stated using a parallel-plate MARKUS chamber. All the details about the ion chamber calibration procedures both for the ^{60}Co and for the 62 MeV proton beams, are elsewhere reported [3, 5]. In fig. 1 the calibration curves obtained in the fully modulated proton beam are shown. In the proton beam the TL response is lower for the batches B and C as compared to the reference one. Anyway the radiation effects on the dosimetry sensitivity are greater than those induced by the thermal stress. Intercomparing the calibration curve in ^{60}Co and clinical proton beam we noted a lower sensitivity for samples B and C after protons irradiations, therefore showing a dependence of the TL response on the LET of the radiation (about 1.5 KeV/μm for protons and 0.25 KeV/μm for ^{60}Co. The difference between ^{60}Co and protons curves is more pronounced for batch C compared to B one so showing again that the radiation damage is more effective than thermal stress in producing higher LET dependence on the TL response [5].

The figure 2 show the TSC response for diamond sample irradiated with protons with doses of 3 Gy, 5 Gy, 8 Gy and 10 Gy. The total charge emitted during the thermal scanning is obtained from current integral dividing by heating rate. The values of collected charge respect to the total absorbed dose from proton and electron beams have been reported in the insert of figure 2. It is evident a good linearity for doses up to 8 Gy. Moreover the measurements with protons seem to show a tendency to saturation at higher doses.

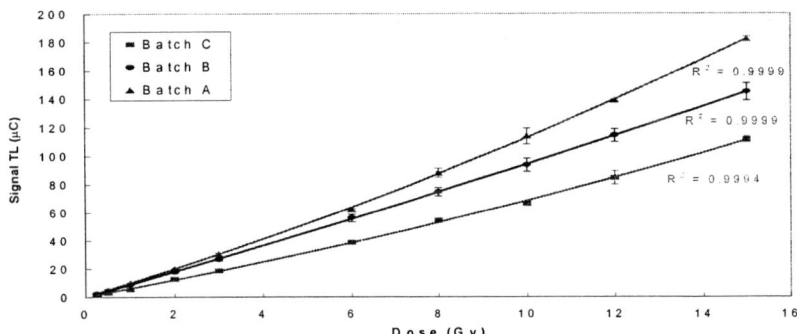

FIGURE 1. TL response curves in 62 MeV fully modulated proton beam.

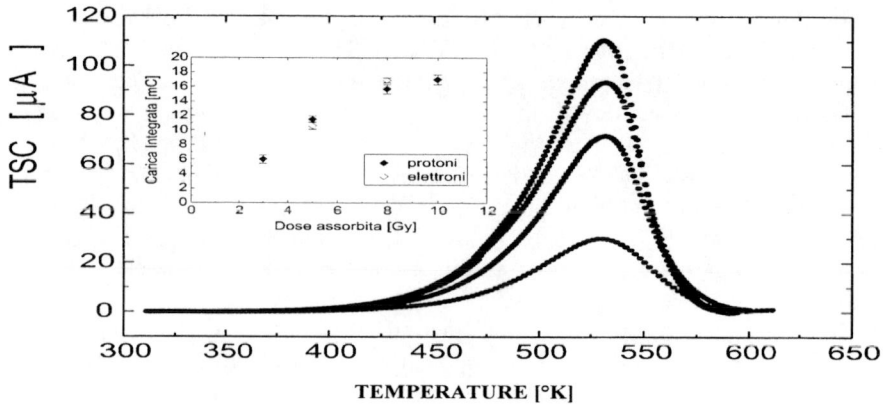

FIGURE 2. TSC response curves for the diamond irradiated with proton beams at various doses. Total charge respect to the absorbed dose is shown in the insert

CONCLUSION

The response of two different solid state dosimeters (TLD and CVD synthetic Diamond) irradiated with proton beams has been studied as a function of the dose. The results obtained show that both systems can be efficiently used for dosimetric applications as in conventional electron beams as in proton ones. Solid state dosimeters seem to have an important role also for protontherapy dosimetry.

REFERENCES

1.Mckeever S.W.S. et al. "Thermoluminescence dosimetry materials: properties and uses" England 1995, Nuclear Technology Publishing.

2.AAPM TG. 21 "A protocol for the determination of absorbed dose from high-energy photon and electron beams" Medical Physics 1983: 10; 741-771.

3. M. Bucciolini et al. *Physica Medica* Vol. XV no. 2 pag. 71-77, 1999.

4. E. Borchi et al. *Solid State Electronics* Vol.423, pp. 428-436, 1988.

5. G. Cuttone et al. *Physica Medica* Vol. XV no. 3 pag. 121-130, 1999.

Relativistic Smearing Detected in the Spectrum of Cyg X-1 in Hard State

Tiziana Di Salvo[1], Luciano Burderi[2], Natale R. Robba[1], Nino La Barbera[1], Chris Done[3]

[1] *Dipartimento di Scienze Fisiche ed Astronomiche, Via Archirafi 36, 90123 Palermo, Italy.*
[2] *Osservatorio di Monteporzio, Roma, Italy.*
[3] *University of Durham, UK.*

Abstract. We report on an observation of the black hole candidate Cyg X-1 performed by the Narrow Field Instruments on board BeppoSAX satellite. During the BeppoSAX observation Cyg X-1 was in its usual low (hard) state with a 0.1-200 keV luminosity of $\sim 2.6 \times 10^{37}$ erg/s, adopting a distance of 2.5 kpc. The broad band spectrum (0.1-30 keV) is well fitted by an absorbed power-law partially reflected by cold matter and a soft excess. The reprocessed component is relativistically smeared and the inferred inner radius of the disc is 20 ± 10 gravitational radii. This inner radius is much smaller than the prediction of the so called Advection Dominated Accretion Flow models: $R_{in} > 200\ R_g$.

INTRODUCTION

Cyg X-1 is the brightest of the persistent galactic black hole candidates (GBHC). The most probable geometry for this system is one in which an outer cold disc (source of soft seed photons) switches into an inner hot region ($T_e \sim 100$ keV and $\tau_T \sim 1$) that is a homogeneous spherical cloud situated around the black hole (probably a torus geometry is more realistic). Successive Compton scatterings (Comptonization) of the soft photons in the hot electron cloud (corona) originate in a power-law continuum, with a cutoff at the electron temperature $T_e \sim 100$ keV. The reflection of this continuum by the accretion disc rises a bump around 20 keV, an iron line at ~ 6.5 keV and an edge at ~ 7 keV. A soft excess observed at energies < 1 keV is interpreted as radiation from the accretion disc (see [1] for a review).

Cyg X-1 was observed by BeppoSAX Narrow Field Instruments (NFI) (see [2] for a description of these instruments) on 1998 May 3 and 4, with a total exposure time of ~ 25 ksec. During the BeppoSAX observation Cyg X-1 was in its usual low (hard) state with a total (0.1-200 keV) luminosity of $\sim 2.6 \times 10^{37}$ erg/s adopting a distance of 2.5 kpc ([3]).

CP513, *Nuclear and Condensed Matter Physics,* edited by A. Messina
© 2000 American Institute of Physics 1-56396-929-7/00/$17.00

RESULTS

To better investigate the shape of the direct component and its reflected continuum we firstly limited the spectral analysis in the 2-30 keV band (MECS and HPGSPC instruments). In this range we can neglect the complexity of the soft excess that is modeled by a disc multitemperature blackbody with the hydrogen column density fixed to the likely interstellar value of 6×10^{21} cm^{-2} ([4]). We used a power-law to describe the direct component (the cutoff at ~ 100 keV is well above the 2-30 keV energy range considered here).

To fit the features due to the reflection we used a Compton reflected continuum and the iron K_α line computed self-consistently for that continuum. The reprocessed component can be smeared to take into account the relativistic and kinematic effects of disc emission (see *e.g.* [5]). The reflected continuum was modeled either by a narrow (not relativistically smeared) component or by a relativistically smeared component (with a little modification of the spectral index from 1.64 ± 0.01 to 1.67 ± 0.01). We obtain a χ^2/d.o.f. value of 395/267 and 338/266 respectively, fixing the iron abundance to the solar value and $\cos i = 0.6$ being i the inclination angle (see [6]).

This demonstrates that a relativistically smeared reflection from cold material gives a better description of the spectrum.

Because a narrow reflection component is also expected from reflection by the companion star and/or the stellar wind and/or the outer flared disc (see e.g. [6]) we added to the previous model another not relativistically smeared and not ionized reflection component, obtaining a χ^2 reduction of $\Delta\chi^2 = 12$. An F-test gives a value of the F function of 9.4, which is significant at more than 99.5% confidence level for one additional parameter.

We then extended the fit to the energy range 0.1-30 keV. In agreement with [7] we found that the soft excess in Cyg X-1 has a more complex shape than a simple blackbody or a disc multitemperature blackbody. To investigate the shape of the soft excess we tried several models. We obtained the smallest value of the χ^2 using the blackbody plus a comptonized model (see [8]) for the soft excess. In Table 1 we report the results of this fit. Figure 1 shows the unfolded spectrum with the used model (top panel) and the residuals (in unit of σ) with respect to the model (bottom panel).

DISCUSSION

The results of our spectral analysis are as follows:
1. The shape of the soft X-ray excess seems to be broader than the disc multitemperature blackbody, as previously suggested by [7]. A model consisting of a blackbody and a comptonization model is preferable.
2. The relativistic smearing of the reflected continuum is unambiguously detected in Cyg X-1, confirming the presence of optically thick material at small radii.

TABLE 1. Results of the fit in the energy range 0.1-30 keV. Uncertainties are at the 90% confidence level for a single parameter.

Parameter	Value
$N_H \times 10^{22}\ cm^{-2}$	0.69 ± 0.13
KT_{BB}	0.13 ± 0.03
KT_{comp}	1.3 ± 0.5
τ_{comp}	10 ± 8
Photon Index	$1.69 + 0.03$
Norm[a]	1.33 ± 0.08
f_{rel}[b]	0.31 ± 0.08
Fe abund	1.0 (fixed)
$\cos i$	0.6 (fixed)
ξ[c]	< 0.3
R_{in}/R_g	20 ± 10
f_{narrow}	0.21 ± 0.08
χ^2/d.o.f.	661/535

[a] Power-law normalization is in unit of 10^{-2} ph keV^{-1} cm^{-2} s^{-1} at 1 keV.
[b] Reflection fraction of the relativistic smeared reflection.
[c] Ionization parameter.

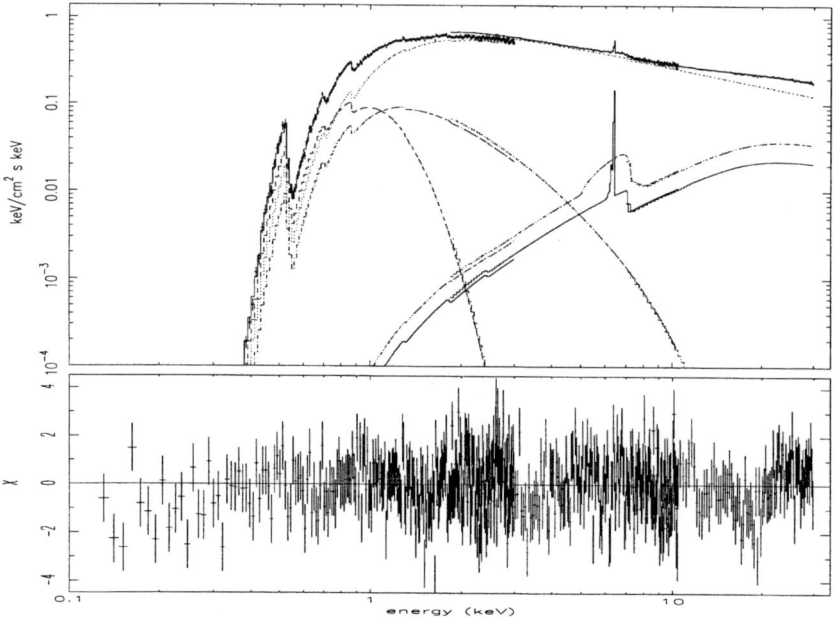

FIGURE 1. Concentration profile climatology and climate model situation.

Although these are preliminary results, seems a robust result that the cold reflecting material does not subtend a large solid angle as viewed by the X-ray source ($\Omega/2\pi = f < 1$) and it is not highly ionized ($\xi << 100$).

3. We also report the presence of a narrow reflection component. The narrow reprocessor subtends a solid angle of $\Omega/2\pi \sim 0.21$. This is somewhat higher than expected from the companion star ($\Omega/2\pi < 0.1$) and is in line with the result of [7], that is $\Omega/2\pi \sim 0.2 - 0.4$. A contribution from the flared outer disc could explain this result. However it could be an artifact of using solar rather than $2\times$ solar abundances.

The result of point 2 indicates that in the low state of Cyg X-1 the spectrum is relativistically smeared and the reprocessing cold matter (probably an accretion disc) extends down to $\sim 20 \ R_g$. This is important to test the so called Advection Dominated Accretion Flow (ADAF). This model could naturally explain the presence of hot, optically thin, geometrically thick matter in the inner region and the observed spectral changes in the GBHCs ([9]). However the inner disc radius we infer is much smaller than their prediction in the low state ($R_{in} > 200 \ R_g$). This model requires the transition from the low to the high state in a very narrow range of luminosity ($0.07 < L_X/L_{Edd} < 0.09$). During the BeppoSAX observation the luminosity was roughly 1% of L_{Edd}. At this luminosity the ADAF models predict that all of the accretion flow should be optically thin. On the contrary our analysis reveals the presence of cold matter at radii as small as $\sim 20 \ R_g$.

REFERENCES

1. Poutanen J., in *Theory of Black Hole Accretion Discs*, Cambridge University Press (1998)
2. Boella G., et al., *A&AS*, **122**, 299 (1997)
3. Liang E. P., and Nolan P. L., *SSRv*, **38**, 353 (1984)
4. Balucinska M., and Hasinger G., *A&A*, **241**, 439 (1991)
5. Zycki P. T., Done C., Smith D. A., *ApJL*, **488**, 113 (1997)
6. Done C., and Zycki P. T., *MNRAS*, in press (1998)
7. Ebisawa K., et al., *ApJ*, **467**, 419 (1996)
8. Titarchuk L., et al., *ApJ*, **434**, 570 (1994)
9. Esin A. A., McClintock J. E., Narayan R., *ApJ*, **489**, 865 (1997)

Neutron Star Spin Evolution: a semi-analytical approach

Antonino La Barbera[1], Luciano Burderi[2], Tiziana Di Salvo[1], Ulrich Kolb[3], Natale Robba[1]

[1] *Dipartimento di Scienze Fisiche ed Astronomiche, Università di Palermo, via Archirafi 36 - 90123 Palermo, Italy*
[2] *Osservatorio di Monteporzio - via di Frascati 33 - 00044 Roma - Italy*
[3] *Open University - Walton Hall - Buckingham - United Kingdom*

Abstract. We report on a semi-analytical model for studying the evolution of the spin period P of a magnetic neutron star as a function of the baryonic mass load M_{acc}. We have taken into account different equations of state and included rotational deformation effects. The presence of a strong gravitational field has also been considered in the context of general relativity. A comparison with numerical fully relativistic codes shows that our description is accurate within 5 %.

I INTRODUCTION

It is currently believed that millisecond radio pulsars are recycled neutron stars (NS) that have been spun up by accretion in binary systems where a donor star pours mass on the NS [1] [2]. The evolution of NS in a binary sistem differs considerably in the case the donor star is a high mass giant star in a HMXB or is a lighter star in a LMXB (see e.g. [3]). It is only in the case of LMXB that the system lives enough time to spin up the NS to millisecond spin period. Till recently a major difficulty of the theory was the undetectability of any spin periodicity in LMXB. The discovery [4] of a 2.49 ms spin period in the soft X-ray transient SAX J1808.4-3658 (a LMXB) is an important advance. In fact SAX J1808.4-3658 could be the connection element between LMXB and millisecond radio pulsar and a confirmation of the recycling scenario.

Even so, anyway, a number of questions remain open. A neutron star can reach a rapid spin rate by accretion provided the magnetic field is low enough ($\leq 10^8$ G). It is largely accepted that, when a NS forms, the magnetic field at the NS surface is very high ($\sim 10^{13}$ G). Some mechanisms of decay of the magnetic field have to be supposed. But at the present moment there is no generally accepted theory. Another problem is the accreted mass in order to achieve millisecond periods. This

CP513, *Nuclear and Condensed Matter Physics*, edited by A. Messina
© 2000 American Institute of Physics 1-56396-929-7/00/$17.00

mass cannot exceed the maximum allowed mass. On the other hand this parameter is stricly depending on the adopted equation of state of the nuclear matter composing the NS. In addition a very important role could be played by accretion rate which depends on the parameters characterizing the binary system.

In order to clear these questions and to study the evolution of the spin of a NS in a binary system a model of a rapidly rotating neutron star in full general relativity is needed. Cook Shapiro & Teukolsky (CST) [5] developed a fully relativistic code to study the spin evolution of an unmagnetized NS accreting from the inner edge of a Keplerian disk with a constant accretion rate. We developed a semi-analytical model which is able to reproduce the results of CST within 5 %.

II   MASS LOADING AND SPIN EVOLUTION

In this section we briefly describe the semi analytical model used to study the relation between P and M_{acc}. The code follows the evolution of a NS wich is accreting mass from a companion donor star. In our model, the gravitational mass M_G and the baryonic mass M_B are connected, at any time, by the relation:

$$M_G = M_B \left(1 + \alpha \frac{M_B}{R_{stat}}\right) \tag{1}$$

where R_{stat} is the circumferential radius for the non-rotating NS, and α is a constant depending on the adopted EOS. Here we are assuming that M_G is independent of the spin P of the NS. In effect the dependence is very weak and the error which we commit is less than 3 %. In addition to this equation we need a relation connecting M_G and R_{stat}. We use an analytical approximation obtained by fitting the results of CST [6]:

$$R_{stat} = R_{stat_{min}} \left\{1 + \left[ln\left(1 - \frac{M_G}{M_{G_{max}}}\right)^{-\frac{1}{\sigma}}\right]^{-\frac{1}{3}}\right\} \tag{2}$$

where $R_{stat_{min}}$ is the minimum radius allowed for the NS when it achieves its maximum static gravitational mass $M_{G_{max}}$ for an adopted EOS. σ is a constant also depending on the EOS. Combining (1) to (2) and deriving respect to time it is possible to bind the accretion rate $\dot{m}_B$ with the increment rate of gravitational mass of the NS $\dot{m}_G$:

$$\dot{m}_G = \phi \dot{m}_B \tag{3}$$

where ϕ is a known function depending essentially on M_G and constants characterizing the adopted EOS.

Of course the accreting mass tranfers angular momentum to the NS. Close to the NS the relativistic corrections to the gravity are important. In this case the specific angular momentum l_{in} differs from the classical newtonian value. So we have considered the Kerr metric for the approximate description of the gravity around the NS.

Supposing during the accretion process the specific angular momentum l_{in} of the accreting matter at the inner radius of the accretion disc is entirely tranfered on the NS, from the angular momentum conservation, we obtain the increment rate of the spin frequency $\dot{\Omega}$ of the NS:

$$\dot{\Omega} = \frac{\dot{m}_B}{I_\Omega} \left(l_{in} - \phi\Omega \frac{\partial I_\Omega}{\partial M_G} \right) \left(1 + \frac{\Omega}{I_\Omega} \frac{\partial I_\Omega}{\partial R_\Omega} \frac{\partial R_\Omega}{\partial \Omega} \right) \tag{4}$$

where I_Ω and R_Ω are the moment of inertia and the circumferencial radius of the neutron star at a given Ω. The rapid rotation inflates and deforms the neutron star surface. In this way I_Ω and R_Ω are significantly altered at high rotational rates. We have found that, for not-supramassive sequences (see [6]) I_Ω and R_Ω can be approximated as:

$$I_\Omega = F(\Omega/\Omega_{max}, M_G)\, I_{stat} \tag{5}$$

$$R_\Omega = F(\Omega/\Omega_{max}, M_G)\, R_{stat} \tag{6}$$

where I_{stat} is the static moment of inertia of the NS and $F(\Omega/\Omega_{max}, M_G)$ is:

$$Log\left(F(\Omega/\Omega_{max}, M_G)\right) = 0.25 \left(\frac{M_G}{1.03 M_{G_{stat}}} - 1 \right)^{50} \left(1 - \left(1 - \frac{\Omega}{1.01\Omega_{max}} \right)^{0.2} \right) \tag{7}$$

Here Ω_{max} is the limiting angular frequency at which the gravitational pull is balanced by the centrifugal forces. We have estimated that the classical expression of Ω_{max} is a valid approximation even in the fully relativistic treatment:

$$\Omega_{max} = \left(\frac{G M_G}{R_{\Omega_{max}}^3} \right)^{\frac{1}{2}} \tag{8}$$

where $R_{\Omega_{max}}$ is the the circumferential radius of the rotating NS when Ω reaches its maximum allowed value Ω_{max}:

$$R_{\Omega_{max}} = F(1, M_G)\, R_{stat} \tag{9}$$

About the static moment of inertia a relativistic expression is needed. We have fitted a result of Ravenhall & Petick [7] obtaining:

$$I_{stat} = 1.82 \left(\frac{M_{G,\odot}}{R_{stat,km}} \right)^{0.62} e^{-6.5 \frac{M_{G,\odot}}{R_{stat,km}}} \frac{M_G R_{stat}^2}{1 - \frac{2 G M_G}{c^2 R_{stat}}} \tag{10}$$

where $M_{G,\odot}$ is the gravitational mass of the NS in solar units and $R_{stat,km}$ is the static radius in kilometres.

III CONCLUSIONS

In this paper we have presented a semi-analytical model to compute P vs M_{acc} relation. We have first derived a simple equation describing the evolution of the gravitational mass as a function of the baryonic load. Then we have introduced the set of equations for the evolution of the rotational frequency of the NS, including rotational effects on the NS radius. The angular momentum tranfer has been computed including the corrections due to a strong gravitational field. Solving for the differential equations (3) and (4), we have determined the NS rotation evolution for stationary accretion. As an example, in figure 1 we show the spin evolution vs the accreted matter M_{acc} of an unmagnetized NS in the FPS-EOS case. Our results are in agreement with the results of the relativistic code of CST [5] within 5 % even in the high spin regime. The related computational effort is greatly reduced with respect to a fully relativistic numerical approach.

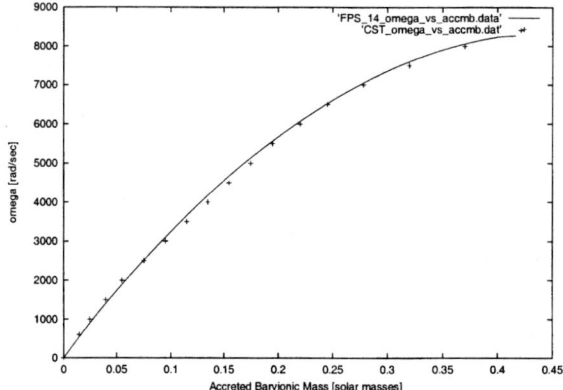

FIGURE 1. Rotation evolution of an unmagnetized NS (initial gravitational mass = 1.4 $M_\odot$) undergoing steady accretion of matter. The Ω vs M_{acc} relation, calculated for FPS-EOS with the semi-analytical model described in the text (solid line) is compared to that from CST

REFERENCES

1. Alpar, M. A., Cheng, A. F., Ruderman, M. A., Shaham, J., *Nature*, **300**, 728 (1982)
2. Bhattacharya, D., van den Heuvel, E. P. J., *Phys. Rep.* **201**,1 (1991)
3. Verbunt, F., *Ann. Rev. Astron. Astrophys.*, **31**, 93 (1993)
4. Wijnands, R., van der Klis, M., *Nature*, **394**, 344 (1998)
5. Cook, G. B., Shapiro, S. L., Teukolsky, S. A., *ApJ*, **423**, L117 (1994)
6. Cook, G. B., Shapiro, S. L., Teukolsky, S. A., *ApJ*, **424**, 823 (1994)
7. Ravenhall, D. G., Pethick, C. J., *ApJ*, **424**, 846 (1994)

Low Power Electronics For NEMO Detector

Domenico Lo Presti[a] for the NEMO Collaboration[b]

[a]*Physics Department, University of Catania and INFN of Catania, Catania, Italy*
[b]*Physics Department, University of Bari and INFN of Bari, Bari, Italy*
Physics Department, University of Bologna and INFN of Bologna, Bologna, Italy
Physics Department, University of Cagliari and INFN of Cagliari, Cagliari, Italy
Physics Department, University of Catania and INFN of Catania, Catania, Italy
Physics Department, University of Messina and INFN of Catania, Messina, Italy
Physics Department, University "La Sapienza" of Roma and INFN of Roma, Roma, Italy
Laboratori Nazionali del Sud (INFN), Catania, Italy
Laboratori Nazionali di Frascati (INFN), Frascati, Italy
Fondazione "Ugo Bordoni", Roma, Italy
"Osservatorio Geofisico Sperimentale", Trieste, Italy
"Istituto per lo studio dell'Oceanografia Fisica" (IOF-CNR), La Spezia, Italy
"Centro Interdip. di Ricerche sulle Coste e sull'Ambiente Marino", Uniersity. Of Cagliari, Italy

Abstract. For the realisation of the submarine detector NEMO it is necessary to design an acquisition system which is able to capture the signals coming from photo-multipliers (PMs) of the optical modules (OMs) and to satisfy several specifications: low power consumption; few submarine interconnections for reliability and simplicity of the deployment; flexibility of the system; minimum dead time; high dynamic range; accuracy in order to have good resolution. Here we present a Very Large Scale Integration full-custom solution for the OMs according to the requirements. It foresees to use a switched capacitor analog memory, a trigger and single photon classification system, a PLL and a control unit able to manage the different operation states of the whole system. For such a system we foresee a power dissipation not higher than 200-300 mV in each OM, 20 bit dynamic range and a dead time of about 0,1 %.

INTRODUCTION

The main constraints for the submarine detector realisation are: very low power because of the distance from the shore; only one submarine interconnecting cable to have the best reliability and to simplify the deployment; flexibility to give the possibility of changing parameters; very small dead-time to get a good detector efficiency; high dynamics range to fit with different kinds of experiments; very good accuracy of experimental data; low costs, if possible!

DESCRIPTION OF THE SYSTEM

CP513, *Nuclear and Condensed Matter Physics*, edited by A. Messina
© 2000 American Institute of Physics 1-56396-929-7/00/$17.00

Signals from the optical modules cannot be transferred to the shore unaltered. It is necessary to transfer a suitable compressed and codified representation from the optical module to a concentrator. An electronic system triggered by the signal holds analog samples taken at very high frequency (200 MHz). It then successively transforms them into a digital code (10 MHz) and applies a first compression algorithm.

The system is timed by a common clock that is sent to all the optical modules from the shore. A slow control system must be provided. Both clock and slow control signals travel in a single electro-optical cable using a self-synchronised serial transmission method. Data flow in both directions is managed using FPGAs and DSPs. The link between an Optical Module and the first concentrator is realised using a coaxial cable.

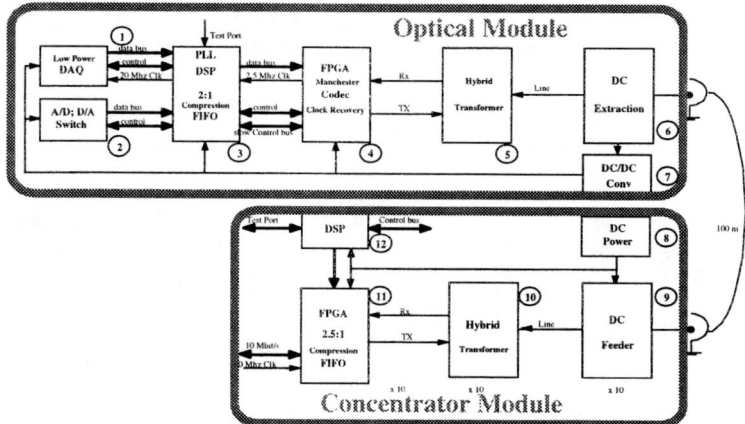

FIGURE 1. Schematic of the Concentrator Module.

A full-duplex bi-directional channel must be used. Because the spectra of the two fluxes are overlapping we must use duplexers, with very high isolation between the two directions, at both ends. The same cable carries the DC power supply for the Optical Module. Devices capable of extracting or injecting DC power in this cable, without attenuating signal levels must be designed.

The concentrator module manages a group of optical modules and performs a further compression. Moreover it provides the temporal alignment of the signals. Finally, a secondary trigger is foreseen in the same module to ameliorate the S/N ratio, ON-LINE. In such a way a dramatic cut is made to the noise coming from the natural disintegration of radioactive Potassium melt in the sea water. In such a way, other, hierarchically higher, concentrator modules, using Artificial Intelligence Techniques, compress data flow further, making it manageable.

These purposes can be reached if several analog, digital and mixed low power ASIC devices are realised. For instance we foresee the design of an ASIC to detect sensor signals, a digital unit to control all the activities in the Optical Module, a Switched Capacitor Analog Memory writing at 200 Mhz and reading at 10 Mhz and a Data Package & Transfer Unit to organise and compress data in the doptical module. The power budget in the optical module is very low, less than 300 mW. In the higher level concentrator we foresee the design of simple low power neural networks like ASICs to perform an increase of S/N ratio.

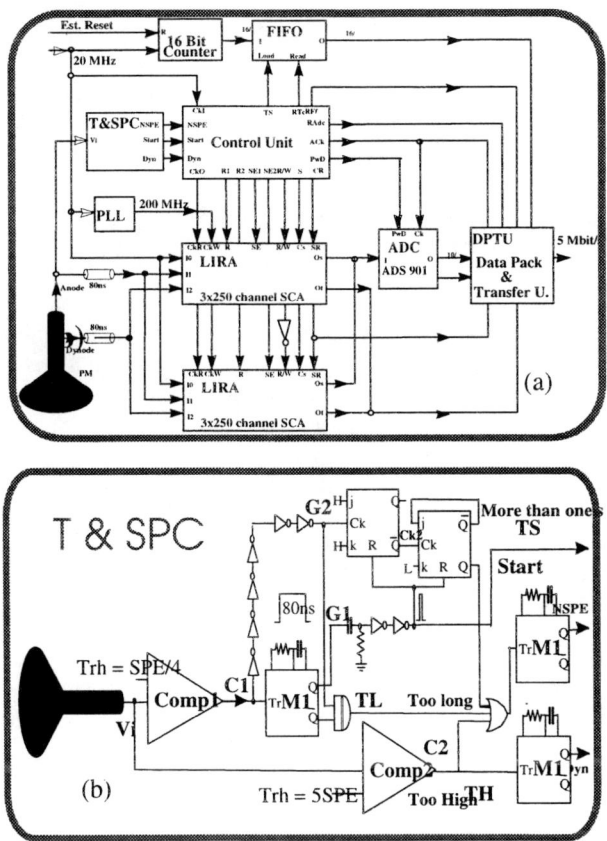

FIGURE 2. (a) Schematic of the whole Photo-multiplier read-out system. (b) Detail of T&PSC.

Two signals are taken from the Photomultiplier. One from the anode, the other one from a dynode. If the anode signal is too high we use sampled dynode signals. A Trigger and Single Photon Classifier chip (T&SPC) is used to take this decision. It, also, classifies input signals as Single PhotoElectron (SPE) or Non Single Photon Electron (NSPE) signals.

In the last case the number of sampling is 100, otherwise it remains the minimum number necessary to sample a whole SPE signal. The sampling is accomplished by a Switched Capacitors Array Analog Memory (LIRA) that consists of 3 channels of 250 cells each. The sampling speed is 200 MHz while the transfer rate is 20 MHz. LIRA samples: anode signal; dynode signal and 20 MHz Master clock. In this way we can record time with 1.4 ns of resolution. We use two LIRA chips that work alternately. While one is in a sampling phase the other one transfers data towards a 20 MHz ADC. The conversion, accomplished only for the samples, is followed by a storage phase of digital data in a Data Package & Transfer Unit (DPTU). The latter transfers data towards a concentrator module.

CONCLUSIONS

The solution we propose has a total power less than 5 kW; not more than 300 mW in each OM. This result can be reached with an extensive use of full-custom VLSI devices both for analog and digital parts in the OM. Ten OMs are connected to a single concentrator module (CM) where we foresee 2 W. In this way it is possible to have about 500 mW per module and about 5 kW in total. Only one cable carries the DC current and, in both directions, the information between OM and CM.

To limit data flow some trigger must be provided. A system having hardware coincidences between two or more OMs lacks flexibility. The interconnecting network complicates deployment and increases costs. A rigid structure must be provided. Dead time can become very large. We foresee the adoption of a software/hardware solution. All the signals exceeding a threshold (about 1/4 p.e.) are sampled and sent to a concentrator module where a software coincidence is made, using the time information contained in the signals.

The foreseen dead time is less than 0.1%. Generally speaking, send a fixed number of samples for each p.e. signal. No oversampling is foreseen. We sample the signals of the PM anode and of a dynode, so more than 14 bit can be achieved for input dynamic range.

We foresee the possibility to achieve a timing accuracy better than 1 ns; the use of only one PM for each OM and a single external interconnecting cable simplifies the mechanical structure and decreases costs both for deployment and maintenance.

REFERENCES

1. D. Lo Presti, S. Panebianco, G.V. Russo, S. Reito, "Switched Capacitor Arrays Analog Memory For Sparse Data sampling", Fourth Workshop on Electronics for LHC Experiments, Cern/LHCC/98-36, 30 Ottobre 1998, pp. 155-159.

2. S. Panebianco, D. Lo Presti, G. V. Russo et al., "Switched Capacitor Arrays Memory For Sparse Data Sampling", Nuclear Instruments & Methods in Physics Research, A 434 (1999) 424-434.

Experimental Study Of The ^{6}Li(d,α)^{4}He Reaction And Its Astrophysical Implications Via The Trojan Horse Method

R.G. Pizzone[1,2], M. Aliotta[1,2], S. Blagus[3], S. Cherubini[4], P. Figuera[2],
M. Lattuada[2,5], M. Milin[3], Đ. Miljanic[3], M.G. Pellegriti[1,2], D. Rendic[3],
S. Romano[2], N. Soic[3], C. Spitaleri[1,2], M. Zadro[3], R.A. Zappalà[6]

[1]Dipartimento di Metodologie Fisiche e Chimiche per l'Ingegneria, Catania, Italy
[2]Laboratori Nazionali del Sud, Catania, Italy
[3]Institut Rudjer Boskovic, Zagreb, Croatia
[4]Institut de Physique Nucleaire, Universitè Catholique de Louvain, Louvain-la-Neuve, Belgium
[5]Dipartimento di Fisica, Università di Catania, Catania, Italy
[6]Istituto di Astronomia, Università di Catania, Catania, Italy

Abstract. The ^{6}Li(d, α)^{4}He reaction, whose astrophysical importance is connected to the primordial nucleosynthesis in the framework of the Inhomogeneous Big Bang, has been studied by using the Trojan Horse Method (THM). We derive and discuss the cross section and the astrophysical S(E)-factor for E_{cm}= 0.025- 0.7 MeV. Results are compared with data from a direct measurement.

INTRODUCTION

Primordial abundances of light elements, such as ^{2}H, ^{3}He, ^{4}He, ^{7}Li, play a crucial role for the determination of the cosmological parameter η (the baryon-to-photon ratio). It can be shown [1] that the adimensional baryon density Ω_b is strictly related to η, but its range of variability is still too large to affirm whether the universe is close or open. Further constraints on η may be derived if primordial abundances and cross sections for other light elements are known. For instance it is possible to obtain limits to the ^{7}Li abundance from that of ^{6}Li [2]. Some effort to measure ^{6}Li primordial abundance and to study the nuclear processes which produce or destroy it. According to the Inhomogeneous Big Bang Model [3], ^{6}Li may be burned via the ^{6}Li(d,α) ^{4}He reaction in neutron-rich regions. We have studied this reaction by means of the Trojan Horse Method in order to extract the S(E)-factor at astrophysical energies. Comparison of our results with direct data is presented whereby hints on the electron screening effect can be obtained.

CP513, *Nuclear and Condensed Matter Physics*, edited by A. Messina
© 2000 American Institute of Physics 1-56396-929-7/00/$17.00

THE METHOD

The basic idea of the Trojan Horse Method consists in the assumption that the quasi-free reaction mechanism is dominant in some particular kinematical conditions [4]. The triple differential cross section $d^3\sigma/d\Omega_1 d\Omega_2 dE$ for the three body reaction $^6Li(^6Li,\alpha\alpha)\,^4He$ can then be related to the $^6Li(d,\alpha)^4He$ nuclear process. The target nucleus ^6Li is assumed to break-up into the clusters α_s and d, respectively spectator and participant to the $^6Li(d,\alpha)\,^4He$ virtual reaction which proceeds within the region of the nuclear interaction if the projectile energy is greater than the Coulomb barrier. Using the Plain Wave Impulse Approximation we express the triple-differential cross section, measured in an $\alpha_1-\alpha_2$ coincidence experiment, through the two-body nuclear cross section $d\sigma^N/d\Omega$, as

$$\frac{d^3\sigma}{d\Omega_1 d\Omega_2 dE} \propto KF \cdot \left|\Phi(\overrightarrow{p_s})\right|^2 \cdot \frac{d\sigma^N}{d\Omega}$$

where KF is a kinematical factor and $\Phi(p_s)$ is the relative momentum distribution of the α_s cluster inside the ^6Li nucleus. Information on the virtual reaction $^6Li(d,\alpha)\,^4He$ can then be achieved from a measurement of the three-body cross section, as:

$$\frac{d\sigma^N}{d\Omega} \propto \frac{d^3\sigma}{d\Omega_1 d\Omega_2 dE_1} \cdot \left[\, KF \cdot \left|\Phi(\overrightarrow{p_s})\right|^2 \,\right]^{-1} \qquad (1)$$

It can be shown [5] that $d\sigma^N/d\Omega$ is related to the laboratory cross section by the relation

$$\frac{d\sigma}{d\Omega} = \sum_l G_l \frac{d\sigma^N}{d\Omega} \qquad (2)$$

where G_l is the transmission coefficient through the Coulomb barrier for the l-th partial wave involved in the process.

EXPERIMENTAL PROCEDURE AND DATA ANALYSIS

The experiment was performed at the 5 MV Van de Graaff Tandem of the Rudjer Boskovic Institute of Zagreb. A $^6Li^{++}$ beam, with $E_b=5.9$ MeV and $i\approx 15$ nA, was sent onto an isotopically enriched $^6Li_2O$ target (125 $\mu g/cm^2$) evaporated on a carbon backing (20 $\mu g/cm^2$). The outgoing α-particles were detected in coincidence by means of two silicon position sensitive detectors, mounted in co-planar geometry at

$\theta_1=60°$ and $\theta_2=73°$, at opposite sides of the beam axis. This geometric choice allowed for investigation of a number of quasi-free angles within the acceptance of both detectors ($\Delta\theta_1=\Delta\theta_2=14°$). Angular calibration of both PSD's was carried out by using grids with 18 equally spaced vertical slits. Due to the high Q-value (20.9 MeV) of the ^{6}Li(^{6}Li,$\alpha\alpha$) ^{4}He with respect to any other reaction occurring in the target, the kinematical locus corresponding to the α–α coincidence could be easily discriminated and no further particle identification was in need. Projections of these selected data in angular correlation show a broad peak around the spectator energy $E_s=0$, its magnitude decreasing as E_s moves away from zero. Such evidence has been taken as a sign for the occurrence of the quasi-free break-up mechanism.

RESULTS AND DISCUSSION

The nuclear cross section $d\sigma^N/d\Omega$ has been obtained within an arbitrary normalization constant. Its value was determined by comparison of the subset of data at $\theta_{cm}=90°\pm 5°$ (where θ_{cm} is the emission angle for the alpha particle in the d-^{6}Li centre of mass frame) with the corresponding data from a direct measurement (fig.1). We shall stress that the nuclear differential cross section $d\sigma^N/d\Omega$, extracted via THM, represents only the contribution of the nuclear cross section: so no correction for Coulomb barrier penetration effect is present. In fig. 2 we show the differential cross section extracted from our data by using (2) compared with direct data [6].

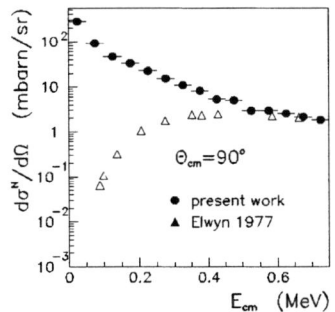
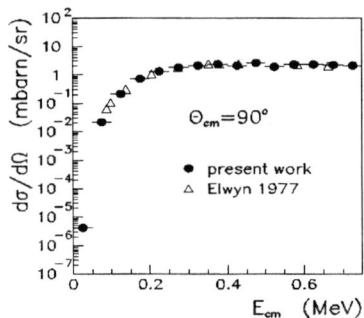

FIGURE 1. Comparison with direct data from ref. 6   **FIGURE 2.** Bare nucleus cross section

We normalize (see fig. 1) the two-body nuclear cross section $d\sigma^N/d\Omega$ to the direct one at $E_{cm}=0.7$, where no barrier penetration effect is expected and the two sets of data should be consistent. It is possible now to extract from our indirect data the values of the astrophysical S(E)-factor through the equation:

$$S(E)= \exp(2\pi\eta)\cdot E\cdot\sigma(E)$$

and compare them with those obtained from direct measurements. In the present case, taking into account the property of the dominant contribution of the s-wave, we can write for the bare cross section $\sigma(E)$:

$$\sigma(E) = G_0 \cdot \sigma^N(E)$$

Figure 3 shows the comparison between the data obtained by means of the THM and the direct ones reported in ref. 7. A good agreement is found in the energy range E_{cm}=0.05-0.7 MeV and in the astrophysical energy region (namely E_{cm}~290 keV for a temperature of 10^9 K), while for E_{cm}< 0.04 MeV the trend of the two data sets appears to be quite different. It is important to note that the extrapolated value at zero energy, obtained by using a third order polynomial fit, $S(0)$=17.6 ± 0.4 keV·barn, appears in very good agreement with that of ref. 7. Therefore the discrepancy in the low energy trend of the two data sets, shown in the figure 3, should be essentially due to the electron screening effect on the directly measured data.

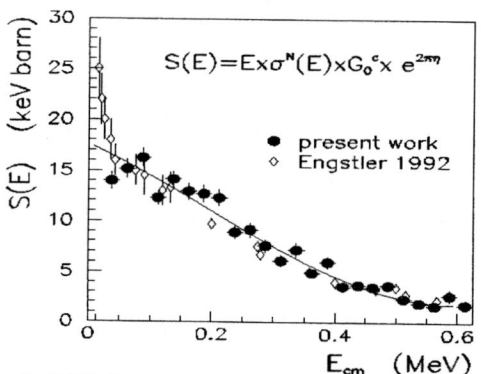

FIGURE 3. The astrophysical S(E) factor obtained in the present measurement; the polynomial third order fit and the direct data (open dots) are also shown.

REFERENCES

1. C.J. Copi et al., *Science.* **627**, 192 (1995)

2. K.M. Nollet et al. *Phys. Rev. C* **56**, 1114 (1997)

3. J.H. Applegate et al., *Astroph. J.,* **329**, 572 (1988)

4. G. Baur, *Phys. Lett. B* **178**, 135 (1986)

5. S. Cherubini et al., *Astroph. J.,* **457**, 855 (1996)

6. A.J. Elwyn et al., *Phys. Rev.,* **C16**, 1744 (1977)

7. S. Engstler et al., *Z. Phys.,* **A342**, 471 (1992)

TECHNICAL AND ORGANIZATIONAL ASPECTS OF PROTECTION FROM IONIZING RADIATIONS WITHIN THE DEFENCE

Dr. Vittorio Sabbatini

CISAM, Triforce Center for Military Application Studies,
Nuclear Office, S.Piero a Grado (Pisa)

Abstract. When the Defence is not interested in the nuclear aspects connected with energy production or basic research, it must feel compelled to follow the nuclear activities for what concerns the nuclear protection needs within the operational forces.
During the years, this has caused the installation and utilization of a nuclear reactor, laboratories specialized in radiological and nuclear matters and the management and utilization of radioactive material and radiogenic machines to satisfy additional requirements.
A specific structure (CISAM) has been created within the Defence for these activities; it is able both to offer a valid protection organisation to the forces assigned to operate militarly and to operate in peace time for the safety of personnel and the protection of environment.
The purpose of this paper is to conduct a quick analysis of the Defence nuclear and radioprotection needs and to illustrate CISAM's function, the technical methods adopted and some specific protection arrangements connected with radiological emergencies.

Defence: Nuclear Energy and Ionizing Radiation

Nuclear and radiological protection for operational forces has gone, hand in hand, with protection from ionizing radiation in peace time,of military and civilian personnel involved in the utilization of radioactive sources and radiogenic machines.

Referring to the the military measures regarding nuclear defence (N), it is important to remember that the mentioned dispositions are organized according to international agreements drawn jointly within NATO.

Indeed the principles governing the N defense may be summarized as follows:

- To preserve in the best way the operational capability of the military units.
- To determine procedures and methods of application of nuclear defence.
- To protect personnel, materiel and equipment from a nuclear threat in operations in which the armed forces are likely to be employed.

The nuclear threat considered is the one deriving from events connected with the following scenarios: employment of nuclear weapons; possible acts of war or terroristic attacks against nuclear installations and industrial risks of various types.

CP513, *Nuclear and Condensed Matter Physics,* edited by A. Messina
© 2000 American Institute of Physics 1-56396-929-7/00/$17.00

The principles of protection and the scenarios mentioned above tell us how deeply committed the Defense has been for the implementation of studies and protective measures dealing with the prevention and management of critical situations.

Two remaining aspects, examined more thoroughly in the paragraphs to follow deal with radiological protection and protection of the environment.

CISAM's Tasks

CISAM is the Defence technical agency called upon to provide solution to nuclear problems and to carry out physical surveillance and environment protection.

The radio-protection organization of the Center has become capable of satisfying the needs of Defence in the following areas:

Operational, health physics checks and Environment surveillance;

Individual dosimetry for X, gamma and neutronic radiation;

Metrology of ionizing radiation;

Disposal of radioactive waste and Radioactive decontamination;

Training of personnel to be assigned to the protection;

Technical Cooperation

CISAM was recognized as Authorized Institution for the physical surveillance by Ministry of Labour and Social Welfare decree.

It has been awarded recognition as secondary calibration center, and entered in the national metrological organization (SIT) for the ionizing radiation fields.

It has participated in comparison campaigns on individual dosimetry (coordinated by ENEA) and on environmental radioactivity (coordinated by ANPA).

It has certified its photographic dosimeter with the ENEA-EDP group.

It has cooperated with AGIP Nucl., FIAT, Pisa University, LENA of Pavia, CNR.

It has made a remarkable contribution in the area of the norms governing radio-protection. Indeed, personnel from the center has participated in UNI groups (dosimetry, waste disposal, filtering systems), ENEA groups (individual dosimetry), ministerial groups; NATO and FINABEL groups.

CISAM's Facilities for the Radiological Protection and for Emergencies

The nuclear plant of CISAM implied the existence of an emergency plan and the creation of a local network for the radioactivity control. Moreover CISAM has been made part of the emergency plans for the presence of nuclear ships in Italian ports.

CISAM carries out the described activities as shown in the following table:

a) environment checks

-sampling of atmospheric dust;

-measuring of total beta radioactivity of fall-out, waters and sediments;

-gamma spectrometric analysis of sediment samples and marine biota;

b)emergency intervention

For possible emergencies, CISAM can count on intervention vehicles (two already available and a third one about to be completed) equipped to operate in areas contaminated by low level radiation (LLR). Moreover, the Center boasts a mobile monitoring station for WBC measurements. Intervention vehicles are kept in efficiency by utilizing them routinely for check and measurement campaigns.

Intervention Vehicle and Mobile Station

As indicated previously the Center is at present equipped with operational vehicles, while the acquisition of an additional, air transportable vehicle is underway. All the mentioned vehicles have been conceived to cover all emergency situations and institutional routine activities. The following is a description of: A-an intervention vehicle; B- a mobile station.

A- The intervention vehicle has been designed to operate mainly in civil type situations. It is capable to perform the following tasks:

- operational checks and radiometric measurements in external structures;
- sample drawing campaigns, treatment and measurement of samples;
- support to decontamination campaigns of material and areas;
- escort to transportation of radioactive material and support to emergency plans.

B-The mobile station has been designed to be used in external areas. It is equipped to measure the internal individual contamination. The mobile station is made of two standard intercommunicating containers:

a) lab for preparation of the patient including an area for calculation and support;

b) revelation lab with screened surfaces up to 3 cm of Pb, containing a hyperpure germanium detector with its screenings, standard chair and various accessories.

The station features a carrying structure with insulated walls, decontaminable surfaces, airtight communication passageways and filtering equipment.

Radiological Emergencies

The radioactive spreading that followed the Chernobyl incident gave the European countries an opportunity to realize the inadequacy of their environment check networks, the gaps in the management of the information sources and in the protective structure.

The Chernobyl experience has tought us to be proactive and set up an adequate organization inside the Defense Administration, whose objective is to cooperate and provide answers for the national organization and to establish a way of conduct to follow in order to safeguard own structures and personnel from the consequences of an external accident.

Organization within the Defence Administration

In the area of radiological emergencies, the Defence Administration is now operating in a coordinate manner, to respond to the following needs:
- definition of Defence tasks within the "National plan of emergencies";
- drawing up a document within Defence "Internal plan for intervention in case of radiological emergencies";
- collaboration to drawing up a document within NATO for military operations in areas with low level radiation.

The occurring of a radiological emergency suggests the necessary setting, within the Defence, of an organization capable to respond to its own specific needs, while demanding the structure responsible for the environment, radiological checks be constantly updated and efficiently kept. The organization must: evaluate information and data (even input from the civil control network); activate its own technical and check units; define the protection measures adopted and activate the territorial operational structure.

Conclusions

The Defence interest in the nuclear sector is strictly connected with its institutional duties, as such it concerns: nuclear defence purely military in character; protection from ionizing radiation of personnel involved in the use of radiological sources and protection of Defence areas; intervention activities in emergency situations.

For the radio-protection, the Defence has issued dispositions concerning the utilization of structures, vehicles and methods in case of emergency, originated by events whose scenarios are those foreseeable both in civilian and military sectors.

The structures and the protective measures contemplated for the Defense lead up to an internal organization which , in a specific plan, contemplates the existence of:
-decision making units, technical units, territorial and support units;
-assignement of duties in case of radiological emergency;
-procedures and methods of communication and information.

In the above context, CISAM is described as a technical triforce organization, providing the ability of its own work force and the scientific equipment of its labs for the performance of its commitments, making use of its own intervention vehicles, performing sampling and identification of radiological agents that allow for rapid intervention in external areas.

Use Of 70 MeV Proton Beam For Medical Applications At INFN-LNS: CATANA Project

M.G. Sabini[a,f], G.A.P. Cirrone[a,f], L. Barone Tonghi[a], A. Bartolotta[e], M. Brai[e], G.Cuttone[a], S. Lo Nigro[d,f], F. Marano[b], G.A. Nicoletti[b], G. Privitera[c], L. Raffaele[a,c], A. Reibaldi[b], N.Romeo[a], A. Rovelli[a],V. Salamone[c], G. Teri[e]

a) *INFN-Laboratori Nazionali del Sud, Catania*
b) *Institute of Ophtalmology -University of Catania*
c) *Institute of Radiology - University of Catania*
d) *Physics Department - University of Catania*
e) *Istituto di Biocomunicazioni - University of Palermo*
f) *Centro Siciliano di Fisica Nucleare e Struttura della Materia, Catania*

Abstract. The project CATANA (Centro di AdroTerapia ed Applicazioni Nucleari Avanzate) is a collaboration between the INFN-Laboratori Nazionali del Sud (LNS), Physics Department, Ophthalmology Institute and Radiology Institute of the Catania University and CSFNSM Catania. The main goal of CATANA is the study and the application of protontherapy for the treatment of shallow tumors (4 cm max) like uveal melanomas and subfoveal macular degenerations.

INTRODUCTION

The increasing interest in the use of protons in external radiotherapy arises from the improvement in the adsorbed dose distributions, as compared to conventional techniques using photon and electron beams; the rationale for using protons in radiotherapy is based on the different modality of interaction with tissue of protons. Bragg peak phenomena, with maximum delivered dose at the end of the proton range, the well defined range, with a sharp fall-off dose, the negligible lateral scattering with associated very small penumbra, enable to confine the high dose region in the target volume with a well defined transversal and longitudinal dose distribution in the tumor volume, realizing the maximum sparing of surrounding normal tissues. At the INFN-LNS a new protontherapy, named CATANA, is under realization mainly dedicated to the treatment of shallow tumors, like those present in the ocular region.

MEDICAL APPLICATIONS

Proton beam irradiation is particularly suitable for various tumors radiotherapy, especially the choroidal melanoma. The development of high precision systems to check the alignment of the patient allowed to localize the position of the tumor into the eyeball by means of tantalum clips and X-ray tubes. All this is necessary to obtain a careful saving of the normal tissues around the tumbrel volume (target) with a precision of about 1 mm [1]. Duration of irradiation is about 30-60 sec. for 4 times, total dose is 60 Gy [2]. Over 3500 cases of choroidal melanoma were treated in some Centers (Harvard-USA, Villigen-Switzerland, etc.) by means of proton beam irradiation. The most important literatures data about the follow-up of these patients are widely reported in literature [1,2].

An other interesting application field of proton beam irradiation in Ophthalmology is the exudative form of age-related macular degeneration (DMLA), which is characterized from a subretinal neovascular membrane (MNVSR) with subretinal exudation and low vision: in some cases, when this membrane is subfoveal, laser photocoagulation is not possible, because it would heavily damage the fovea. Age-related macular degeneration (ARMD) is currently the most common cause of blindness in patients over 65 year of age. A pilot study to demonstrate the feasibility of proton beam irradiation, the efficacious dose, the modality of irradiation, and the eventual side effects of the treatment of CNV in patients affected with age-related macular degeneration will be carried out at LNS.

RADIOTHERAPY FACILITY

The accelerators in operation at LNS are a 15 MV Tandem and a K=800 Superconducting Cyclotron (CS). An external source for the CS axial beam injection has been installed. In this configuration the CS will permit also to have proton beams at an energy between 45 and 100 MeV, with an intensity particularly suited for therapy (10-20 nA). In order to realize the clinical proton beam, a dedicated beam line in air has been realized. The beam dimensions will be defined by the first collimator (10-15 mm in diameter). A second collimator placed after 40-50 cm stops a considerably fraction of the diffused beam to obtain a beam distribution centered with respect to the collimator. A third collimator will be placed 10 cm before the isocentre position having a cross section different for each patient according to the shape of the tumor to be treated. A rotating wheel will be used as a diffuser spreading out the Bragg peak in order to have a flat dose distribution at the entrance. For the control of the dose during irradiation we will use a monitor chamber constituted by two independent transmission ion chambers calibrated with respect to the reference one and by a four sector transmission chamber to check the beam displacement. The positioning system is consisting of a chair with five independent motions and fully computer controlled by using stepping motors and absolute encoders. The patient positioning, with respect to the metallic clips inserted in the eye, is obtained using two orthogonal X-ray tubes. The eye motion during the treatment will be checked using a CCD camera and a frame grabber for image digitalization. The treatment plan will be studied using EYEPLAN. In case of ARMD treatment the positioning verification will be different considering that metallic clips will be not inserted. For this reason a double low power laser beam will be used, checking the alignment of the eye with respect to the beam axis and its orthogonal direction.

DOSIMETRY

The dosimetry of the clinical proton beam represents one of the main research and development goals of this project. The main requirements are the absolute and relative dose determinations. Our interest has been drawn in particular to the use of Ionisation Chambers (IC). In order to check the proton dose calibration methods two dosimetry intercomparisons were carried out at the 62 MeV clinical proton beam OPTIS line of the Paul Scherrer Institute (PSI) and at the Clatterbridge Centre for Oncology (CCO), using the 60 MeV clinical proton beam. For absolute dosimetry measurements in proton beams, we used two different types of air-filled IC; an Exradin T1 cylindrical IC (0.05 cm^3) made of A-150 tissue-equivalent plastic, and a MARKUS-PTW plane-parallel IC (0.056 cm^3). The use of A-150 tissue-equivalent thimble IC, calibrated in terms of air-kerma at ^{60}C0, is recommended in the ECHED Code of Practice for Clinical Proton Dosimetry [3] and in the ECHED Supplement to the Code of Practice [4]. The absorbed dose to water (D_w) when using Exradin T1 thimble IC is determined according to the ECHED recommendations. Recently the IAEA has published a new Code of Practice [5] on the use of plane parallel ICs in high-energy electron and photon beam. Extrapolating this formalism to proton beams, the absorbed dose to water at the reference point of the plane parallel Markus IC is given by the following equation:

$$D_w^Q (P_{eff}) = M_Q \cdot N_{D,air,Q0}^{pp} \cdot \left[(W_{air})_Q / (W_{air})_{Q0} \right] \cdot (S_{w,air})_Q \cdot p_Q$$

where Q and Q$_0$ mean the actual proton beam quality and the reference beam quality respectively; the beam quality specifier for the proton beam is the effective energy of protons ($E_{P,eff}$) at the IC measurement point, determined as recommended by the ECHED reports, Q$_0$ denotes the calibration beam quality used for the experimental evaluation of the absorbed dose to air chamber factor of the MARKUS chamber ($N^{pp}_{D,air,Q0}$), M$_Q$ is the MARKUS chamber reading (C) in the proton beam corrected for atmospheric factors, recombination and polarity effects, $[(W_{air})_Q/(W_{air})_{Q0}]$ is the ratio of average energy required to produce one ion pair in air in the user's proton beam quality Q, and in the calibration beam quality Q$_0$, $(S_{w,air})_Q$ is the water-to-air mass stopping power ratio for protons of quality Q, and p$_Q$ corrects for perturbations introduced by the IC. As recommended by Medin et al. [5], we determined directly the value of N $^{pp}_{D,air,Q0}$ for the parallel-plate MARKUS chamber, following the recommendations of the IAEA 1997 Code [6] and of the AAPM TG-39 by experimental intercomparison in an high energy electron beam with a reference Farmer-type IC having a known $N^{ref}_{D,air}$ ($4.64 \cdot 107$ Gy/C). A dosimetry intercomparison was carried out at the Douglas Cyclotron Unit of CCO (Clatterbridge Centre for Oncology, UK), in order to assess consistency of clinical proton beam calibrations between the two institutions, expecially when the absorbed dose is measured by using the plane-parallel MARKUS. For the purpose of the intercomparison the CCO used a FWT IC-18 0.1 cm^3 thimble IC constructed from A-150 tissue equivalent plastic, with an air-kerma calibration factor N_k measured in a ^{60}Co beam against a secondary standard chamber. Measurements were performed at the therapy position of a patient eye (isocenter), located 70 mm from the final collimator (Φ=25 mm). The results of the intercomparison, assuming as reference the CCO FWT-IC18 chamber, indicate a difference between the maximum and minimum dose values for all three chambers of 1.6% and 2.1% in modulated and unmodulated beam respectively, below the intercomparison tolerance level of 3% suggested in the ECHED code of practice .The results are satisfactory, considering that different procedures for the IC calibrations and different calibration factors were used. A second intercomparison was performed at the PSI using only the MARKUS chambers. The PSI MARKUS chamber was originally calibrated by intercomparison with the Harvard cyclotron in Boston; presently the calibration is repeated yearly in terms of N_k (^{60}Co) in an accredited dosimetry laboratory. MARKUS chamber intercomparison was performed in the modulated beam ($E_{p,incident}$=60 MeV) at the center of SOBP, corresponding to an effective proton energy of about 40 MeV. The percentage difference in measured dose for the two institutions was of 1.5%. More details about the dosimetric intercomparisons are elsewhere reported [7]. The facility so far described is now under development. The first proton beams is expected at the end of 1999. In 2000 the dosimetric beam characterization will be and the positioning system will be installed in its final position in the end of 1999. We expect to have first patient irradiation in 2000.

REFERENCES

1. Gragoudas E.S., et al. "Charged particle irradiation for choroidal melanoma" S.J.Ryan, ed. Retina, C.V. Mosby Company, vol. 1, cap. 45, 703-711, 1989

2. Zografos L., et al. "Le traitment des tumeurs oculaires par faisceau de protons acceleres. 7 ans dexperience" Klin. Monatsbl. Augenheilkd., 200:431-435, 1992

3. Vynckier S, Bonnett DE, Jones DTL "Code of practice for clinical proton dosimetry" Radiotherapy and Oncology 1991: 20; 53-63.

4. Vynckier S, Bonnett DE, Jones DTL "Supplement to the code of practice for clinical proton dosimetry" Radiotherapy and Oncology 1994:32; 174-179.

5. IAEA. The use of plane parallel ionization chambers in high energy electron and photon beams. 1997:Technical Report Series No.381

6. Medin J, Andreo P, et al. "Ionization chamber dosimetry of proton beams using cylindrical and plane parallel chambers. N_w versus N_k ion chamber calibrations" Physics in Medicine and Biology 1995: 40; 1161-1176.

7. G. Cuttone et al. " First Dosimetry Intercomparison Results for the CATANA Project" Physica Medica , Vol. XV, N. 3, July-September 1999

The α-disk instability in the radiation pressure dominated zone

V. Teresi, D. Molteni, M.A. Valenza

Department of Physical and Astronomical Sciences, Palermo University
Via Archirafi 36, Palermo 90123 Italy

Abstract. It is well known from analytical work that the accretion disk model stated by Shakura and Sunyaev (Shakura and Sunyaev, 1973) shows, in the radiation pressure dominated zone, called the A zone, an instability, called the 'Shakura-Sunyaev instability'.

This work shows the results of some simulations of α-disks in the conditions that the analytical theory states as instability conditions.

Our simulations are two dimensional (r-z) and time dependent. We use the Smoothed Particle method in cylindric coordinates.

The simulations of the A zone follow the time evolution of the disks. It appears that convection is very important to stabilize the flow.

INTRODUCTION

The question about the 'Shakura-Sunyaev instability' arises from the problem of the occurance (in the reality) of the equality between Q_+ and Q_-, i.e. the energy emitted per unit area of the disk and the rate of energy- generation in the disk. If the equality $Q_+ = Q_-$ is imposed, the stationary accretion (in the A zone) is possible only if the value of the viscosity is $\eta_c = \frac{4m_p}{9\sigma_T c}$, a value too big for the accretion disks. This situation can be explained only by assuming a difference between Q_+ and Q_- and/or a violation of stationarity.

Shibasaki and Hochi (1975) and Shakura and Sunyaev demonstrated the existence of thermal instabilities connected with a difference between Q_+ and Q_-.

Formally, the way to study the problem is the following.

An exact thermal equation, i.e. an energy equation that doesn't assume the equality $Q_+ = Q_-$, is considered together with the dynamic equations.

The system of equations so obtained is closed, and its solutions can be studied with respect to the problem of the stability against perturbations, whose radial length scale (called wavelength) Λ satisfies the inequality $H_0 < \Lambda < R$.

This condition on Λ allows to consider negligible, in a perturbative approach, the terms of order $(H_0/R)^2$ and $H_0^2/R\Lambda$ when compared with terms of order $(H_0/\Lambda)^2$.

CP513, *Nuclear and Condensed Matter Physics*, edited by A. Messina
© 2000 American Institute of Physics 1-56396-929-7/00/$17.00

By applying this procedure, what is obtained is that exist perturbations which grow with time and therefore give rise to disk instabilities.

The two fundamental variables that caracterize these instabilities are the wavelength and the growth rate of the perturbation.

First, we will consider the case of gas pressure totally negligible.

In the A zone, if the wavelength Λ satisfies the condition $2H_0 < \Lambda < 4H_0$, the perturbation takes the form of a ring moving across the surface of the disk.

Corrispondingly, the growth rate Ω is $\Omega = 0$ when $\Lambda = 2H_0$ and $\Omega = \alpha\omega/10$ when $\Lambda = 4H_0$ (ω is the angular velocity of the disk at the radius considered).

If $\Lambda > 4H_0$ there are two branches.

One of them is caracterized by a growth rate decreasing asymptotically to zero with Λ, while the other one by a growth rate increasing asymptotically to a maximum value with Λ.

The physical nature of these two types of instability is conceptually different. In fact, the 'decreasing' branch, for big values of Λ, gives a situation in which Q_+ and Q_- are nearly equal, and therefore doesn't correspond to a thermal instability, but a dynamic one. This instability, in this limit, is the instability discovered by Lightman and Eardley. Instead, the 'increasing' branch is caracterized by the fact that Q_+ and Q_- are significantly different. So, this branch must be considered a description of the dependence of Ω on Λ for instabilities of thermal nature.

When the gas pressure isn't totally negligible, an important parameter of the problem is

$$\beta_r = \frac{p_r}{p_r + p_g} \tag{1}$$

that is the fraction of the total pressure represented by the radiation pressure.

When β_r decreases from the value 1, that represents the case of gas pressure totally negligible, there are yet two branches, but they don't begin from the wavelength $4H_0$: the 'bifurcation' point corresponds to a greater Λ, that increases with decreasing β_r.

Moreover, the 'increasing' branch has a maximum value that decreases with decreasing β_r.

Finally, it must be considered that, before the 'bifurcation' point, the growth rate increases from zero with increasing Λ, with a value of Λ for which $\Omega = 0$, that we will name $\Lambda_{\min}$, which increases with decreasing β_r.

The minimum value of $\Lambda_{\min}$ is, as we said before, $2H_0$, value which corresponds to $\beta_r = 1$.

When β_r tends to $3/5$, $\Lambda_{\min}$ tends to infinity. This means that, at this value of β_r, there are no perturbations which grow with time. In fact, calculation gives the value $\Omega = 0$, for this β_r.

For $\beta_r < 3/5$, $\Omega < 0$, i.e. we have perturbations that decay with time.

THE Z-STRUCTURE

To simulate the time evolution of an α-disk, we need to know exactly the equilibrium configuration of the disk, to use it as the initial condition for the simulation. The density and temperature distributions in r are standard (Shakura and Sunyaev already calculated them in 1973), but, as regards the distributions in z, there are some aspects (particularly about the role of the convection) that need a detailed mathematical analysis.

We have performed this analysis and here we explain the followed method.

The z-structure equations in stationary regime are:

$$-\frac{1}{\rho}\frac{\partial P}{\partial z} = \omega^2 z \tag{2}$$

$$\frac{\partial}{\partial z}\left(\frac{16\sigma T^3}{3k_r\rho}\frac{\partial T}{\partial z}\right) + \eta\left(r\frac{d\omega}{dr}\right)^2 = 0 \tag{3}$$

with P total pressure, η dynamical viscosity (set proportional to the total pressure), ω angular velocity (set equal to the keplerian one) of the gas rotation around the black hole, k_r the total opacity of the gas, due to the electron scattering and the free-free absorption.

Therefore we solve a system composed by the equation of the hydrostatic vertical equilibrium (eq. (2)) and the equation (the (3)) that binds the energy produced by the viscous dissipation to the variation with z of the radiated electromagnetic energy.

The system composed by the equations (2) and (3) has been solved through a numerical method (finite difference integration).

The density always increases as a function of z.

The temperature, instead, always decreases as a function of z.

We have simulated the evolution of these configurations.

They show a z-structure instability, that, obviously, must not be confused with the Shakura-Sunyaev instability.

In order to have a stable initial configuration (as regards the z-structure) whose evolution can be simulated to investigate whether the Shakura-Sunyaev instability develops, we have tried to find the physical phenomenon that stabilizes the z-structure.

In reality, the neglected phenomenon in the whole previous analysis is the phenomenon of the convective heat transfer just along the z-direction.

We have included this phenomenon by modifying the equation (3) in order to consider the role of the heat convective flux F_{conv}, given by the following formulas:

$$F_{conv} = \frac{8k_B\rho T H\omega}{m_H\beta_g^{5/2}}\left(\nabla - \nabla_{ad}\right)^{3/2} \tag{4}$$

with:

$$\nabla - \nabla_{ad} = \frac{\beta_g}{16} \left(\frac{12 - 10.5\beta_g}{4 - 3\beta_g} - \frac{d \ln \rho}{d \ln T} \right) \tag{5}$$

and $\beta_g = P_{gas}/(P_{gas} + P_{rad})$.

The inclusion of the convective flux in the eq. (3) gives rise to the following equation:

$$-\frac{\partial}{\partial z} \left(F_{rad} + F_{conv} \right) + \eta \left(r \frac{d\omega}{dr} \right)^2 = 0 \tag{6}$$

with F_{rad} given, as before, by $F_{rad} = -\frac{16\sigma T^3}{3k_r \rho} \frac{\partial T}{\partial z}$

Solving this system we have obtained configurations that always show a decreasing part in the density profile.

Depending on the values of the accretion rate and of α one can then have or not, close to the disk surface, an increasing part of the density profile.

When there isn't this part, and therefore the density profile is monotonically decreasing, the configuration, as results from the simulations, is stable.

The temperature profile, taking into account the convection, is increasing up to a given value of z and then decreasing.

CONCLUSIONS

The results of our simulations of the time evolution of α-disks, with a large portion in radiation pressure dominating conditions, show clearly enough that no instability develops. This fact is particularly interesting, if we consider that the local analysis performed by Shakura and Sunyaev demonstrates, for a range of values of the perturbation wavelength and the pressure rate β_r, that both thermal and secular instabilities are present.

We suppose that the reason for which we see no instabilities can be:

1. the small extension of the A zone, in the sense that the radiation pressure dominated region isn't large enough to enable the disk to produce a perturbation of the necessary wavelength;

2. the presence of convection, that isn't considered in the Shakura and Sunyaev analysis, but is naturally simulated by our program.

We will make investigations about the problem through new simulations.

REFERENCES

1. Shakura, Sunyaev: 1973, A.A., **24**, 337.

2. Shakura, Sunyaev: 1976, M.N.R.A.S., **175**, 613.

Optically Stimulated Luminescence Dating Of Sediments

S.O.Troja[*], C.Amore[†], G.Barbagallo[*], G.Burrafato[*], R.Forzese[*],
F.Geremia[†], A.M.Gueli[*], F.Marzo[*], D.Pirnaci[*], M.Russo[*], E.Turrisi[*]

[*]Dipartimento di Fisica and Centro Siciliano di Fisica Nucleare e Struttura della Materia and
[†]Dipartimento di Scienze Geologiche, Università di Catania, Corso Italia 57, 95129 Catania, Italy

Abstract. Optically stimulated luminescence (OSL) dating methodology was applied on the coarse grain fraction (100÷500 μm thick) of quartz crystals (green light stimulated luminescence, GLSL) and feldspar crystals (infrared stimulated luminescence, IRSL) taken from sections at different depths of cores bored in various coastal lagoons (Longarini, Cuba, Bruno) in the south-east coast of Sicily. The results obtained give a sequence of congruent relative ages and maximum absolute ages compatible with the sedimentary structure, thus confirming the excellent potential of the methodology.

INTRODUCTION

Coastal wetlands are transitional ecosystems between terrestrial and aquatic systems, where the land is covered by shallow water and the factors determining the sedimentary characteristics are mainly mechanical, due to action of waves, littoral currents and wind. In south-eastern Sicily, however, these environments, made up of sedimentary deposits situated near the coast, have, or have had, a significant interaction with the marine environment, through a limited communication with the open sea, accumulating fine-grained sediment of terrestrial provenance (1). The Vendicari region is one of the most important of these coastal lagoon systems and includes, among others, the Longarini, Bruno and Cuba lagoons. The coastal lagoon of Longarini communicates with the sea through two artificial canals, partially occluded, whereas the Cuba and the Bruno lagoons are completely separated from the sea by a 5 m high ridge of sandy dunes.

The sediment samples taken away from the bottoms of the coastal lagoons were studied in terms of texture, composition and fauna, allowing the preparation of thematic maps from which the sedimentary evolution of the area can be reconstructed. The results obtained from absolute dating methodologies applied to the sediment cores allow, under suitable conditions, a reconstruction of the chronology of the sedimentary sequences and provide information regarding the rate of sediment deposition and the temporal evolution of the coastal lagoons.

CP513, *Nuclear and Condensed Matter Physics,* edited by A. Messina
© 2000 American Institute of Physics 1-56396-929-7/00/$17.00

EXPERIMENTAL METHODS

The OSL analysis was conducted on the sediment hand-cores taken from Longarini, Cuba and Bruno lagoons. The sediment cores were divided vertically into two parts, one to give the annual dose for radioactive measurements, and the other for the determination of the paleodose for luminescence measurements. The hand cores were analysed using different procedures: for the Cuba and Bruno lagoons the measurements were performed along the entire length of the cores, on sections of 1 cm, collected from the central part of sedimentary intervals showing different textural characteristics. For the Longarini lagoon, the measurements were effectuated on crystals, sampled from sections of 1 cm at different depths, in 5 cm steps, independently from textural considerations.

The GLSL measurement was carried out separating the quartz crystalline fraction with a granulometry of 100-300µm. Some IRSL measurements on feldspars were performed in some particular cases. The usual separation procedures of sieving, etching with acids and separation of the mineral phases with variable density liquids were used (2). The "purity" of the crystalline phases obtained was verified by X-ray diffraction measurements. A Risø TL/OSL-DA-12 system able to supply the green or infrared radiation and artificial radioactive irradiation (calibrated β source ^{90}Sr-^{90}Y) was used for all the luminescence measurements. The paleodose was obtained applying the S.A.R.A. procedure (3). The annual-dose was evaluated using the values of concentration of the U, Th and K chains obtained by high resolution gamma spectroscopy (HPGe). The evaluation of the annual-dose involves only the dose components which come from β and γ radiation of the sand sample, considered in this case to be homogeneous with the environment (2). A constant dose-rate contribution of 90µGy/year was considered for the component due to cosmic rays.

RESULTS

Tab.I reports the final results of OSL dating for the samples measured, with the data of the dose essential for the calculation. Samples in which the necessary characteristics of linearity, reduced dispersion in the intensity of luminescence or fading were not verified, were rejected and considered not datable. The data obtained on the different intervals of the sediments from Cuba and Bruno lagoons show, in both cases, well differentiated sequences of ages. An irregularity is found for Longarini lagoon, with ages clearly lower than others, which cannot apparently be attributed to an inversion of the layers. Before making further hypotheses, therefore, a feldspars fraction was separated from the same layers for IRSL dating.

TABLE 1. Results of OSL dating obtained from different samples analysed

Sample	Fractions Separated	Paleodose (Gy)	Annual Dose (μGy/a)	Age (BP)	Error (%)
Cuba Coastal Lagoon					
Cuba A	Q	0.59	990	550	6
Cuba B	Q	1.06	840	1250	14
Cuba C	Q	2.07	860	2400	5
Bruno Coastal Lagoon					
Bruno A	Q	0.23	870	220	11
Bruno B	Q	1.34	810	1610	8
Bruno C	Q	1.52	820	1810	4
Longarini Coastal Lagoon					
Long4	Q	Not linear	810	Not datable	
Long9	Q	Not linear	810	Not datable	
Long14	Q	0.09	810	65	39
Long19	Q	Not linear	810	Not datable	
Long24	Q	0.35	790	430	8
Long29	Q	0.39	790	450	22
Long34	Q	0.71	790	850	19
Long39	Q	0.64	790	760	10
Long44	Q	0.78	790	940	6
Long49	Q	0.34	790	380	23
	FK	0.74	790	890	7
Long54	Q	0.38	790	440	17
	FK	0.76	790	920	8
Long59	Q	0.87	790	1050	20
	FK	1.05	790	1280	8
Long64	Q	1.47	790	1820	10

The measurement on feldspars were performed only on those samples where the quartz presented an anomalous behaviour, providing dating values outside the relative sequence. Fig.1 shows the trend of the ages checked for the sections of Longarini lagoon where the solid squares refer to quartz analysis of the regular sequence, the open squares to quartz analysis related to crystals with anomalous behaviour and the solid triangles refer to feldspars analysis.

CONCLUSION

The results obtained applying the optically stimulated luminescence dating methodology to the coarse grain fraction of quartz and/or feldspar crystals extracted from sediments confirm the capacity of the methodology which is proposed for the dating of sedimentary environments where other methodologies are not suitable. The age values appear congruent as a sequence of relative and absolute ages.

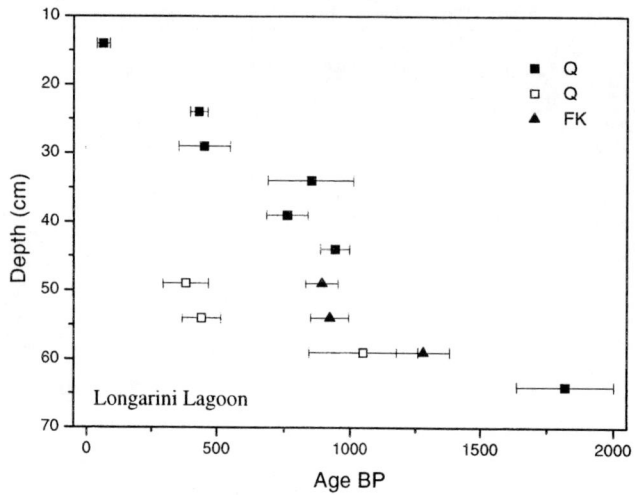

FIGURE 1. Trend of the ages checked for the sections of Longarini marsh.

This appears evident in the sequence on the sediments of Longarini lagoon where, however, an anomalous behaviour was noted compared to the predicted sequence. The measurements on feldspars, however, confirmed the sequence. To obtain a reliable confirmation, the measurement was also performed on the first two near layers which had provided regular dating in quartz fraction analysis. The confirmation of these ages indicates that the problem is directly linked to the quartz fraction. The result on feldspars, in fact, establishes the principal sequence. The results show that greater reliability can only be obtained from a series of data regarding a lagoon rather than by considering as a single age measurement.

ACKNOWLEDGMENTS

This work was supported by the Consiglio Nazionale delle Ricerche, Progetto Finalizzato Beni Culturali, Italy.

REFERENCES

1. Amore, C., *et al.*, *Rivista Italiana di Paleontologia e Stratigrafia* **103**, 3-14 (1997).

2. Aitken, M.J., *Thermoluminescence dating*, Oxford: Academic Press Inc., 1985.

3. Aitken, M.J., *An Introduction to Optical Dating: The Dating of Quaternary Sediments by the Use of Photon-stimulated Luminescence*, Oxford: Oxford University Press, 1998.

LNS: Present and Future

D.Vinciguerra

Dipartimento di Fisica, Università, Catania,
Laboratori Nazionali del Sud, INFN, Catania

Abstract. The status of the Laboratori Nazionali del Sud and of their scientific activities is briefly described. The main projects concerning the acquisition of new facilities and techniques and the opening of new fields of research are also presented.

RESEARCH AT LNS

The study of the atomic nucleus dates back to the beginning of the century. Yet many essential aspects of its structure, and of the ways in which nuclei interact, have not been unveiled. This is due to the fact that nuclei are made up by a number of nucleons which is too large to allow for a description in terms of a two- or few-body interaction, and too small to permit easily the use of statistical approaches. Moreover new techniques allow nowadays the study of unstable nuclei with very short life-times.

At the Laboratori Nazionali del Sud (LNS), which are one of the four large facilities which the Istituto Nazionale di Fisica Nucleare (INFN) has built in Italy (1), research is mainly devoted to nuclear physics at low and intermediate energy (experimental and theoretical) and to applied physics with nuclear techniques (solid state, cultural heritage). Experiments are carried on by teams which are partially or entirely formed by researcher belonging to other Italian or foreign Institutions.

MAIN EXPERIMENTAL EQUIPMENT

Two large heavy ion accelerators are operational at LNS, together with complex experimental set-ups for particle and γ-ray detection. Some of the most relevant pieces of equipment will be briefly described in the following.

Accelerators

A Tandem van de Graaff with a 15 MV maximum terminal voltage and a K=800 cyclotron with superconducting coils (CS) are installed at LNS. The heavy ions are ionized, pre-accelerated in a 450 kV injector which has been designed and built at

CP513, *Nuclear and Condensed Matter Physics*, edited by A. Messina
© 2000 American Institute of Physics 1-56396-929-7/00/$17.00

LNS, and accelerated by the Tandem. If a higher energy is required, CS is used as a booster to accelerate the ions up to several tens of MeV/nucleon

By the end of 1999 there will be a change in the mode of operating CS. The injection will take place axially and the ions will be delivered by an efficient superconducting ECR source, already operational at LNS, which to date has the best performances in the world. This new arrangement will have important consequences: the beam current can increase by three orders of magnitude, with an increase also in the maximum energy of the heaviest ions. Moreover the two accelerators can run in an independent way.

Detection Systems

When a nuclear reaction takes place at CS energies, a large number of nuclear fragments and of γ-rays can be produced in a single event. It is then important to use multi-detectors which cover as much as possible of 4π and have a large granularity and a high detection selectivity.

Only MEDEA and CHIMERA will be described here. MEDEA consists of 180 scintillators, for γ-ray and proton detection, which are now coupled with a forward wall of charged particle detectors and a superconducting solenoid which focuses charged fragments, emitted at very small angles, 15 m downstream for TOF measurements. CHIMERA is more apt to detect and identify charged particles with a high granularity. It consists of about 700 triple detectors, which will be 1192 in the final configuration. Other multi-detectors in use are HODO, ARGOS, TRASMA, NEUTRONI,...

Finally let us mention MAGNEX, a magnetic spectrometer which is under construction and is presented elsewhere at this Conference (2).

MAIN LNS PROJECTS

The daily activity which makes use of the LNS equipment – accelerators, detection set-ups and ancillary devices – does not exhaust the work of the physicists, engineers and technicians. A large effort is actually devoted to the definition and realization of new projects, which represent an investment for the future of the laboratory. Only four of them will be described in the following.

The EXCYT Project

Research in nuclear physics with accelerators has been carried out till recent years by using beams and targets made of stable nuclei. In some cases beams or targets are made of radioactive nuclei with a half-life long enough to perform experiments.

In recent years a new technique, called ISOL, has been developed to produce radioactive ion beams (RIB) with half-life as short as a few ms, which can interact with stable targets. At LNS, CS will be used as a primary accelerator to produce fragments in the interaction with a thick target. The produced particles will be selected and accelerated by the Tandem so to have the desired RIB. By the end of 2000 the system will be operational and the first RIB will be delivered in 2001. Note that while the stable or long-lived nuclei are about 2000, this technique allows in principle to study about 4000 other nuclei, thus opening new and challenging frontiers for research in nuclear physics and related fields.

The CATANA Project

Protons or heavier ions can be used in the treatment of localized tumors, their advantage with respect to electrons, X- and γ-rays being the possibility of releasing a dose at the desired emplacement, exploiting the Bragg peak characteristics through the use of suitable diaphragms and absorbers. Once the axial injection will be installed, the cyclotron will be able to deliver protons with an energy of 70 MeV which corresponds to a range of about 4 cm in human tissue. The CATANA Project concerns the treatment of eye deseases like uveal melanoma and macular degeneration with 70 MeV protons. All the techniques concerning beam transport, local dosimetry and computerized treatment plans have already been developed. This project has been made possible by a fruitful collaboration with the Ophtalmology and the Radiological Institutes and the Physics Department of the Catania University, which has received a support from the EC to buy the irradiation chair and the ancillary control equipment. The treatment room has already been prepared at LNS and in a few months we should be ready to start operating the first hadrotherapy facility in Italy.

Accelerator Development

The availability of two heavy ion accelerators allows for a variety of possible researches, as we have seen. However higher beam current and better performances are demanded to enlarge the fields of interest. This can be done either by upgrading CS or by installing a new cyclotron. Both solutions are under study. Among the advantages of the new installation are the possibility of producing radioisotopes for industry and medicine, and the Phase II of EXCYT, with CS accelerating the radioactive ions produced by the new cyclotron, thus producing RIB's at intermediate energy with the ISOL technique. This will be a facility unique in the world.

The NEMO Project

It concerns a very large detection system which has to be installed in deep sea water (3 – 4 km). The water will have the double task of shielding the detectors from cosmic rays, and to fire them through Cerenkov light. This subject is treated in greater detail elsewhere in this conference (3). It is worth mentioning here the interdisciplinary potentialities of such a project, the role LNS is playing in activating a test site off the Catania harbour, and the role it can have once a site off the Sicilian coast is chosen for the final detector.

THE ROLE OF THE INSTITUTIONS

The birth of LNS dates back to 1976 when an agreement was signed between INFN, the University of Catania and the Regione Siciliana through CSFNSM. Since then, all three Institution have played a role, though at different levels, in the life of LNS.

The buildings of LNS lay on the University campus, but they have been built with the INFN financial support. The whole of instrumentation is of INFN property. The entire technical and administrative personnel (80 persons) belongs to INFN, with one exception, while researchers are both from INFN (18) and University (17). Moreover about 40 fellows and students participate to the LNS activities. The annual budget is entirely INFN, with a small University contribution.

The Regione Siciliana has contributed significantly to LNS at the very beginning with a determinant contribution of 1 billion lire (1976). From that date on, only the annual support of CRRNSM has very marginally helped the research at LNS as in other research institution of Sicily. However, in spite of its relatively small amount, it is worth stressing the positive impact the regional support has had through the assignment of dedicated fellowships. Not to mention more than one example, this is the case of the CATANA activities, which have recently benefit from the regional support, through the fellowships of CSFNSM.

ACKNOWLEDGEMENTS

It is a pleasure and a duty to acknowledge the effort, the dedication and the scientific and professional skill of all those physicists, engineers and technicians who make an advanced laboratory like LNS run at a highly competitive level.

REFERENCES

1. Finocchiaro, P., and Vinciguerra, D., *Laboratory Portrait: LNS*, Nucl. Phys. News Int., NuPECC, 1998, vol.8 n. 1, pp. 4-16 and references therein.

2. Cappuzzello, F. , *contribution to this Conference.*

3. Randazzo, N. , *contribution to this Conference.*

LIST OF PARTICIPANTS

Agnello Simonpietro	sagnus@yahoo.com
Albergo Sebastiano	sebastiano.Albergo@ct.infn.it
Andaloro Giuseppina	bellomon@fisica.unipa.it
Arena Antonella	aarena@ortica.unime.it
Barbera Marco	barbera@oapa.astropa.unipa.it
Bartolotta Antonio	djrba@tin.it
Bartolotta Antonio	bartolot@hpits1.its.me.cnr.it
Bassani Franco	bassani@sns.it
Bellomonte Leonardo	bellomon@fisica.unipa.it
Benedetti Armando	armando.Benedetti@cisam.it
Boscaino Roberto	boscaino@fisica.unipa.it
Brai Maria	mbrai@unipa.it
Branca Caterina	branca@dsme01.messina.infm.it
Burderi Luciano	burderi@gifco.fisica.unipa.it
Burlon Riccardo	burlon@www.unipa.it
Caccamo Carlo	caccamo@vulcano.unime.it
Calareso Carmen	calareso@ortica.unime.it
Cannas Marco	cannas@fisica.unipa.it
Cannistraro Giuseppe	giaconia@unipa.it
Caponetto Luigi	luigi.Caponetto@ct.infn.it
Cappuzzello Francesco	cappuzzello@lns.infn.it
Carini Giuseppe	carini@imeuniv.unime.it
Carollo Angelo	carollo@isiosf.isi.it
Castiglia Francesco	castiglia@din.din.unipa.it
Chiorboli M.	
Cirone Markus	cirone@ifata1.deaf.unipa.it

Cirrone Giuseppe	cirrone@lns.infn.it
Compagno Giuseppe	compagno@fisica.unipa.it
Cordone Lorenzo	cordone@fisica.unipa.it
Costa Dino	tritone@unime.it
Cottone Grazia	grazia@ns.iaif.pa.cnr.it
Cubiotti Gaetano	cubiotti@vulcano.unime.it
Cunsolo Angelo	cunsolo@lns.infn.it
Cupane Antonio	cupane@fisica.unipa.it
Cutroni Maria	cutroni@dsme01.messina.infm.it
Cuttone Giacomo	cuttone@lns.infn.it
D'Angelo Giovanna	dangelo@dsme01.messina.infm.it
Della Sala Dario	dario.dellasal@casaccia.enea.it
Di Maio Pietro	vella@din.din.unipa.it
Di Salvo Tiziana	disalvo@gifco.fisica.unipa.it
Di Stefano Omar	distef@ortica.unime.it
Donato Mariagrazia	mdonato@vulcano.unime.it
Donato Dorina Ines	donato@unipa.it
Emanuele Antonio	antonio.emanuele@iaif.pa.cnr.it
Falci Giuseppe	gfalci@cdc.unict.it
Falciglia Filippo	falciglia@dmfci.ing.unict.it
Faraone Antonio	afaraone@dsme01.messina.infm.it
Fauci Francesco	faucipro@gifco.fisica.unipa.it
Fazio Enza	enza_f@hotmail.com
Ferrante Gaetano	ferrante@unipa.it
Figuera Pierpaolo	figuera@lns.infn.it
Foti Antonino	foti@ct.infn.it
Foti Gaetano	gaetano.foti@ct.infn.it
Frisone Fulvio	frisone@ct.infn.it

Gallo Carrabba Daniele	dgc99@fastlink.it
Giaconia Carlo	giaconia@unipa.it
Gelardi Franco	gelardi@fisica.unipa.it
Geraci Elena	geraci@lns.infn.it
Gerardi Gaetano	gerard@gifco.fisica.unipa.it
Giuliano Sandro	esg@ginestra.unime.it
Grasso Vincenzo	grasso@ortica.unime.it
Gueli Anna	anna.gueli@ct.infn.it
Iacono Manno Marcello	iacono@lns.infn.it
Iatì Maria Antonia	iati@ortica.unime.it
Immè Josette	imme@ct.infn.it
Insolia Antonio	antonio.insolia@catania.infn.it
La Barbera Antonino	nino@gifpa2.fisica.unipa.it
Lattuada Marcello	lattuada@lns.infn.it
Leone Claudio	leone@unipa.it
Leone Maurizio	leone@fisica.unipa.it
Librizzi Fabio	dottorat@fisica.unipa.it
Lillo Fabrizio	lillo@mari.deaf.unipa.it
Lipari Eugenio	euro@tao.it
Li Vigni Maria	livigni@fisica.unipa.it
Lo Nigro Salvatore	lonigro.salvatore@ct.infn.it
Magazù Salvatore	magazu@dsme01.messina.infm.it
Maisano Giacomo	maisano@dsme01.messina.infm.it
Majolino Domenico	majolino@dsme01.messina.infm.it
Mandanici Andrea	andrea@nucleo.unime.it
Maniscalco Sabrina	messina@fisica.unipa.it
Manno Mauro	mauro@iaif.pa.cnr.it
Mastellone Andrea	andrea@femto.if.ing.unict.it

Melita Augusto Luciano	melita@lns.infn.it
Messina Antonino	messina@fisica.unipa.it
Mezzasalma A.	
Migliardo Federica	migliardo@alpme2.me.infn.it
Migliardo Placido	migliard@dsme01.messina.infm.it
Migliore Rosanna	messina@fisica.unipa.it
Militello Valeria	militelo@fisica.unipa.it
Molteni Diego	molteni@gifco.fisica.unipa.it
Mondio Guglielmo	mondio@ortica.unime.it
Montereali Rosa Maria	montereali@frascati.enea.it
Morelli Daniela	environ@ct.infn.it
Moya-Cessa Hector	hector@camcat.unicam.it
Napoli Anna	messina2@fisica.unipa.it
Neri Fortunato	neri@ortica.unime.it
Nociforo Chiara	nociforo@lns.infn.it
Oliveri Maria Elisabetta	meolive@dmfci.ing.unict.it
Paladino Elisabetta	paladino@alp1ct.ct.infn.it
Palma G.Massimo	palmagm@mildred.physics.ox.ac.uk
Palma M.Ugo	palma@iaif.pa.cnr.it
Pappalardo Lorenzo	lorenzo.Pappalardo@ct.infn.it
Patti Fabrizio	patti@ifata1.deaf.unipa.it
Pellegriti Maria Grazia	pellegriti@lns.infn.it
Pellicane Giuseppe	pellican@vulcano.unime.it
Persano Adorno Dominique	dpersano@ifata1.deaf.unipa.it
Petta Catia	catia.Petta@ct.infn.it
Piccolo Antonio	piccolo@nucleo.unime.it
Pizzone Rosario	rgpizzone@lns.infn.it
Potenza Renato	renato.potenza@ct.infn.it

Privitera Giuseppe	giuseppe.Privitera@dmfci.ing.unict.it
Puccia Daria	leone1@fisica.unipa.it
Randazzo Nunzio	nunzio.randazzo@ct.infn.it
Rausei Magda	cupane@fisica.unipa.it
Reale Fabio	reale@oapa.astropa.unipa.it
Robba Natale	robba@gifco.fisica.unipa.it
Russo Giovanni Valerio	valerio.russo@ct.infn.it
Sabbatini Vittorio	Vittorio.Sabbatini@cisam.it
Saija Rosalba	rosalba@ortica.unime.it
Sanfratello Vincenzo	leone1@fisica.unipa.it
Savasta Salvatore	savasta@ortica.unime.it
Serio Salvatore	serio@oapa23.astropa.unipa.it
Silipigni Letteria	silipign@ortica.unime.it
Spagnolo Bernardo	spagnolo@unipa.it
Teresi Vincenzo	teresi@gifpa2.fisica.unipa.it
Teri Giovanni	gteri@unipa.it
Tricomi Alessia	alessia.Tricomi@ct.infn.it
Triglia Antonino	atriglia@dmfci.ing.unict.it
Triolo Roberto	triolo@unipa.it
Tripodo Gaspare	tripodo@dsme01.messina.infm.it
Troja Sebastiano	olindo.troja@ct.infn.it
Trusso Sebastiano	nelloòhpits1.its.me.cnr.it
Tudisco Salvatore	tudisco@lns.infn.it
Tumino Aurora	tumino@lns.infn.it
Tuvè Cristina	cristina.tuve@ct.infn.it
Vaiana Sara	sara@iaif.pa.cnr.it
Valenti Davide	valentid@locascio.fisica.unipa.it
Vella Giuseppe	vella@din.din.unipa.it

Venuti Valentina	vvenuti@dsme01.messina.infm.it
Villari Valentina	villari@dsme01.messina.infm.it
Vinciguerra Domenico	vinciguerra@lns.infn.it
Vitrano Eugenio	vitrano@fisica.unipa.it
Vittorelli Beatrice	vittorel@iaif.pa.cnr.it
Zarcone Michelangelo	zarcone@unipa.it

Author Index

A

Adorno, D. P., 202
Agliolc Gallitto, A., 3
Agnello, S., 7, 63
Agozzino, P., 110
Agudov, N. V., 11
Aiello, S., 257
Albergo, S., 326
Aliotta, M., 261, 298, 385
Allegrini, M., 15
Aloisi, A., 19
Amore, C., 409
Amorini, F., 278
Anastasi, M., 333
Andaloro, G., 333
Annino, A., 337
Anzalone, A., 257
Arena, A., 15
Arrighi, V., 234

B

Baldo, M., 257
Ballone, P., 90, 114
Barbagallo, G., 409
Barnà, R., 257
Barone, G., 341
Barraco, R., 333
Barreca, F., 23
Bartolotta, A., 27, 31, 35, 75, 401
Basile, S., 35
Bellomonte, L., 333
Belozyorov, A. V., 270
Betts, D., 234
Biagi, A., 345
Bivona, S., 39
Blagus, S., 385
Boemi, D., 326
Boffi, A., 218
Bonvicini, V., 306
Borghese, F., 43
Boscaino, R., 7, 63
Bradfield-Smith, W., 278
Brai, M., 31, 35, 333, 401
Branca, C., 47, 118, 146, 250

Bruno, G., 35
Bruzzi, M., 369
Bucciolini, M., 369
Buemi, A., 389
Bulone, D., 214
Burderi, L., 349, 353, 373, 377
Burlon, R., 39
Burrafato, G., 409

C

Cabibbo, M., 278
Caccamo, C., 51, 55, 90
Caccia, Z., 326
Calareso, C., 59
Campisi, M. G., 257
Cannas, M., 7, 63
Cannistraro, G., 67, 357, 361
Caponetto, L. M., 71
Cappuzzello, F., 266, 270
Cardella, G., 257, 278
Carini, G., 27, 75
Carollo, A., 79
Carrotta, R., 83
Castiglia, F., 286
Cavallaro, S., 257
Cefali, E., 15
Celata, G. P., 286
Chen, C.-X., 326
Cherubini, S., 261, 298, 385
Cirone, M. A., 365
Cirrone, G. A. P., 369, 401
Compagno, G., 87, 242
Cordone, L., 83, 130
Costa, D., 51, 90
Costa, S., 326
Cramer, C., 150
Crawford, H. J., 326
Cronqvist, M., 326
Crupi, V., 94, 246
Cumo, M., 286
Cunsolo, A., 266, 270
Cupane, A., 174, 210, 218
Cutroni, M., 98, 150
Cuttone, G., 35, 369, 401

Tuvè, C., 19, 326
Typel, S., 261

U

Urso, S., 345

V

Vacchi, A., 306
Vaiana, A. C., 238
Vaiana, S. M., 214
Valenti, D., 242
Valenza, M. A., 405
Valvo, G., 306
Vasi, C., 23
Vella, G., 274
Venuti, V., 94, 246
Verde, G., 290
Vidiella-Barranco, A., 182
Villari, V., 47, 118, 146, 246, 250, 393
Vinciguerra, D., 270, 322, 413
Vitrano, E., 130

W

Waddington, C. J., 326
Webber, W. R., 326
Wefel, J. P., 326
Weiss, U., 194
Winfield, J. S., 266, 270
Wolter, H. H., 261, 266
Woods, P. J., 278

Y

Yeremin, A., 270

Z

Zabalza, R. Rey, 35
Zadro, M., 298, 385
Zappalà, R. A., 298, 385
Zarcone, M., 202
Zeilinger, A., 79
Zetta, L., 257
Zummo, G., 286

6-4 D